Studies in Neuroscience, Psychology and Behavioral Economics

Series Editors

Martin Reuter, Rheinische Friedrich-Wilhelms-Universität Bonn, Bonn, Germany
Christian Montag, Institut für Psychologie und Pädagogik, Universität Ulm, Ulm,
Germany

More information about this series at http://www.springer.com/series/11218

Harald Baumeister · Christian Montag

Editors

Digital Phenotyping and Mobile Sensing

New Developments in Psychoinformatics

 Springer

Editors
Harald Baumeister
Department of Clinical Psychology
and Psychotherapy
Institute of Psychology and Education
Ulm University
Ulm, Germany

Christian Montag
Department of Molecular Psychology
Institute of Psychology and Education
Ulm University
Ulm, Germany

ISSN 2196-6605   ISSN 2196-6613   (electronic)
Studies in Neuroscience, Psychology and Behavioral Economics
ISBN 978-3-030-31622-8   ISBN 978-3-030-31620-4   (eBook)
https://doi.org/10.1007/978-3-030-31620-4

This Springer imprint is published by the registered company Springer Nature Switzerland AG
The registered company address is: Gewerbestrasse 11, 6330 Cham, Switzerland

Foreword for Digital Phenotyping and Mobile Sensing

It is an axiom in the business world that you can't manage what you can't measure. This principle, usually attributed to the business guru Peter Drucker, is equally true in medicine. Imagine managing diabetes without HbA1c, hypertension without blood pressure readings, or cancer diagnosis without pathology. Perhaps the foundational measure in medicine was thermometry. The discovery that our subjective sense of being "chilled" accompanied the objective evidence of body temperature rising and our subjective sense of "hot" matched a fall in body temperature eliminated forever the idea that we could manage in medicine without objective measurement. We need objective data to understand and interpret subjective experience.

Unfortunately, the field of mental health has failed to benefit from the kinds of measurement that revolutionized business and medicine. While it is true that we have a century of psychological research on objective tests of cognition, mood, and behavior; little of this science has translated into clinical practice. Over the last half-century, biologically oriented researchers have followed the medical model, exploring blood, urine, and cerebrospinal fluid in the hope of finding the equivalent of HgA1c or some circulating marker of mental illness. EEG readings have been mined like EKG tracings. Brain scans and protocols for brain imaging have become ever more sophisticated in the hopes of finding the engram or some circuit dysregulation or a causal lesion. And more recently, genomics seemed a promising path to finding a biomarker for mental illness. In oncology, the most clinically useful genetic signals have proven to be somatic or local mutations in tumors, not germ line genetic variants found in blood cells which is the basis of psychiatric genetics. Nevertheless, we continue to seek causal signals in circulating lymphocytes assuming these will reflect the complex genomics of brain.

This half-century search for biological markers for mental states has been, for patients, a roller coaster of hype followed by disappointment. Thus far, science has not delivered for patients with mental illness the kind of measurement that has transformed care for people with diabetes, hypertension, or cancer. There are many potential reasons why we have failed to discover objective markers. The most common explanation is that brain and behavior are more complicated than glucose

v

regulation or vascular tone or uncontrolled cell division. Finding the EEG signal for psychosis or the brain signature of depression will take longer. I accept this excuse, but there are three other explanations that are worth our consideration.

First, most clinicians rightly value the subjective reports of their patients as the most critical data for managing mental illness. They point out that the subjective experience of pain, anxiety, or despair is the hallmark of a mental disorder. They are not looking for quantitative, objective measures. Instead, clinicians hone skills of observation to translate their patients' reports into something more objective, usually defined by clinical terms if not a clinical numerical score. Master clinicians base their assessments not only on what they observe in the patient but on their own subjective experience, which they have learned to use as a barometer of paranoia or suicidal risk. While this approach, combining the subjective reports of the patient with the subjective experience of the clinician, might work for the provider, patients are increasingly expecting something better. Many patients realize, just as we learned from thermometry, that they cannot trust their subjective experience. Just as people with diabetes learn that every moment of lethargy is not hypoglycemia and people with hypertension learn that every headache does not mean elevated blood pressure, people with mental disorders are asking for something more objective to help them to manage their emotional states, distinguishing joy from the emergence of mania and disappointment from a relapse of depression.

A second reason for our failure to develop objective markers is that we lack a ground truth that can serve as the basis for qualifying a measurement as accurate. This is one reason why it took 200 years for thermometry to become a standard for managing an infectious disease—we had no simple proof of the value of body temperature, especially when the measure did not conform with subjective experience. Much of the clinical research on measuring biological features of mental illness has tried to validate the measure against a diagnosis. If only 40% of patients with Major Depressive Disorder had abnormal plasma cortisol levels, then measuring cortisol could have little value as a diagnostic test. The problem here is that Major Depressive Disorder does not represent ground truth. It is simply a consensus of master clinicians who voted that five of nine subjective symptoms constituted Major Depressive Disorder. And none of those symptoms, including sleep disturbance and activity level, are actually measured.

For me, the most important shortcoming in our approach to measurement is that we have put the cart before the horse: we are attempting to find biological correlates of cognition, mood, and behavior before we have better objective measures of cognition, mood, and behavior. Our measures, when we make them, are usually at a single point in time (generally during a crisis), captured in the artificial environment of the lab or clinic, and represent a burden to both the patient and the clinician. Ideally, objective measures would be captured continually, ecologically, and efficiently.

That ideal is the promise of mobile sensing, which has now become the foundation for digital phenotyping. As described in detail in this volume, wearables and smartphones are collecting nearly continuous, objective data on activity, location, and social interactions. Keyboard interactions (i.e., reaction times for typing and tapping) are being studied as content-free surrogates for specific cognitive domains, like executive function and working memory. Natural language processing tools are transforming speech and voice signals into measures of semantic coherence and sentiment. Of course, the rich content of social media posts, search queries, and voice assistant interactions can also provide a window into how someone is thinking, feeling, and behaving. Digital phenotyping uses any or all of these signals to quantify a person's mental state.

While most of the focus for digital phenotyping has been on acquiring these signals, there is a formidable data science challenge to converting the raw signals from a phone or wearable into valid, clinically useful insights. What aspects of activity or location are meaningful? How do we translate text meta-data into a social interaction score? And how to define which speech patterns indicate thought disorder or hopelessness? As you will see in the following chapters, machine learning has been employed to solve these questions, based on the unprecedented pool of data generated. But each of these questions requires not only abundant digital data, we need some ground truth for validation. Ground truth in academic research means a clinical rating, which we know is of limited clinical value. Ground truth in the real world of practice is functional outcome, which is difficult to measure.

It's useful to approach digital phenotyping or, as it is called in some of these chapters, psychoinformatics, as a work in three parts. First, we need to demonstrate the feasibility. Can the phone actually acquire the signals? Will people use the wearable? Will there be sufficient consistent data to analyze? Next, we have the validity challenge. Does the signal consistently correlate with a meaningful outcome? Can the measure find valid differences between subjects or is it only valid comparing changes across time within subjects? Can this approach give comparable results in different populations, different conditions, different devices? Finally, we face the acid test: is the digital measure useful? Utility requires not only that the signals are valid but that they inform diagnosis or treatment in a way that yields better outcomes. Patients will only use digital phenotyping if it solves a problem, perhaps a digital smoke alarm that can prevent a crisis. Providers will only use digital phenotyping if it fits seamlessly into their crowded workflow. As a chief medical officer at a major provider company said to me, "We don't need more data; we need more time".

Mastering feasibility, validity, and utility will also require engaging and maintaining public trust. Trust is more than ethics, but certainly the ethical use of data, consistent protection of privacy, and full informed consent about the phenotyping process are fundamental. Trust also involves providing agency to users, so that they are collecting their data for their use. There may be technical assets that can help. For instance, processing voice and speech signals internally on the phone might prove useful for protecting content privacy. The use of keyboard interaction signals, which consist of reaction times and contain no content, might be more trustworthy for some users. But it is unlikely tech solutions will be sufficient to overcome the appearance of and very real risk of surveillance. It is important, therefore, as you read the following chapters that you distinguish between the use of this new technology in a medical setting where consenting patients and families can be empowered with information versus the use of this technology in a population where monitoring for behavioral or cognitive change can be a first step down a slippery slope toward surveillance.

If we can earn public trust, there is every reason to be excited about this new field. Suddenly, studying human behavior at scale, over months and years, is feasible. Recent research is proving out the validity of this approach, already in thousands of subjects for some measures. We have yet to see clinical utility but there is every reason to expect that in the near future, digital technology will create objective, effective measures. Finally, in mental health, we may be able to measure well and manage better. Patients are waiting.

Thomas R. Insel, MD
Mindstrong Health, Mountain View, CA, USA
e-mail: tom@mindstronghealth.com

Contents

Digital Phenotyping and Mobile Sensing In Psychoinformatics—A Rapidly Evolving Interdisciplinary Research Endeavor

Harald Baumeister and Christian Montag

Many scientists are currently considering whether we are seeing a paradigm shift in the psychosocial and behavioral health sciences from narrow experimental studies to ecological research driven by big data. At the forefront of this trend is the implementation of smart device technologies in diverse research endeavors. This enables scientists to study humans in everyday life on a longitudinal level with unprecedented access to many relevant psychological, medical, and behavioral variables including communication behavior and psychophysiological data. Although the smartphone without doubt presents the most obvious "game changer" (Miller 2012), it only represents a small part of a larger development toward the Internet of Things, where everything from household machines to the car will be connected to the Internet (Montag and Diefenbach 2018). Therefore, human interaction with all these Internet-connected devices will leave digital traces to be studied by scientists in order to predict bio-psycho-social variables ranging from personality to clinical variables including states of physical and mental health (Markowetz et al. 2014; Montag and Elhai 2019).

The present volume gives an overview on current developments in this area, looking at digital phenotyping and mobile sensing as two prominent approaches in *Psychoinformatics*, i.e., the research field that combines innovative technological attempts with the psychosocial and behavioral health science traditions (Montag et al. 2016). Digital phenotyping extends the construct of phenotypes as the observable (biological) traits of organism to digital traces of people in a digital era of mankind (Jain et al. 2015; Insel 2017). Given almost omnipresent human–machine interactions, people's digital traces might allow for diagnostic, prognostic, and intervention activities in different areas of life such as predicting product needs (e.g., by GPS tracking or bio-sensing approaches; even microtargeting (see Matz et al. 2017)), estimating personality traits, attitudes and preferences (e.g., predicting people's political orientation by social–network interaction (Kosinski et al. 2013)) or improving patients health care (e.g., estimating disease and treatment trajectories based on ecological momentary assessment data). Mobile sensing is the most prominent driver of this new approach, given the already substantial penetration of smartphones around the world (at the time of writing 3.3 billion smartphone users have been estimated (Statista.com 2019)).

Systematizing this dynamic and fast developing field of research is challenging, as technological development in diverse and unconnected research areas might already have stimulated the next wave of innovations prior to this book being published. However, while the specific approaches might vary, a first framework for using digital phenotyping and mobile sensing in psychosocial and behavioral health sciences can be proposed, as is depicted in Fig. 1.

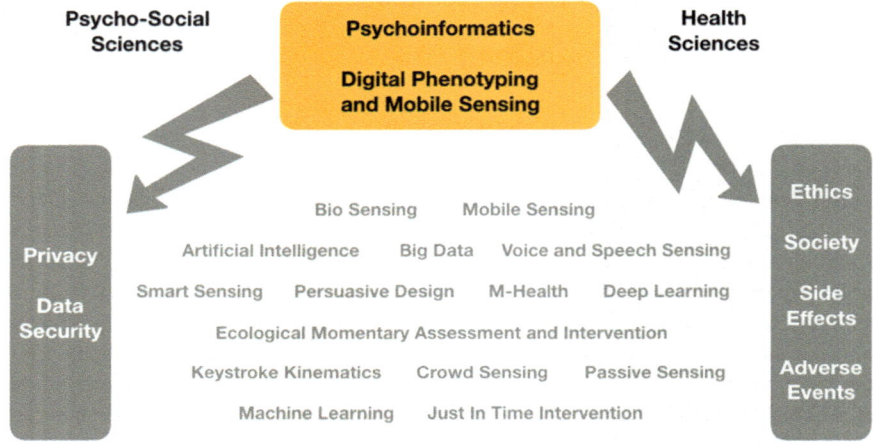

Fig. 1 Digital phenotyping and mobile sensing conceptual framework

A multitude of buzzwords such as machine and deep learning, big data, crowdsensing, bio-sensing, EMA, and EMI (ecological momentary assessment/intervention) are frequently used to express the impact of digitalization on people's lives and on societies as a whole. Note that we use the term *digitalization* here, because it describes how the use of digital technologies shapes society (or here scientific research), whereas digitization refers to the mere process of transforming analog into digital data.

Mobile sensing is often specified by the device providing the mobile sensing data, i.e., smart sensing devices such as smartphone, smartwatch, and smart-wearables, and by the data a smart sensing device is tracking, e.g., voice and speech sensing, bio-sensing, passive sensing, crowdsensing, and ecological momentary assessment or facial emotion recognition sensing. Using these terms already stimulates our imaginations regarding both risks and potentials associated with this still new technology, capable of altering peoples´ and societal life to a degree that has not yet been fully grasped.

Digital phenotyping constitutes one relevant application in the area of mobile sensing, with the potential to substantially improve our knowledge in the realm of latent constructs such as personality traits and mental disorders. A better understanding of these variables is of great relevance, because personality traits such as conscientiousness are good predictor for a healthy living style (Bogg and Roberts 2004) and mental disorders are a tremendous source of individual suffering and high costs for society (Trautmann et al. 2016). At the same time, digitally supported health care offers seemingly are capable of improving mental and behavioral health (Ebert et al. 2017; Bendig et al. 2018; Ebert et al. 2018; Paganini et al. 2018).

Our life has become digital and this digital image of our lives and persons can be ephemeral or used to provide the data basis necessary to estimate people's traits, states, attitudes, cognitions, and emotions (Montag et al. 2016; Martinez-Martin et al. 2018; Lydon-Staley et al. 2019). What does your smartphone usage pattern tell us about you and your state of mind? What does your vacation, social, and work life pictures posted in social networks tell us about your happiness and your attitude toward work, holiday time, and your spouse? Would we recognize a change in mental state when comparing voice recordings from today and 5 years ago? What will your lunch look like tomorrow? You might not know it, but maybe your bio-sensing signals will tell us now already. Two advances enable us to provide increasingly sophisticated digital phenotyping estimates: Big Data and machine/deep learning approaches.

Big Data constitutes a precondition for digital phenotyping based on a granular matrix of our digital traces, consisting of a multitude of (longitudinally) assessed variables in large cohorts coming at different degrees of velocity (speed), variety (data format), and volume. In short, Big Data can be described by varying degrees of VVVs (Markowetz et al. 2014). Once those databases have been established, we are confronted with a large set of complex data for which established statistical methods are often not the best fit.

Machine learning and deep learning (ML/DL) are the buzzwords that promise to make sense out of the big data chaos (Lane and Georgiev 2015). While some rightly argue for a more well-thought through scientific basis in the current machine learning hype (Kriston 2019), these analytical approaches are undoubtedly the key for the last puzzle of digital phenotyping and mobile sensing covered in the present volume: artificial intelligence (AI). This said, ML as an integral part of AI comes with many problems such as a lack of understanding of what kind of patterns the computer actually recognizes or learns when predicting a variable such as a tumor from an MRI scan (Mohsen et al. 2018). Aside from this, a demanding topic is that of programming ethics into deep learning algorithms (the field of machine ethics, see in, e.g., Moor and James 2006; Brundage 2015).

Artificial intelligence (AI) might become central to several fields of application, by pattern recognition using deep learning algorithms (Ghahramani 2015; Topol 2019). For instance, in a first development step, users of AI-based medical programs will be supported in interpreting diagnostic results and receiving AI-based prognostic feedback regarding the current treatment course of their patients. Predicting economic or environmental impacts of political decisions and tailoring product placement according to peoples' personality based on AI algorithms are further likely fields of application. Once developed, these artificial intelligence applications might again use mobile sensing techniques to further improve their prognostic power by means of a deep-learning-based self-improvement cycle (see Fig. 2).

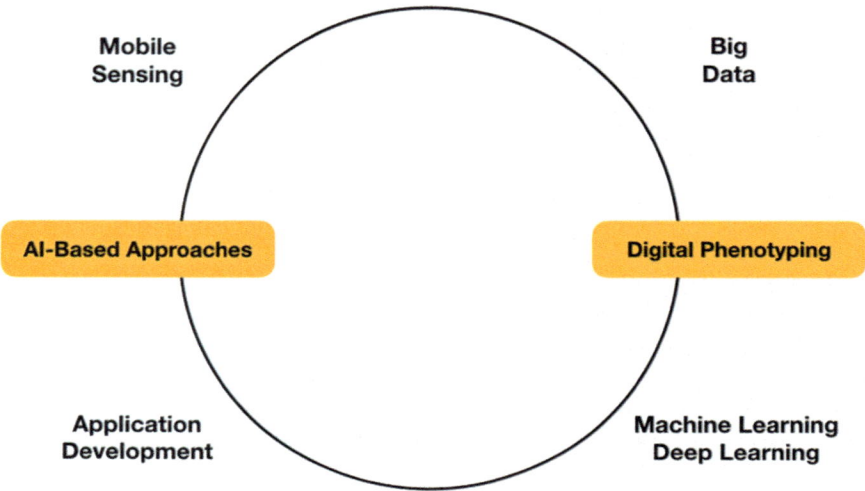

Mobile Sensing   **Big Data**

AI-Based Approaches   **Digital Phenotyping**

Application Development   **Machine Learning Deep Learning**

Fig. 2 Mobile sensing, digital phenotyping, and artificial intelligence life cycle

The chapters of this book provide a snapshot of what is already possible and what science might allow in the near future. Thus, most chapters not only focus on the areas of applications and the potentials that come along with these approaches but also on the risks that need to be taken into account, principally in terms of data privacy and data security issues as well as ethical and societal considerations and possible side effects of mobile sensing approaches. Regarding the risks, chapters by Kargl and colleagues (Chap. 1), who provide an overview of privacy issues as inherent aspect of mobile sensing approaches, and by Dagum and Montag (Chap. 2), who reflect on ethical implications of digital phenotyping, consider the ethical boundaries of our actions. Examples of unintended de-anonymization of data summarized by Kargl et al. (Chap. 1) shows how easily supposedly anonymized data can quickly become person identification data when combined with the almost infinite information on everything and everyone in our Internet of Things world. Answers on these ethical questions need a scientific discourse on how to exploit the potential of mobile sensing in an ethical way but also a societal discourse on what we are willing to accept in light of the conveniences mobile sensing approaches provide (e.g., accepting that Google knows where we are as a trade-off for using Google maps to navigate through traffic or find our ways in unknown places).

With these privacy and ethics boundaries in mind, readers of this book are provided with a look into the future that is already happening.

Several chapters provide exemplary research and conceptualization frameworks on mobile sensing approaches across the psychosocial and behavioral health sciences fields. Digital phenotyping, mental health prediction models, ecological momentary assessments, and academic performance estimates (Cao, Gao, and Zhou: Chap. 8; Kubiak and Smyth: Chap. 12; Marengo and Settanni: Chap. 7; Rozgonjuk, Elhai and Hall: Chap. 11; Sariyska and Montag: Chap. 4; Schlee et al.: Chap. 13; Vaid and Harari: Chap. 5) are only some of the possibilities in the realm of mobile sensing discussed in this book. The chapters range from established fields of mobile sensing that can already draw on substantial evidence (Saeb et al. 2015; Sariyska et al. 2018; Montag et al. 2019), such as predicting personality or mental and behavioral health status by means of smartphone usage patterns to less established fields such as the potential of bio-sensing, which might allow in future such things as physiologically delineated measures to improve health (e.g., cortisol implants measuring stress-related symptoms that could inform a mobile health application to provide a just-in-time intervention).

A second group of chapters focuses on the potential for machine learning and deep learning approaches. Geiger and Wilhelm (Chap. 3), for example, illustrate the research potentials of combining mobile devices with face recognition software allowing for immediate facial emotion expression recognition based on machine learning algorithms. Similarly, Hussain and colleagues (Chap. 10) reflect on the potential of machine-learning-based keyboard usage and corresponding typing kinematics and speech dynamics analysis for predicting mental health states. Regarding speech-analysis-based machine learning approaches, Cummins and Schuller (Chap. 9) present a selection of already available open-source speech analysis toolkits along with a discussion on their potentials and limitations.

Finally, four chapters provide frameworks and insights into how these fields of research can be used and combined to inform complex intervention developments in order to further improve people's health and living. While Messner and colleagues (Chap. 15) report on the current state of research regarding mHealth Apps and the potential that comes with new sensing and AI-based approaches, Pryss and colleagues (Chap. 16) present a framework for chatbots in the medical field. While most of the current chatbot approaches are based on a finite amount of answer options the chatbot can access (Bendig et al. 2019), the framework presented in this chapter looks at how the expert knowledge database necessary for complex communication situations such as psychotherapy can be generated and iteratively improved until a truly artificial intelligent chatbot therapist is at place. Baumeister et al. (Chap. 17), with a focus on persuasive-design-based intervention development, and Rabbi et al. (Chap. 18), with a model for just-in-time interventions, provide further details on how technology can be used to further improve existing interventions, enhance intervention uptake and adherence, and ultimately increase effectiveness by exploiting the full potential of mobile sensing.

Writing these few paragraphs on the content of our book fills us with excitement about the perspectives digital phenotyping and mobile sensing offer for research and practice. At the same time, however, we feel uneasy in light of the obvious risks for individuals and society at large. Researchers should not usually argue based on their emotions, but in this case these two emotions—positive and negative—might guide the next steps by a development process for future innovations that is ethical and informed on issues of privacy. It therefore seems that the development of new technical solutions will take place anyway given their potential economic value, leaving research to establish conceptual frameworks and guidelines to set the guardrail. Focusing on the example of artificial intelligence, large-scale companies will probably revolutionize the product market with ever more intelligent systems, increasing the convenience of consumers and at the same time reducing human workforce needs. However, these companies will probably not provide the urgently needed answers on how to develop and implement such innovations in a way that benefits society (Russell et al. 2015) considering all the scenarios relating to potentially malevolent AI (Pistono and Yampolskiy 2016). Broadening the focus again, we need to establish good scientific practice for mobile sensing in order to exploit its full potential. First, scientists currently discuss whether the paradigm shift postulated at the beginning of this editorial needs to undergo fine-tuning, setting the ecological correlation research approach into the context of explorative and explanatory research paradigms. Exploratively fishing for hypotheses is the beginning and not the end of methodologically sound psychosocial and behavioral health science (Kriston 2019).

This said, at the end of this short introduction we want to express our gratitude to all our authors for their important chapters in this book. They all invested a lot of their time and energy to provide insights into their different research perspectives.

We have not mentioned so far that Thomas Insel, former director of the National Institute of Mental Health (NIMH) in the USA and a prominent advocate of the digital phenotyping movement, was kind enough as to provide us with his thoughts on this relevant research area. His efforts are also much appreciated.

References

Bendig E, Bauereiß N, Ebert DD, Snoek F, Andersson G, Baumeister H (2018) Internet- and mobile based psychological interventions in people with chronic medical conditions. Dtsch Aerzteblatt Int 115:659–665

Bendig E, Erb B, Schulze-Thuesing L, Baumeister H (2019) Next generation: chatbots in clinical psychology and psychotherapy to foster mental health—a scoping review. Verhaltenstherapie 1–15. 10.1159/000499492

Bogg T, Roberts BW (2004) Conscientiousness and health-related behaviors: a meta-analysis of the leading behavioral contributors to mortality. Psychol Bull 130:887–919. 10.1037/0033-2909.130.6.887

Brundage M (2015) Limitations and risks of machine ethics. In: Risks of artificial intelligence. Chapman and Hall/CRC, pp 141–160

Ebert DD, Cuijpers P, Muñoz RF, Baumeister H (2017) Prevention of mental health disorders using internet-and mobile-based interventions: a narrative review and recommendations for future research. Front Psychiatry 8. 10.3389/fpsyt.2017.00116

Ebert DD, Van Daele T, Nordgreen T, Karekla M, Compare A, Zarbo C, Brugnera A, Øverland S, Trebbi G, Jensen KL, Kaehlke F, Baumeister H (2018) Internet-and mobile-based psychological interventions: applications, efficacy, and potential for improving mental health: a report of the EFPA E-Health taskforce. Eur Psychol 23:167–187. 10.1027/1016-9040/a000318

Ghahramani Z (2015) Probabilistic machine learning and artificial intelligence. Nature 521:452–459. 10.1038/nature14541

Insel TR (2017) Digital phenotyping. JAMA 318:1215. 10.1001/jama.2017.11295

Jain SH, Powers BW, Hawkins JB, Brownstein JS (2015) The digital phenotype. Nat Biotechnol 33:462–463. 10.1038/nbt.3223

Kosinski M, Stillwell D, Graepel T (2013) Private traits and attributes are predictable from digital records of human behavior. Proc Nat Acad Sci 110:5802–5805. 10.1073/pnas.1218772110

Kriston L (2019) Machine learning's feet of clay. J Eval Clin Pract jep.13191. 10.1111/jep.13191

Lane ND, Georgiev P (2015) Can deep learning revolutionize mobile sensing? In: Proceedings of the 16th international workshop on mobile computing systems and applications—HotMobile'15. ACM Press, New York, New York, USA, pp 117–122

Lydon-Staley DM, Barnett I, Satterthwaite TD, Bassett DS (2019) Digital phenotyping for psychiatry: accommodating data and theory with network science methodologies. Curr Opin Biomed Eng 9:8–13. 10.1016/J.COBME.2018.12.003

Markowetz A, Błaszkiewicz K, Montag C, Switala C, Schlaepfer TE (2014) Psycho-informatics: big data shaping modern psychometrics. Med Hypotheses 82:405–411. 10.1016/j.mehy.2013.11.030

Martinez-Martin N, Insel TR, Dagum P, Greely HT, Cho MK (2018) Data mining for health: staking out the ethical territory of digital phenotyping. npj Digit Med 1:68. 10.1038/s41746-018-0075-8

Matz SC, Kosinski M, Nave G, Stillwell DJ (2017) Psychological targeting as an effective approach to digital mass persuasion. Proc Natl Acad Sci U S A 114:12714–12719. 10.1073/pnas.1710966114

Miller G (2012) The smartphone psychology manifesto. Perspect Psychol Sci 7:221–237. 10.1177/1745691612441215

Mohsen H, El-Dahshan E-SA, El-Horbaty E-SM, Salem A-BM (2018) Classification using deep learning neural networks for brain tumors. Futur Comput Informatics J 3:68–71. 10.1016/J.FCIJ.2017.12.001

Montag C, Baumeister H, Kannen C, Sariyska R, Meßner E-M, Brand M, Montag C, Baumeister H, Kannen C, Sariyska R, Meßner E-M, Brand M (2019) Concept, possibilities and pilot-testing of a new smartphone application for the social and life sciences to study human behavior including validation data from personality psychology. J 2:102–115. 10.3390/j2020008

Montag C, Diefenbach S (2018) Towards homo digitalis: important research issues for psychology and the neurosciences at the dawn of the Internet of Things and the digital society. Sustainability 10:415. 10.3390/su10020415

Montag C, Duke É, Markowetz A (2016) Toward psychoinformatics: computer science meets psychology. Comput Math Methods Med 2016:1–10. 10.1155/2016/2983685

Montag C, Elhai JD (2019) A new agenda for personality psychology in the digital age?. Personality Individ Differ 147:128–134. 10.1016/j.paid.2019.03.045

Moor JH, James H (2006) The nature, importance, and difficulty of machine ethics. IEEE Intell Syst 21:18–21. 10.1109/MIS.2006.80

Paganini S, Teigelkötter W, Buntrock C, Baumeister H (2018) Economic evaluations of internet-and mobile-based interventions for the treatment and prevention of depression: a systematic review. J Affect Disord 225:733–755

Pistono F, Yampolskiy RV (2016) Unethical research: how to create a malevolent artificial intelligence. In: Proceedings of ethics for artificial intelligence workshop (AI-Ethics-2016). New York

Russell S, Dewey D, Tegmark M (2015) Research priorities for robust and beneficial artificial intelligence. AI Mag 36:105. 10.1609/aimag.v36i4.2577

Saeb S, Zhang M, Karr CJ, Schueller SM, Corden ME, Kording KP, Mohr DC (2015) Mobile phone sensor correlates of depressive symptom severity in daily-life behavior: an exploratory study. J Med Internet Res 17:e175. 10.2196/jmir.4273

Sariyska R, Rathner E-M, Baumeister H, Montag C (2018) Feasibility of linking molecular genetic markers to real-world social network size tracked on smartphones. Front Neurosci 12:945. 10. 3389/fnins.2018.00945

Statista.com (2019) Number of smartphone users worldwide from 2016 to 2021 (in billions). https://www.statista.com/statistics/330695/number-of-smartphone-users-worldwide. Accessed 16 Oct 2019

Topol EJ (2019) High-performance medicine: the convergence of human and artificial intelligence. Nat Med 25:44–56. 10.1038/s41591-018-0300-7

Trautmann S, Rehm J, Wittchen H (2016) The economic costs of mental disorders. EMBO Rep 17:1245–1249. 10.15252/embr.201642951

Part I
Digital Phenotyping and Mobile Sensing: Privacy and Ethics

Chapter 1
Privacy in Mobile Sensing

Frank Kargl, Rens W. van der Heijden, Benjamin Erb and Christoph Bösch

Abstract In this chapter, we discuss the privacy implications of mobile sensing and modern psycho-social sciences. We aim to raise awareness of the multifaceted nature of privacy, describing the legal, technical and applied aspects in some detail. Not only since the European GDPR, these aspects lead to a broad spectrum of challenges of which data processors cannot be absolved by a simple consent form from their users. Instead appropriate technical and organizational measures should be put in place through a proper privacy engineering process. Throughout the chapter, we illustrate the importance of privacy protection through a set of examples and also technical approaches to address these challenges. We conclude this chapter with an outlook on privacy in mobile sensing, digital phenotyping and, psychoinformatics.

1.1 Introduction

While mobile sensing provides substantial benefits to researchers and practitioners in many fields including psychology, the data collected in the process is often sensitive. The data collected by smartphones and other devices with sensors, such as fitness trackers, is clearly related or relatable to persons. Therefore, the researcher or practitioner that collects, processes, and stores such data has moral and legal obligations to handle this data responsibly. This is especially important if the data is related to health or mental disorders of a person.

F. Kargl (✉) · R. W. van der Heijden · B. Erb · C. Bösch
Institute of Distributed Systems Ulm University, Albert-Einstein-Allee 11,
89081 Ulm, Germany
e-mail: frank.kargl@uni-ulm.de

R. W. van der Heijden
e-mail: rens.vanderheijden@uni-ulm.de

B. Erb
e-mail: benjamin.erb@uni-ulm.de

C. Bösch
e-mail: christoph.boesch@uni-ulm.de

© Springer Nature Switzerland AG 2019
H. Baumeister and C. Montag (eds.), *Digital Phenotyping and Mobile Sensing*,
Studies in Neuroscience, Psychology and Behavioral Economics,
https://doi.org/10.1007/978-3-030-31620-4_1

The right to protection of personal data has been recognized as a central human right and is, for example, embedded in the European Charter on Human Rights.[1] In this chapter we want to raise awareness of the importance of privacy and data protection. To this end, we introduce privacy from a legal, technical, and applied perspective, as well as discussing some of the associated challenges. In particular, we would like to dispel the myth that a consent form from a study participant relieves the researcher from legal obligations. Beyond legal obligations, we discuss some evidence that a lack of privacy may negatively affect the participants' or patients' trust in systems or procedures. Finally, we will provide a positive outlook on how both the legal and ethical obligations could be achieved by proper privacy engineering and the application of privacy-enhancing technologies (PETs).

Privacy protection has already been recognized as an important issue within the psycho-social research community, after controversial incidents such as the Tearoom Trade study by Humphreys. The name of the study refers to male-male sexual behavior in public bathrooms. In his work, Humphreys not only surveyed unwitting subjects in extremely private and intimate situations (sexual intercourse in public bathrooms) without their consent, he also collected personally identifiable data (license plates) to later de-anonymize the subjects and visit their homes under false pretenses for follow-up interviews. This study demonstrated the fatal effects when personal data is collected in studies without regard to privacy (Kelman 1977).

One of the first to investigate the anonymization of health data was Latanya Sweeney, who showed that anonymous hospital discharge records contained sufficient information to de-anonymize 87% of the US population by matching the zip code, gender, and date of birth information of the records to US census data (Sweeney 2002). Many more examples have led to the conclusion that proper anonymization becomes extremely hard if the opponent has sufficient context knowledge.

De-anonymization is also an issue for location privacy and other data collected by mobile devices such as smartphones and fitness trackers. For example, the company Strava released data from its fitness tracking service, where many people uploaded GPS traces from their daily runs. People analyzing this massive dataset quickly found out that it contained runs from soldiers from supposedly undisclosed military bases in Afghanistan and elsewhere.[2] Similarly, anonymous trip records published from New York taxis have been used to identify trips from celebrities and find out whether they tipped the driver or not.[3] The extent by which seemingly innocent data gives away our most intimate information is best illustrated by an example reported in the New York Times where shopping data from customers was analyzed to learn that a high-school girl was pregnant even before her own family knew.[4] Such examples have led to the conclusion that good anonymization is really hard and the category

---

[1] http://www.europarl.europa.eu/charter/pdf/text_en.pdf.

[2] https://www.theguardian.com/world/2018/jan/28/fitness-tracking-app-gives-away-location-of-secret-us-army-bases.

[3] https://research.neustar.biz/2014/09/15/riding-with-the-stars-passenger-privacy-in-the-nyc-taxicab-dataset/.

[4] https://www.nytimes.com/2012/02/19/magazine/shopping-habits.html.

of personal identifiable information needs to be broadened up substantially. This is also reflected in the modern understanding of privacy and current lawmaking, such as the European General Data Protection Regulation (GDPR).

1.2 Privacy as a Multifaceted Concept

1.2.1 Privacy as a Legal Concept

Many countries have regulated different aspects of privacy in their laws. Legal protection of personal data is termed data protection. One of the most holistic data protection frameworks is the European General Data Protection Regulation (GDPR), which went into force throughout Europe in May 2018, unifying the different data protection regimes throughout Europe.

While it is beyond the scope of this chapter to provide a complete overview of the GDPR, we still use it here to illustrate major concepts of data protection. GDPR, or "Regulation (EU) 2016/6791" as it is officially named, regulates the *processing* of *personal data* relating to a natural person in the EU, by an individual, a company or an organization. As described in Article 4, *personal data* includes data that is indirectly related to a person, while *processing* has to be understood to include activities such as collection or storage of personal data.

The GDPR places many requirements on anyone that either conducts data processing (termed "data processor") or is responsible for data processing (termed "data controller"). In addition, "data subjects", which are the natural person(s) to whom the data relates, are given a broad set of rights, such as the right to be informed or the right to erasure of the processed data. It is important to note that any processing of personal data by default requires *informed consent* of the data subject—exceptions being, e.g., if that data is required for legal reasons or to fulfill a contract with that person. In psychological studies, such informed consent is typically the basis for processing of personal data. However, such consent does not free the data controller or processor from the broad set of obligations that come with the right to process personal data (Schaar 2017). These obligations include the information rights of data subjects, such as the right to be informed, the right to access such data in a portable format, and the right to object to data processing, even after the fact. This obviously clashes with some obligations of researchers on research data management.

In this chapter, we want to focus on yet another aspect of the GDPR: Privacy by Design (PbD) and Privacy by Default. Especially when processing sensitive data or in high-risk cases, the GDPR requires any system that processes personal data to be designed as privacy-friendly from the ground up. This is done by following a PbD design and development process based on a Data Protection Impact Assessment (DPIA) that investigates potential privacy issues right from the start. The GDPR further requires that state-of-the-art technical and organizational measures (TOMs) be integrated into the foundations of every system. However, it is currently still open how courts will decide on the definition of "state-of-the-art".

This will have substantial implications on how psychological research can be conducted. In the remainder of this chapter, we will first illustrate these consequences and then address how such privacy and data protection may be achieved. Finally, we will provide an outlook on how modern privacy engineering might also be applied to conduct research in psycho-social research in a compliant and responsible way.

1.2.2 Privacy as a Technical Concept

In the technical literature, privacy is often defined in a quantifiable way, which greatly simplifies the analysis of technical solutions employed to protect privacy. The simplest conception of privacy in the technical literature is anonymity, which means that "a subject is not identifiable within a set of subjects, referred to as the anonymity set" (Pfitzmann and Hansen 2010). The anonymity set of a subject is the set of subjects that have the same properties (e.g., age category, gender, disease characteristics) so that one cannot distinguish one particular person from the others in the group. If the anonymity set is a database published as research data, removing names or pseudonyms from the subjects may not be enough to achieve anonymity, since each subject may have other unique properties. Another related but distinct concept is that of linkability, which is defined between "two or more items of interests (e.g., subjects, messages, actions, …) […], the attacker can sufficiently distinguish whether these are related or not". This is a stronger requirement than that of anonymity, since knowing whether two messages originate from the same source does not (necessarily) identify the source. However, in many cases, linkability of messages implies the possibility of de-anonymization. For example, in the location privacy example of New York taxis above, linkability of locations led to de-anonymization of individuals, as explained (Douriez et al. 2016).

To design and validate practical privacy enhancing technologies (PETs), many technical articles also quantify the privacy associated with a specific system. In particular, many different metrics exist (Wagner and Eckhoff 2018) to quantify exactly how anonymous a subject is within their anonymity set. The fundamental challenge is that many de-anonymization attacks typically work using external information sources. This led to the development of many classes of metrics, the most well-known of which are data similarity (where k-anonymity is a widely-known example) and indistinguishability (where differential privacy is a prominent example). We refer interested readers to Wagner and Eckhoff (2018) for a detailed survey of these metrics. But how does this all apply to psychological research?

1.2.3 Privacy as a Concept in Research Studies and Treatments

In the context of research studies, experiments, and treatments related to healthcare and psychological well-being, privacy is particularly relevant from two perspectives. First, the protection of data from participants or patients raises strong obligations for researchers or therapists, as the ethical principles in these professions go well beyond the legal requirements. Second, the participants' or patients' perception of their privacy influences their trust in the procedure, which can potentially even negatively affect the results or the outcome of the treatment.

Organizations such as the American Psychological Association (APA) take into account privacy obligations as part of their ethical principles. For instance, section four of the APA Ethics Code specifically addresses privacy and confidentiality, requiring that the confidentiality of collected information is maintained, as well as the minimization of privacy intrusions. Best practices have been established to adhere to these principles in traditional settings (e.g., usage of pseudonymization codes). However, novel approaches such as mobile sensing and smartphone-based data collection, entail new threats to privacy and confidentiality, which cannot be addressed with existing practices alone. In fact, an increasing technologization of experiments and treatments requires equal advancements in the safeguarding and (technical) implementation of ethical principles.

Orthogonal to the ethical considerations, addressing the privacy concerns of participants and patients has a positive impact on the procedure outcome. According to a model of Serenko and Fan (2013), informational privacy (i.e., information acquisition and ownership) has the strongest influence on a patient's privacy perception in the healthcare context. The level of perceived privacy is associated with the level of trust of the patient in the treatment. Again, this trust level is associated with the behavioral intentions of the patient such as commitment, adherence, and compliance with the treatment. Furthermore, anonymity and confidentiality in studies work against the social desirability bias (Krumpal 2013)—a tendency to give answers that are considered to be favorably by other peers. Joinson (1999) found similar effects in early, web-based questionnaires.

On the other hand, recent developments in psycho-social research have shown many results were not reproducible by further studies. This has given rise to the open science movement, whose primary goal is to improve reproducibility through data availability. A wide variety of data management platforms, such as the open science framework (https://osf.io) and Zenodo (https://zenodo.org), have risen in this context, whose primary aim is to widely disseminate research data. However, sensitive personal data clearly cannot be published in this fashion under the GDPR regime; technical solutions are required to ensure that the data is used only in correspondence with the consent provided by the user, under the obligations posed by the GDPR.

As we have shown, privacy is a complex and challenging concept to data processing in psycho-social research from many perspectives. We now continue to refine these challenges in more detail before discussing possible technical solutions.

1.3   Challenges in Privacy

While mobile sensing enables new forms of participatory research and discovery by collecting almost endless amounts of sensor data, such systems create distributed and massive databases of individuals' data and maybe even of unrelated surrounding people. This collection and processing of large amounts of sensor data leads to a unique source of information about, e.g., environmental conditions and changes, user activity patterns and health, and behavioral habits. By collecting this various data about the human body as well as its direct and extended environment, it is possible to create precise personal profiles including massive amounts of sensitive information. This data can often even be used to make behavioral predictions about an individual. Thus, protecting this data from unauthorized access during the whole data lifecycle, i.e., during its collection, transfer, processing, storage, potential release or final deletion, is a major challenge.

Even anonymization of data often mentioned in this context cannot convincingly eliminate this risk. Often one can infer identities from anonymous data when linking it to other (public) data. An example of such indirect leakage is data regarding location information. Tracking a person's location inevitably implies collecting and processing data about the whereabouts—and thus the behavior—of persons. This might infer sensitive information about users' activities, inter alia, sexual habits/orientation, drinking and social behavior, physical or mental health, religious and/or political beliefs. A famous example of the analysis of driving behavior is the 2012 Uber blog-post "Rides of Glory",[5] in which Uber showed how to spot candidates for one-night stands among its riders. While the risks from location-based attacks are fairly well understood given years of previous research, our understanding of the dangers of other modalities (e.g., activity inferences, social network data) are less developed.

For some data, the impact on an individual's privacy appears insignificant at first glance, but contains sensitive information that can be derived or inferred from the data. Most users are not aware of the amount and extent of data as well as the expressiveness and significance of the collected data. Since Sweeney's work (Sweeney 2002) on de-anonymization through seemingly anonymized, innocent data, many other examples of de-anonymization attacks have been reported. This includes those of the web search queries of over half-million America Online (AOL) clients (Barbaro and Zeller 2006) and the movie reviews of a half-million Netflix subscribers (Narayanan and Shmatikov 2008).

In AOL's case, user IDs of search queries were replaced with unique identifiers per user to allow researchers using the data to access the complete list of a person's search queries. Unfortunately, the complete lists of search queries were so thorough that individuals could be de-anonymized simply based on the contents of search queries. In the Netflix case, Narayanan and Shmatikov used a different approach and re-identified several Netflix users by correlating the available research data with publicly available movie ratings data. In addition, they were able to infer more sensitive information from the available data since "[m]ovie and rating data contains

---

[5]https://web.archive.org/web/20140828024924/http://blog.uber.com/ridesofglory.

information of a more highly personal and sensitive nature. The member's movie data exposes a Netflix member's personal interest and/or struggles with various highly personal issues, including sexuality, mental illness, recovery from alcoholism, and victimization from incest, physical abuse, domestic violence, adultery, and rape."[6]

These examples have shown that the concept of personally identifiable information (PII) is a challenging concept that is not as straight-forward as it appears. Due to the diversity and efficiency of modern de-anonymization algorithms (Narayanan and Shmatikov 2008, 2009), it is often possible to re-identify an individual, even in the absence of personally identifiable information. While some attributes in personal data may not be uniquely identifying on their own, almost all information can be personal or identifiable when combined with enough other relevant context information (Narayanan and Shmatikov 2010).

Furthermore, an intentional release or unintended leak of personal data brings new challenges, since a data set cannot be protected to preserve privacy once it is public. A major challenge is thus to determine whether the data stored or to be released is sufficiently and adequately anonymized. There is a growing number of examples and techniques for reconstruction attacks, where data that may look safe and innocuous to an individual user may allow sensitive information to be reverse-engineered. This also means that your data may have to be regularly reconsidered for their privacy sensitivity.

Starting from a background of legal and ethical obligations to safeguard privacy, we have illustrated how difficult and complex the notion of anonymization and privacy protection is. It becomes evident how privacy and data protection need to become essential elements in mobile sensing. Motivated by data protection regulation such as the GDPR, one needs to actively consider how to limit the collection, flow, use/processing, storage, release, and deletion of personal data in own research. We will next illustrate what role Privacy Enhancing Technologies (PETs) can play in this protection and how they can be embedded in a privacy engineering process to design mobile sensing systems.

1.4 Privacy Protection

We will start our discussion on how to properly protect privacy in mobile sensing with a recap of some GDPR principles as listed in Article 5. It already foresees legitimate purposes related to scientific research and statistics, for which specific rules are outlined in Article 89. There is also an exemption that states personal data may be archived for these purposes (which would normally violate the storage limitation principle). However, these processing and archiving of personal data for scientific research is subject to the condition that technical *and* organizational measures (TOMs) are taken to protect the data. As discussed earlier, Article 25 of the GDPR specifies that data protection should be done by design and by default

---

[6]http://www.wired.com/threatlevel/2009/12/netflix-privacy-lawsuit.

but it leaves open what this could mean in practice. In this section, we now discuss examples how one can practically implement this requirement.

Hoepman (2014) introduced a set of *strategies* to improve privacy during the design process. They provide overarching categories to then derive more specific design patterns that can be used to finally select appropriate technical and organizational measures to protect personal data. Hoepman's strategies are called *Minimize, Separate, Aggregate, Hide, Inform, Control, Enforce*, and *Demonstrate*. Taking *Aggregate* as an example, this strategy states that "Personal data should be processed at the highest level of aggregation and with the least possible detail in which it is (still) useful". For publication and archiving of scientific data, this strategy is often used when dealing with particularly sensitive information, such as personal profiling or health-related data. On the technical side, privacy design strategies are used to classify privacy patterns, the goal of which is to provide proven best practices for some settings and enable "off-the-shelf" usage of PETs. As discussed by Hoepman (2014) and others, not all strategies have received sufficient attention from research, and the technological maturity varies greatly between PETs. However, recent years have seen significant improvements in this regard and we highly recommend the use of these strategies and review of available privacy patterns and PETs when considering processing of personal data, especially in the context of mobile sensing or big data.

Returning to the fitness tracker example from the introduction, we now describe the practical application of the *Aggregate* strategy. For the purpose of monitoring fitness activity and overall health, a fitness tracker collects information such as the average heart rate, as well as the minimum and maximum rates during a sport session. Rather than collecting each pulse measurement and sending it directly to a server where it is stored, the aggregate strategy recommends local aggregation of this information in a way that is consistent with application goals. For example, to monitor overall health and track fitness, the average, maximum and minimum heart rates over the past minute can be computed in the device and transmitted. This information reveals a lot less about activities compared to per-second measurements. In this specific example, note how the aggregate strategy has benefits beyond privacy; removing unnecessary data from the collection process reduces the bandwidth, storage, and processing time required to analyze the data, illustrating that privacy and data processing are not always a zero-sum game. Similar aggregation can be performed by discretization of continuous data (e.g., only collecting heart rate only as low, medium, and high heart); however, the information loss associated with this process may not be suitable in every application. For this reason, the choice of PETs is inherently dependent on what the data is used for; this is one reason why the GDPR requires a specification of purpose for data processing and entails that domain experts and privacy experts need to work hand in hand in such projects.

Another relevant example is the use of analysis techniques from "big data". The purpose of these techniques is to extract useful patterns from large volumes of data. In such settings, the volume of data is too large to be analyzed on a regular computer, and thus computations are performed on the data by the database or a cluster. In medical and psychological applications, the interesting patterns are typically correlations between observable behaviors and markers on the one hand and associated conditions on the other. These can be computed by the database directly, therefore removing the need for direct access to the data, although this access is often still available. Cryptographic research has dedicated significant effort (Lindell and Pinkas 2002) to the design of *privacy-preserving data mining* techniques, whose goal is to extract such patterns reliably, while being able to provably *hide* the individual inputs and thus protect them from misuse. This privacy design strategy particularly suitable for situations where aggregation or minimization are difficult to apply. In those situations, privacy-preserving data mining techniques offer a compromise between privacy and data access: a researcher can run certain analyses on encrypted data, but not retrieve an individual's data from the dataset. For example, using suitable protocols like Secure Multi-Party Computation or Secure Function Evaluation, it is possible to calculate certain averages on data without learning any individual data items anywhere in the system. However, this often requires to carefully think about the analysis one wants to conduct before collection of data, since the modes of analysis are necessarily restricted to preserve privacy. Here, careful compromises between privacy and use of the data need to be found. Solutions that limit data access to policy-compliant operations like shown by Kargl et al. (2010) and SGX-based solutions like the one from Al-Momani et al. (2018) may allow more flexible analysis but provide different and maybe weaker privacy guarantees. Another aspect to consider is the technological maturity of these approaches is still often limited and constrained by high computation overhead that some of these solutions still imply.

1.5 Conclusion and Outlook

With this chapter we aimed to raise the awareness of privacy in mobile sensing. We outlined how legal and ethical considerations require every researcher applying mobile sensing to carefully consider the privacy implications of this data collection, and to apply a privacy by design and default process to come up with the best protection possible while still keeping the data useful for the researchers.

Current legal frameworks for privacy and data protection clearly state that collecting a consent form from study participants is not sufficient in most cases. Additional technical and organizational measures should be put in place to increase privacy protection. This typically involves collaboration with privacy experts with appropriate legal and technical background. As privacy is an inherently interdisciplinary topic, privacy researchers are always eager to find interesting, challenging use cases and data to which their technologies may be applied. The appropriate and visible application of such technologies can lead to an increase in user trust, while also reducing the impacts of specific biases, such as social desirability bias.

While not all technologies are readily available, the field is evolving rapidly, and there is an increasing interest in the research community to apply these new technologies in challenging applications, including mobile sensing, digital phenotyping, and psychoinformatics. Envisioning global Internet of Things and sensors, it is extremely important that privacy becomes a core value embedded into its foundations.

References

Al-Momani A, Kargl F, Schmidt R, Bösch C (2018) iRide: a privacy-preserving architecture for self-driving cabs service. In: 2018 IEEE vehicular networking conference (VNC), pp 1–8. https://doi.org/10.1109/VNC.2018.8628378

Barbaro M, Zeller T (2006) A face is exposed for AOL searcher no. 4417749. New York Times

Douriez M, Doraiswamy H, Freire J, Silva CT (2016) Anonymizing nyc taxi data: does it matter?. In: 2016 IEEE international conference on data science and advanced analytics (DSAA), pp 140–148. https://doi.org/10.1109/DSAA.2016.21

Hoepman JH (2014) Privacy design strategies. In: IFIP international information security conference. Springer, pp 446–459

Joinson A (1999) Social desirability, anonymity, and internet-based questionnaires. Behav Res Methods Instrum Comput 31(3):433–438

Kargl F, Schaub F, Dietzel S (2010) Mandatory enforcement of privacy policies using trusted computing principles. In: Intelligent information privacy management symposium (Privacy 2010) AAAI. Stanford University, USA

Kelman HC (1977) Privacy and research with human beings. J Soc Issues 33(3):169–195

Krumpal I (2013) Determinants of social desirability bias in sensitive surveys: a literature review. Qual Quant 47(4):2025–2047. https://doi.org/10.1007/s11135-011-9640-9

Lindell Y, Pinkas B (2002) Privacy preserving data mining. J Cryptol 15(3):177–206

Narayanan A, Shmatikov V (2008) Robust de-anonymization of large sparse datasets. In: 2008 IEEE symposium on security and privacy (S&P 2008), 18–21 May 2008. California, USA, IEEE Computer Society, Oakland, pp 111–125

Narayanan A, Shmatikov V (2009) De-anonymizing social networks. In: 30th IEEE symposium on security and privacy (S&P 2009). California, USA, IEEE Computer Society, Oakland, pp 173–187, 17–20 May 2009

Narayanan A, Shmatikov V (2010) Myths and fallacies of "personally identifiable information". Commun ACM 53(6):24–26

Pfitzmann A, Hansen M (2010) A terminology for talking about privacy by data minimization: Anonymity, unlinkability, undetectability, unobservability, pseudonymity, and identity management (v0.34). http://dud.inf.tu-dresden.de/Anon_Terminology.shtml

Schaar K (2017) Anpassung von Einwilligungserklärungen für wissenschaftliche Forschungsprojekte. Die informierte Einwilligung nach den Vorgaben der DS-GVO und Ethikrichtlinien. Zeitschrift für Datenschutz 5:213–220

Serenko N, Fan L (2013) Patients' perceptions of privacy and their outcomes in healthcare. Int J Behav Healthc Res 4(2):101–122

Sweeney L (2002) k-anonymity: a model for protecting privacy. Int J Uncertain Fuzziness Knowl-Based Syst 10(5):557–570. https://doi.org/10.1142/S0218488502001648

Wagner I, Eckhoff D (2018) Technical privacy metrics: a systematic survey. ACM Comput Surv 51(3):57:1–57:38. https://doi.org/10.1145/3168389

Chapter 2
Ethical Considerations of Digital Phenotyping from the Perspective of a Healthcare Practitioner

Paul Dagum and Christian Montag

Abstract In this chapter we introduce digital phenotyping and its applications to healthcare. Despite the promise of this new form of clinical diagnosis in medicine and psychiatry, use of digital phenotyping raises several ethical concerns. We use insights derived from a clinical case study to frame these different ethical questions. We discuss how current healthcare practice and privacy policies address these questions and impose requirements for non-healthcare scientists and practitioners using digital phenotyping. We emphasize that this chapter frames the discussion from the perspective of the healthcare practitioner. We conclude by briefly reviewing more strongly theoretically based discussions of this emerging topic.

2.1 Digital Phenotyping

The term *digital phenotyping* was first introduced in 2015 by Jain et al. (2015) as a modern-day adaptation of evolutionary biologist Richard Dawkins' 1982 concept "extended phenotype". Dawkins argued that phenotypes should not be limited to observable characteristics of biological processes but should be extended to include all observable characteristics including behavior. A digital phenotype was conceptualized as an extended phenotype that used online activity and behavior data that could be linked to clinical data to improve diagnosis and prognosis. While the original concept of digital phenotyping was largely behavioral motivated, prior research on clinical measurements from wearable sensors (Elenko et al. 2015) to cognitive function and mood (Dagum 2018; Kerchner et al. 2015) embraced this nomenclature. A more recent discussion of digital phenotyping by Torous et al. (2017) extends the concept to include digital self-reports and physiological measures from sensors. Insel

P. Dagum
Mindstrong Health, 303 Bryant St., Mountain View, CA, USA

C. Montag (✉)
Institute of Psychology and Education, Ulm University, Ulm, Germany
e-mail: mail@christianmontag.de

© Springer Nature Switzerland AG 2019
H. Baumeister and C. Montag (eds.), *Digital Phenotyping and Mobile Sensing*,
Studies in Neuroscience, Psychology and Behavioral Economics,
https://doi.org/10.1007/978-3-030-31620-4_2

(2017) positions digital phenotyping in a clinical context of measurement-based care enabled by digital measures of behavior, cognition and mood.

Digital phenotyping broadly encompasses two distinct areas: *behavioral phenotyping* and *digital biomarkers*. Behavioral phenotyping more closely resembles the original definition of digital phenotyping. Behavioral phenotyping is the use of digital information such as location tracking, motion sensors and email, text or call activity to infer behavior-expressed symptoms (Chittaranjan et al. 2013; Montag et al. 2014, 2019; Stachl et al. 2017). For example, symptoms of anhedonia may manifest as a decrease in locations visited or email and text activity (Markowetz et al. 2014; Saeb et al. 2015). Behavioral phenotyping also encompasses the use of social media "likes", post content or search terms to infer changes in behavior and mood (Kosinski et al. 2015).

Digital biomarkers purport to measure trait and state changes in neuropathology that can be indicative of disease risk, disease onset, disease progression or recovery. The Biomarkers Definitions Working Group defines a biomarker as "a characteristic that is objectively measured and evaluated as an indicator of normal biologic processes, pathogenic processes, or pharmacologic responses to a therapeutic intervention" (Biomarkers Definitions Working Group 2001). HbA1c is an example of a serum-measurable biomarker used to manage diabetes. Another example of a biomarker is the interval between the onset of the Q wave and the conclusion of the T wave in the heart's electrical cycle, known as the QT interval. There is an association with QT interval prolongation and fatal cardiac arrhythmias, making QT a widely used safety biomarker in drug development.

Digital biomarkers are examples of biomarkers extracted from the data created by the use of digital devices such as smartphones and wearable devices. Despite considerable recent interest in this field there are no examples today of a digital biomarker that would fully satisfy the Biomarkers Definitions Working Group criteria. Possibly the best example is the early work that validated digital biomarkers of cognitive function against gold-standard psychometric instruments (Dagum 2018). While these early results are promising, further work is needed in this exciting area.

2.2 Digital Phenotyping in Healthcare

In healthcare delivery, digital biomarkers are emerging as a clinical measure used by care providers for clinical decision making. Mindstrong Health Services, a California based healthcare provider for individuals with a serious mental illness uses an application running on a member's smartphone to create digital biomarkers and to provision telehealth services (mindstronghealth.com).

To make ethical considerations of digital phenotyping concrete, we describe how Mindstrong's digital biomarker application and healthcare services are used in crisis prevention. The continuum of crisis services is a cycle of prevention, intervention, response and stabilization. Unfortunately, patients with serious mental illness experience primarily respond-and-stabilize services. The goal of crisis services should be to preempt and intervene before a crisis occurs. In the early period of clinical decompensation low acuity interventions delivered through telehealth can be sufficient. The challenge in achieving this goal has been the lack of valid, objective and ecological measures of worsening function to alert the provider. Digital biomarkers have the potential to detect clinical deterioration experienced early but not reported by the patient.

2.2.1   Case Report: Patient AB[1]

We consider the following case report of patient AB in her 40s with a history of bipolar disorder and psychosis. AB was enrolled in the Mindstrong Health service. Enrollment involved downloading the Health by Mindstrong application (app) on her smartphone. The app captures the data needed to create five digital biomarkers. The biomarkers are computed daily from the AB's normal use of her smartphone by measuring response times to different patterns of tapping and swiping that she used repeatedly throughout the day. These biomarkers are *transdiagnostic* meaning that they are not specific to any particular diagnosis. Each biomarker measures state-dependent changes in cognition or mood that may be indicative of illness relapse.

Figure 2.1 illustrates the digital biomarker data available to AB in the Mindstrong app on her smartphone. Shown in the figure is the biomarker for cognitive control over several days, with each "dot" representing a day. On a smartphone, cognitive control can be measured from a subject's tap reaction times and the variability in reaction times to certain events that are aggregated throughout the day during the subject's normal phone use. During this period AB's daily cognitive control is within the white zone set at ±1.5 standard deviations. The light pink zone covers ±2 standard deviations and values outside that range will alert both AB and the clinician. Figure 2.1 shows a period of good cognitive control whereas Fig. 2.2 shows a period of erratic swings in cognitive control outside of the ±2 standard deviation zone. Note the break in data between September 20th and September 25th. With the Mindstrong app on her smartphone, AB has 24/7 access to Mindstrong Health Services licensed providers via secure messaging and private videoconferencing. Figure 2.3 shows messages exchanged during the period of erratic variability just before the break in the data. When the digital biomarkers are outside the ±2 standard deviation zone, AB and her Mindstrong Health Services care manager receive alerts. Figure 2.4 shows an expanded view of the cognitive control biomarker and working memory biomarker during the period of erratic variability. During the four-day break in the data, AB was hospitalized and treated for relapse. Post-discharge from the hospital we notice significant improvement in ABs digital biomarkers. Immediately following discharge AB communicated to her Mindstrong Health Services care manager the following:

---

[1] We obtained the patient's informed consent to publish this material. The patient's initials were changed to protect her privacy.

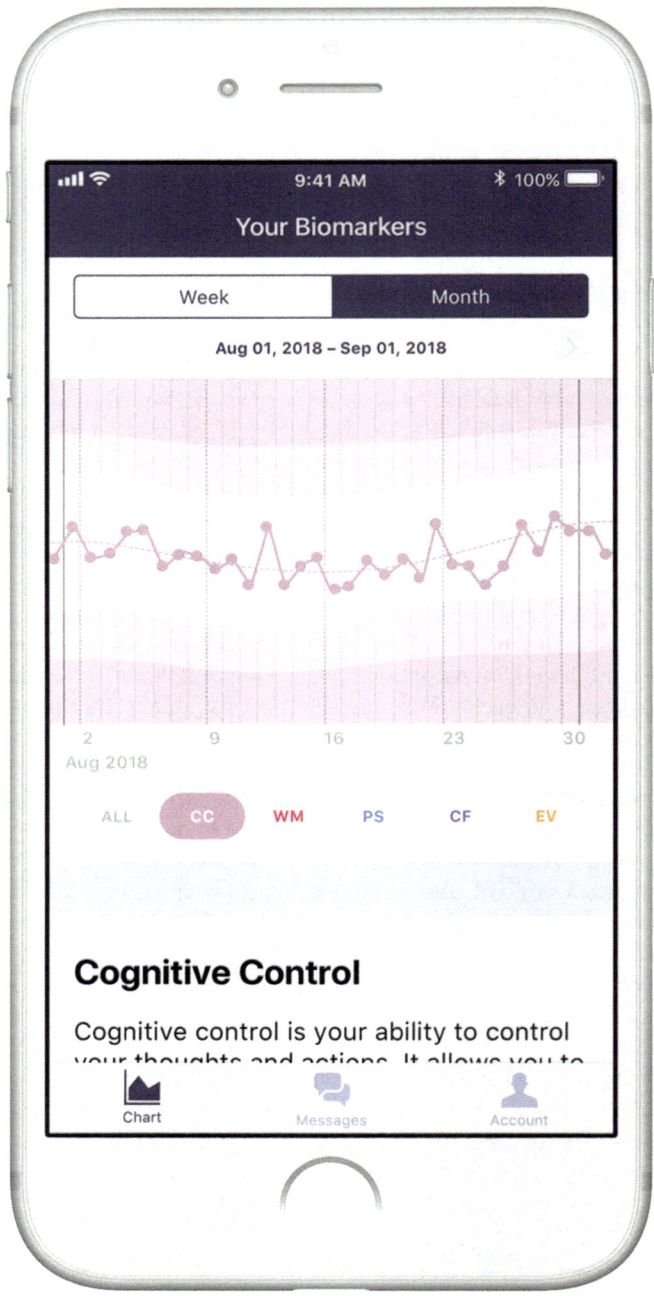

Fig. 2.1 Shown is AB's cognitive control (CC) digital biomarker in August 2018. Each filled circle corresponds to a daily measurement. The white zone spans ±1.5 standard deviations and the light pink zone spans ±2 standard deviations. During August we note that AB's cognitive control remained in the white zone

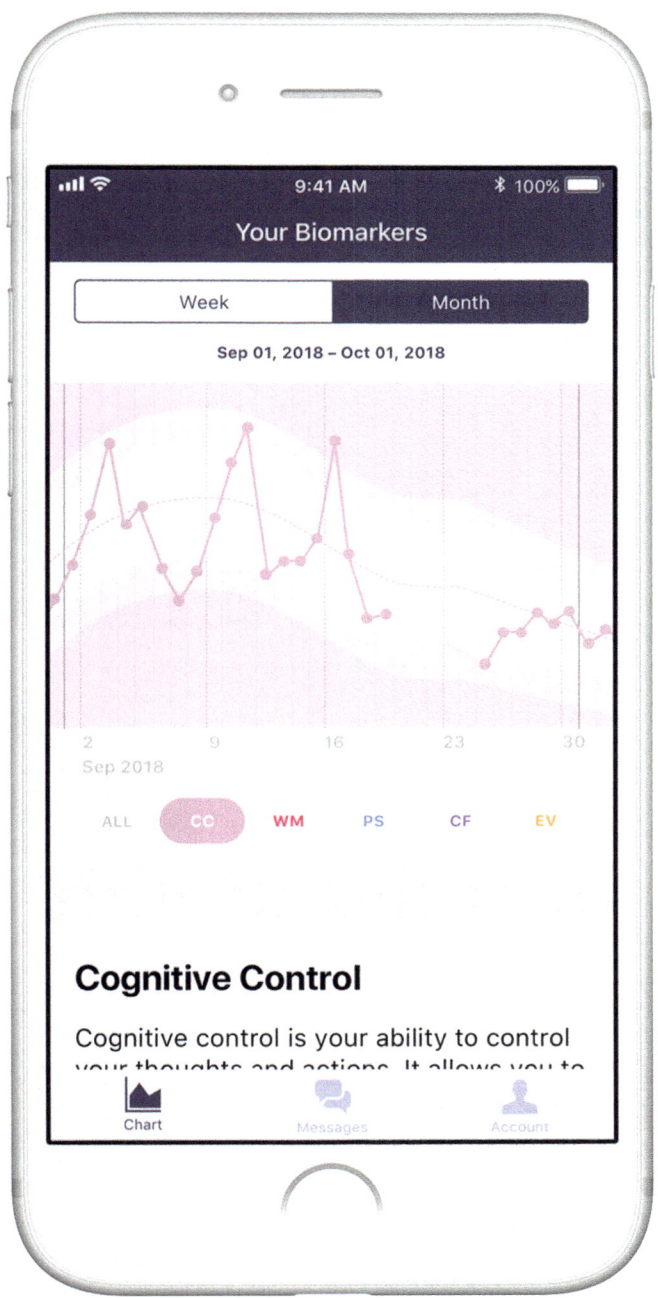

Fig. 2.2 Shown is AB's cognitive control biomarker in September 2018. During this period, we note erratic variability in her biomarker with swings into the light pink and dark pink zones that exceed ±2 standard deviations. We also see a break in the data beginning September 20 and ending September 24 when she was hospitalized

Fig.2.3 Shown is AB's communication with her Mindstrong Health Services care manager prior to her hospitalization. AB had 24/7 access to her Mindstrong care manager and both her and the care manager received automated alerts when AB's biomarkers were outside of the ±2 standard deviation zone. The names of AB and her Mindstrong care managers were changed to protect privacy

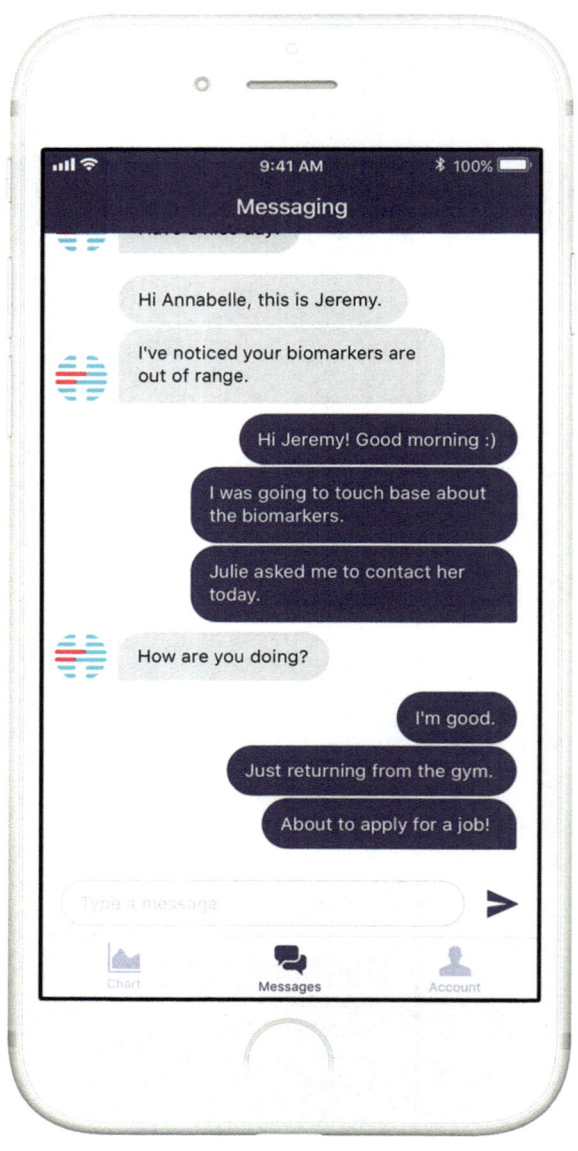

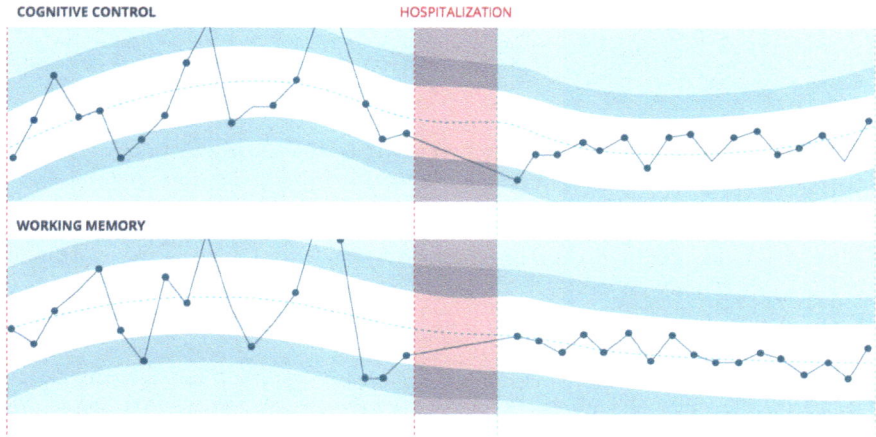

Fig. 2.4 Shown are AB's digital biomarkers for cognitive control and working memory pre-hospitalization and post-discharge. AB is successfully stabilized during her hospital stay. AB's cognitive control and working memory biomarkers return to their healthy baseline following discharge

> I'm doing a lot better. I was experiencing a lot of auditory hallucinations. They made it difficult to sleep which made things progressively worse.

> I checked myself into the hospital. They adjusted my medications, gave group therapy, and monitored me. I believe I slept for 12 hours each night 3 days in a row. What a relief! The hallucinations finally subsided.

AB's transparency following her hospital admission about her symptom burden and clinical decline was markedly distinct from pre-hospitalization. Agency is a recurring theme in mental health. AB's need to stay in control of her illness was evident in retrospect. Safety is the other recurring theme in mental health and represents an individual's desire to be alerted to impeding risks in their condition early while they still have agency.

Figure 2.5 illustrates the user and provider journey through crisis prevention beginning with increased symptom burden and functional decompensation detected by changes in digital biomarkers captured from the user's smartphone. A care manager or other care professional contacts the user through secure messaging or videoconferencing to clinically evaluate the individual and confirm that the observed changes in digital biomarkers represent a clinical decompensation. On confirmation, the patient is escalated to a licensed clinical psychologist for evidence-based telehealth therapeutic interventions and re-evaluated regularly for improvement. If the patient does not improve or continues to decompensate, care is escalated to a telepsychiatric consultation or a clinic consultation.

AB's case report highlights a number of ethical areas that require consideration.

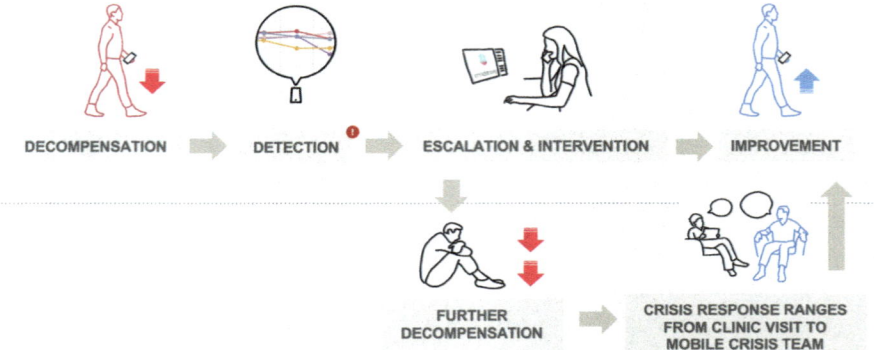

Fig. 2.5 An illustration of a user journey through decompensation, early intervention and crisis intervention while on the Mindstrong Health app and receiving Mindstrong Health Services

1. Does the informed consent properly explain the relevant risks and benefits to AB, the range of data collected, and the digital phenotype findings that are derivable from that data?
2. Does the informed consent properly disclose who has access to her data and digital phenotyping findings that can be derived from her data?
3. Did AB consent to receive evidence-based telehealth care based on the digital phenotype findings?
4. Did AB and her provider understand the intended use of digital phenotyping?
5. Was AB informed about the security of her data and restrictions to its access?

2.2.2 *Informed Consent*

In healthcare, digital phenotyping can be viewed as a risk indicator of an adverse outcome and can inform clinical decision making. In the United States, informed consent that properly explains the relevant risks and benefits to the patient is necessary and sufficient to receive healthcare services. Like diagnostic tests that require tissue or blood samples from which multiple test can be run, the data samples required for digital phenotyping can also be analyzed differently to reveal new findings. It is worth noting that in European countries, under GDPR informed consent may not be sufficient (see Chap. 1 by Kargl et al.).

Digital phenotyping in healthcare qualifies as protected health information. State and federal regulations govern the conditions of access to protected health information (PHI). The Health Insurance Portability and Accountability Act (HIPAA) of 1996 in the United States provides data privacy and security provisions for safeguarding medical information. The HIPAA Privacy Rule allows HIPAA-covered entities,

such as healthcare providers and health plans, to use and disclose individually identifiable PHI without an individual's consent for treatment. In all cases, sharing of
individually identifiable PHI must be limited to the minimum necessary information
to achieve the purpose for which the information is disclosed.

If protected health information needs to be disclosed for one of the following
purposes, a HIPAA release form must be obtained from the patient:

- Disclosure of PHI to a third party for reasons other than the provision of treatment,
  payment or other standard healthcare operations. For example, disclosing data to
  an insurance underwriter
- Use of PHI for marketing or fund-raising purposes
- Disclosure of PHI to a research organization
- Disclosure of psychotherapy notes
- Sale or sharing of PHI for remuneration.

2.2.3 Intended Use

Clear language must accompany the intended use of digital phenotyping in healthcare. For example, the Mindstrong Health application is intended for adjunctive
remote health management of patients under the care of a behavioral health provider
for one or more serious mental illnesses, including schizophrenia, bipolar disorder,
and major depressive disorder.

2.2.4 Security and Access

When working with PHI, a systematic approach to managing this information is
necessary to ensure it remains secure. The approach which includes people, processes
and IT systems, should be independently audited on a yearly basis. The International
Organization of Standardization (ISO) and The Health Information Trust Alliance
(HITRUST) are two certification organizations.

ISO is an independent, non-governmental international organization founded
in 1946 and headquartered in Geneva. ISO has members from 164 countries and
786 technical committees and subcommittees involved in maintaining and improving standards developments (ISO—International Organization for Standardization
2019). The ISO 27000 family of standards certifies an organization's policies, procedures and controls for securing information assets. HITRUST is a private certification standard in the U.S. that like ISO 27001 certifies organizations on their security
framework for the creation, access, storage or exchange of sensitive information. ISO
27001 or HITRUST certification is often required before sharing sensitive information such as PHI between organizations.

The ethical areas discussed here overlap with the work of Fuller et al. (2017) that focuses on medical research. The authors reflect on ethical implications in the context of location and accelerometer measurement and point to four areas of relevance, namely consent, privacy and security, mitigating risk, and consideration of vulnerable populations (page 87). In Table 1 on page 87 of their work they summarize their position that consent should be renewed during the clinical study after participants are provided their actual collected data.

2.3 Digital Phenotyping in Medical Research and Drug Discovery

New clinical tests and measurements create new opportunities for discovery in medicine. Digital phenotyping is not an exception. Pharmaceutical companies have been among the first to embrace the potential of digital biomarkers (Dorsey et al. 2017; Madrid et al. 2017; Rodarte 2017; Smith et al. 2018). Patient stratification using daily, ecological, multidimensional measures of cognition and mood may be indicative of who will respond to a new drug or intervention. Digital phenotyping can provide new digital endpoints to clinical trials that are proximal to patient function, improving on existing symptom reported outcomes.

Medical research on human subjects needs to be reviewed and approved by an independent ethics committee known as an institutional review board (IRB). The purpose of the IRB is to ensure that the rights and welfare of humans participating as subjects in a research study are protected. IRBs provide scientific, ethical, and regulatory oversight of human subject research. Part of the review is the informed consent process which must clearly explain to the individual the risk and benefits from the proposed research. Human subject research that involves digital phenotyping must follow these established guidelines.

More research is necessary to understand how participants perceive their anonymity in study designs using digital phenotyping and how it is explained in the informed consent. We need to consider new methods to ensure subject privacy, a topic that is further developed in the privacy Chap. 1 of this book by Kargl et al. and in recent empirical work by Nebeker et al. (2016).

Fuller et al. (2017) ask if researchers need to protect subjects from imminent harm in a study. The authors discuss if GPS data of a study should be used to find a lost child. The authors take a stance that researchers should not be burdened by these requests and this also needs to be made clear in the consent form. In contrast, incidental findings on a brain scan in a study that is using for example functional magnetic resonance imaging would ethically need to be reviewed by a licensed radiologist and communicated to the subject. Other considerations include whether researchers should be obliged to report illegal activity detected in data patterns to law officials.

Finally, we need to consider vulnerable groups such as children who need stronger protection and vulnerable groups that may be systematically excluded because of limited access to the internet and mobile devices (Hokke et al. 2018). In this context, researchers recently discussed the importance of several prerequisites to be able to use the internet and mobile devices including technical, reading and writing competence (Rubeis and Steger 2019). Such skills need to be considered when designing mental health applications.

2.4   Uses of Digital Phenotyping Outside of Healthcare

Non-clinical uses of digital phenotyping are varied. Examples of non-clinical uses include government, employers, education, direct-to-consumer marketing and insurance risk stratification. Non-clinical use of health information that in a clinical context would be regulated, strains existing ethical governance rules. Government today performs DNA tests on prisoners willingly or under court mandate. The DNA test can be used as forensic evidence but can also provide other information about the subject. For neuropsychological testing of prisoners, ethical guidelines exist (Vanderhoff et al. 2011) and should be used for digital phenotyping.

In the workforce, behavioral phenotyping from posts in social media sites can influence hiring decisions (CareerBuilder 2018). This practice is mostly used to screen-out applicants but on occasion can be helpful if an applicant has contributed a well-written post relevant to their career. Facebook's use of data to predict voter preferences or suicide risk (Eichstaedt et al. 2018; Matz et al. 2017; Reardon 2017) represent other non-healthcare examples of behavioral phenotyping. The connection between a user's posts and their future behavior is nuanced and requires the use of advanced machine learning algorithms trained on vast amounts of data. In these cases, the user has less control or transparency of how their data and predictions of preferences and behavior are used.

Direct-to-consumer marketing has long exploited digital data to discover consumer preference to create an asymmetric commercial advantage. Both behavioral phenotyping and digital biomarkers have the potential to sharpen this asymmetric advantage against the consumer. Behavioral phenotyping in the form of emotional response through Facebook likes or facial expression captured from a smartphone camera are in use today to market to consumers (Settanni et al. 2018; Wang and Kosinski 2018). In all these cases users should explicitly opt-into the use of their digital data for digital phenotyping. Informed consent would need to disclose all possible health-related findings that would be derived from the data captured and how that information would be used. For example, ads for therapists or psychiatrists targeting a user who is deemed to be depressed would feel invasive to most people.

The preceding practices highlight several ethical questions that are reminiscent of a HIPAA release form for PHI data.

- Who owns the user data?
- Who owns what can be inferred from the user data?
- Who decides what can be done with the user data?

- Can the data be used for marketing?
- Can the data be disclosed to other organizations?
- Can the data be sold or shared?

In this context, it would seem that the aforementioned uses of digital phenotyping should require explicit informed consent.

Insurance companies have increasingly used personal data to set premiums. Personal attributes such as occupation, level of education or credit score can significantly impact home and auto insurance premiums (National Association of Insurance Commissioners 2012). This practice has come under closer scrutiny with legislators in some states striking back at the use of such criteria claiming they unfairly discriminate against those who lack a college degree or work in certain industries. Several states have banned the use of credit information to set premiums. New York state went further and banned insurance companies from using occupation and education related criteria to set rates (Scism 2017).

Insurance companies claim that using these factors improves rating accuracy and avoids cross-subsidization, or the charging of higher prices for one group to lower the price for others. Cross-subsidization results in good drivers paying for the behavior of bad drivers.

It is interesting to contemplate whether digital phenotyping could improve rating accuracy without unfairly discriminating based on social economic status. Digital biomarkers that measure cognitive control, or impulsivity, are likely good predictors of risk. Similarly, measures of processing speed or attention should be indicative of higher or lower accident risk. Age is a strong covariate of these cognitive measures, showing a declining performance with advancing age. Auto insurance premiums also show an increase with advancing age beyond the 6th decade (Cobb and Coughlin 1998), possibly reflecting a higher risk with declining cognitive function. But using a population measure such as age to adjust premiums is very different from using an individual clinical measure such as a person's cognitive control biomarker. Data from onboard diagnostics in a car can determine the driver's gender, another attribute used in setting premiums and recently banned in California (Leefeldt 2019; Stachl and Bühner 2015). The question is whether insurance companies can use health information to set premiums.

Following the 2010 passing of the Affordable Care Act (ACA) in the United States, healthcare insurance companies where barred from using pre-existing health condition to increase a person's premium. Similarly, the Americans with Disabilities Act (ADA) of 1990 made it illegal for insurance companies to charge an individual a higher rate because of a physical or mental disability. The ADA is a civil rights law that prohibits discrimination against individuals with disabilities in all areas of public life, including jobs, schools, transportation.

Both the ACA and ADA establish an ethical and legal framework that limits an insurance company's ability to adjust premiums based on a pre-existing health condition or disability. The ADA further extends that framework to employers and education. The ADA defines a person with a disability as someone who has a physical or mental impairment that substantially limits one or more major life activities. Could certain behavioral phenotypes qualify as a pre-existing condition or a disability? Could digital biomarkers of cognitive function qualify as a pre-existing condition or a disability? Today they do not, but if that were to change, uses of digital phenotyping outside of healthcare would be limited and regulated by existing laws.

2.5 The Future of Digital Phenotyping

For over a century we have failed to develop objective clinical measures of mental health or sensitive ecological markers of cognitive function. The brain is a complex organ and unlike other organs, normal physiology and pathology cannot be measured through blood tests or physiological markers. Nonetheless, the paper-and-pencil gold-standard neuropsychological tests have demonstrated that we can create sensitive indicators of neural circuit dysfunction from repeated simple timed challenges. Human-computer interfaces on digital devices present parallel repeated simple timed challenges. These repeated timed challenges create a time-series of measurements because they are repeated frequently throughout the day during normal digital device use. The timings of those events, and the many features that can be extracted from those timings, can be used to reproduce the gold-standard tests objectively and are representative of the user in their ecological environment. Thus, one might argue that the adoption of ubiquitous human-computer interfaces was a prerequisite for the development of ecological and objective clinical measures of the brain.

Digital phenotyping emerged from the ubiquitous adoption of digital devices and explosion of online user data. This data is collected by the websites and telecommunication companies that provide service to the devices. Online data has been used to influence user behavior since 2000. Privacy policies emerged during this period, but they addressed direct marketing uses of the data. Beginning in 2013, early work at Mindstrong demonstrated that the HCI data from smartphones could be converted into validated clinical measures of cognitive function. Subsequent findings demonstrated that posts used in Facebook could be used to predict depression three months before a formal diagnosis (Eichstaedt et al. 2018). The misuse of this information and data privacy sparked ethical concerns (Martinez-Martin et al. 2018).

Ethical considerations of digital phenotyping can proceed along two directions. We can choose to categorize digital phenotyping as protected health information. This approach is pragmatic and allows us to leverage an established and proven framework for individual consent, governance and control of the data. That has been the direction taken at Mindstrong. As future research continues to validate clinical use of digital biomarkers, considering digital phenotyping as protected health information is appropriate.

Alternatively, we can choose to construct a de novo regulatory framework for digital phenotyping. This path is appropriate for digital phenotypes that are not clinical measures or when health information will not be identified from the data. For example, certain behavioral phenotypes are expressions of preferences and may not qualify as health information. But a de novo regulatory framework may not be a pragmatic solution. Such a framework would need to anticipate all future possibilities and uses of the data.

Human-computer interfaces will continue to evolve from touch-screen interfaces to augmented reality to direct brain-computer interfaces. Throughout this evolution, the data available to create digital biomarkers and phenotypes will continue to improve as will prognosis and prediction of brain-related disorders. To safeguard a better future, we must collaborate with policy makers, regulatory bodies, insurance companies, healthcare systems and technology companies to agree on proper governance and use of this data.

Conflict of Interest   Christian Montag mentions that he currently receives funding from Mindstrong Health for a project on molecular genetics and digital phenotyping. Of importance, his views on ethics as presented in this work have not been influenced by this financial support for his research.

References

Biomarkers Definitions Working Group (2001) Biomarkers and surrogate endpoints: preferred definitions and conceptual framework. CPT Pharmacomet Syst Pharmacol 69(3):89–95. https://doi.org/10.1067/mcp.2001.113989

CareerBuilder (2018) More than half of employers have found content on social media that caused them not to hire a candidate, according to recent CareerBuilder survey. PRN Newswire. https://www.prnewswire.com/news-releases/more-than-half-of-employers-have-found-content-on-social-media-that-caused-them-not-to-hire-a-candidate-according-to-recent-careerbuilder-survey-300694437.html. Accessed 8 Jul 2019

Chittaranjan G, Blom J, Gatica-Perez D (2013) Mining large-scale smartphone data for personality studies. Pers Ubiquit Comput 17(3):433–450. https://doi.org/10.1007/s00779-011-0490-1

Cobb RW, Coughlin JF (1998) Are elderly drivers a road hazard?: Problem definition and political impact. J Aging Stud 12(4):411–427. https://doi.org/10.1016/S0890-4065(98)90027-5

Dagum P (2018) Digital biomarkers of cognitive function. NPJ Digit Med 1(1):10. https://doi.org/10.1038/s41746-018-0018-4

Dawkins R (1982) The extended phenotype. Oxford University Press, Oxford

Dorsey ER, Papapetropoulos S, Xiong M, Kieburtz K (2017) The first frontier: digital biomarkers for neurodegenerative disorders. Digit Biomark. https://doi.org/10.1159/000477383

Eichstaedt JC, Smith RJ, Merchant RM et al (2018) Facebook language predicts depression in medical records. Proc Natl Acad Sci USA 115(44):11203–11208. https://doi.org/10.1073/pnas.1802331115

Elenko E, Underwood L, Zohar D (2015) Defining digital medicine. Nat Biotechnol 33(5):456–461. https://doi.org/10.1038/nbt.3222

Fuller D, Shareck M, Stanley K (2017) Ethical implications of location and accelerometer measurement in health research studies with mobile sensing devices. Soc Sci Med 191:84–88. https://doi.org/10.1016/j.socscimed.2017.08.043

Hokke S, Hackworth NJ, Quin N et al (2018) Ethical issues in using the internet to engage participants in family and child research: a scoping review. PLoS ONE 13(9):e0204572. https://doi.org/10.1371/journal.pone.0204572

Insel TR (2017) Digital phenotyping: technology for a new science of behavior. JAMA Netw 318(13):1215–1216. https://doi.org/10.1001/jama.2017.11295

ISO—International Organization for Standardization (2019) ISO—International Organization for Standardization. http://www.iso.org/cms/render/live/en/sites/isoorg/home.html. Accessed 8 Jul 2019

Jain SH, Powers BW, Hawkins JB, Brownstein JS (2015) The digital phenotype. Nat Biotechnol 33(5):462–463

Kerchner GA, Dougherty RF, Dagum P (2015) Unobtrusive neuropsychological monitoring from smart phone use behavior. Alzheimers Dement 11(7):272–273. https://doi.org/10.1016/j.jalz.2015.07.358

Kosinski M, Matz SC, Gosling SD et al (2015) Facebook as a research tool for the social sciences: opportunities, challenges, ethical considerations, and practical guidelines. Am Psychol 70(6):543–556. https://doi.org/10.1037/a0039210

Leefeldt E (2019) California bans gender in setting car insurance rates. CBS NEWS. https://www.cbsnews.com/news/car-insurance-california-bans-gender-as-a-factor-in-setting-rates/

Madrid A, Smith D, Alvarez-Horine S et al (2017) Assessing anhedonia with quantitative tasks and digital and patient reported measures in a multi-center double-blind trial with BTRX-246040 for the treatment of major depressive disorder. Neuropsychopharmacology 43:372–372

Markowetz A, Błaszkiewicz K, Montag C et al (2014) Psycho-Informatics: big data shaping modern psychometrics. Med Hypotheses 82(4):405–411. https://doi.org/10.1016/j.mehy.2013.11.030

Martinez-Martin N, Insel TR, Dagum P et al (2018) Data mining for health: staking out the ethical territory of digital phenotyping. npj Digital Med 1(1):68. https://doi.org/10.1038/s41746-018-0075-8

Matz SC, Kosinski M, Nave G, Stillwell DJ (2017) Psychological targeting as an effective approach to digital mass persuasion. Proc Natl Acad Sci USA 114(48):12714–12719. https://doi.org/10.1073/pnas.1710966114

Mindstrong (2019). Mindstrong Health. https://mindstronghealth.com/

Montag C, Baumeister H, Kannen C et al (2019) Concept, possibilities and pilot-testing of a new smartphone application for the social and life sciences to study human behavior including validation data from personality psychology. J 2(2):102–115. https://doi.org/10.3390/j2020008

Montag C, Błaszkiewicz K, Lachmann B et al (2014) Correlating personality and actual phone usage: evidence from psychoinformatics. J Individ Differ 35(3):158–165. https://doi.org/10.1027/1614-0001/a000139

National Association of Insurance Commissioners (2012) Credit-based insurance scores: how an insurance company can use your credit to determine your premium. https://www.naic.org/documents/consumer_alert_credit_based_insurance_scores.htm. Accessed 8 Jul 2019

Nebeker C, Lagare T, Takemoto M et al (2016) Engaging research participants to inform the ethical conduct of mobile imaging, pervasive sensing, and location tracking research. Behav Med Pract Policy Res 6(4):577–586. https://doi.org/10.1007/s13142-016-0426-4

Reardon S (2017) AI algorithms to prevent suicide gain traction. Nature News

Rodarte C (2017) Pharmaceutical perspective: how digital biomarkers and contextual data will enable therapeutic environments. Digit Biomark. https://doi.org/10.1159/000479951

Rubeis G, Steger F (2019) Internet- und mobilgestützte Interventionen bei psychischen Störungen. Nervenarzt 90(5):497–502. https://doi.org/10.1007/s00115-018-0663-5

Saeb S, Zhang M, Karr CJ et al (2015) Mobile phone sensor correlates of depressive symptom severity in daily-life behavior: an exploratory study. J Med Internet Res 17(7):e175. https://doi.org/10.2196/jmir.4273

Scism L (2017) New York car insurers could soon be banned from asking about your education. The Wall Street Journal. https://www.wsj.com/articles/new-york-car-insurers-could-soon-be-banned-from-asking-about-your-education-1494952287

Settanni M, Azucar D, Marengo D (2018) Predicting individual characteristics from digital traces on social media: a meta-analysis. Cyberpsychology Behav Soc Netw 21(4):217–228. https://doi.org/10.1089/cyber.2017.0384

Smith DG, Saljooqi K, Alvarez-Horine S et al (2018) Exploring novel behavioral tasks and digital phenotyping technologies as adjuncts to a clinical trial of BTRX-246040. International Society of CNS Clinical Trials and Methodology

Stachl C, Bühner M (2015) Show me how you drive and I'll tell you who you are recognizing gender using automotive driving parameters. Procedia Manuf 3:5587–5594. https://doi.org/10.1016/j.promfg.2015.07.743

Stachl C, Hilbert S, Au J-Q et al (2017) Personality traits predict smartphone usage: personality traits predict smartphone usage. Eur J Pers 31(6):701–722. https://doi.org/10.1002/per.2113

Torous J, Onnela J-P, Keshavan M (2017) New dimensions and new tools to realize the potential of RDoC: digital phenotyping via smartphones and connected devices. Transl Psychiatry 7(3):e1053. https://doi.org/10.1038/tp.2017.25

Vanderhoff H, Jeglic EL, Donovick PJ (2011) Neuropsychological assessment in prisons: ethical and practical challenges. J Correct Health Care 17(1):51–60. https://doi.org/10.1177/1078345810385914

Wang Y, Kosinski M (2018) Deep neural networks are more accurate than humans at detecting sexual orientation from facial images. J Pers Soc Psychol 114(2):217–228

Part II
Digital Phenotyping and Mobile Sensing in Psycho-Social Sciences

Chapter 3
Computerized Facial Emotion Expression Recognition

Mattis Geiger and Oliver Wilhelm

Abstract Facial emotion expressions are an important gateway for studying human emotions. For many decades, this research was limited to human ratings of arousal and valence of emotional expressions. Such ratings are very time-consuming and have limited objectivity due to rater biases. By exploiting improvements in machine learning, the demand for a swifter and more objective method to assess facial emotional expressions was met by a plethora of software. These novel approaches are based on theories of human perception and emotion and their algorithms are often trained with massive and almost-generalizable data bases. However, they still face limitations such as 2D recognition and cultural biases. Nevertheless, the accuracy of computerized emotion recognition software has surpassed human raters in many cases. Consequently, such software has become instrumental in psychological research and has delivered remarkable findings, e.g. on human emotional abilities and dynamic expressions. Furthermore, recent developments for mobile devices have introduced such software into daily life, allowing for the immediate and ambulatory assessment of facial emotion expression. These trends provide intriguing new opportunities for studying human emotions, such as photograph-based experience sampling, incidental or implicit data recording in interventions, and many more.

3.1 Introduction

From planning behavior, over automated decision-making, to actual behavior such as communication—emotions shape our everyday life. It comes as no surprise that emotions are among the most researched topics in psychology and related disciplines. When it comes to social interaction, the communicative function of emotions is of key interest. That is, emotions elicit automatic expression and influence controlled

M. Geiger (✉) · O. Wilhelm
Department of Individual Differences and Psychological Assessment, Faculty of Engineering,
Computer Science and Psychology, Institute of Psychology and Education, Ulm University, 89069
Ulm, Germany
e-mail: mattis.geiger@uni-ulm.de

© Springer Nature Switzerland AG 2019
H. Baumeister and C. Montag (eds.), *Digital Phenotyping and Mobile Sensing*,
Studies in Neuroscience, Psychology and Behavioral Economics,
https://doi.org/10.1007/978-3-030-31620-4_3

communication behavior. Typical emotion communication behavior includes vocal or facial expressions, such as crying or smiling. It serves as a fast and automatic pathway for the distribution of information into our social environment, e.g. about dangers or requests for help, and thereby facilitates survival (Darwin 1987). Throughout the history of emotion research, there has been a demand for the systematic assessment of emotion communication. Initially, this demand was met by having human raters go through records of emotional behavior. In the last decade, there has been a rise in the development and use of facial emotion expression recognition software.

In this chapter, we focus on facial emotion expression and its recognition by humans or machines. We will begin by introducing major theories of emotion expression and the tools developed to systematically assess facially expressed emotions with human raters. Next, we will describe contemporary approaches towards machine-based classifications of facially-expressed emotions, including machine learning. Limitations of current methods and perspectives to overcome them will be provided next. Furthermore, we present examples from basic and applied research to show how computerized emotion expression recognition software can facilitate and initiate research in fields that could not be tapped hitherto. We introduce the concept of socio-emotional abilities and present how emotion recognition software expands this field to emotion expression and emotion regulation abilities. Finally, we provide a number of examples to illustrate how mobile sensing of facially-expressed emotions can dramatically change research approaches towards emotion and how it can be useful in a variety of novel applied settings.

3.2   Human Facial Emotion Expression Recognition

Emotions are latent states and therefore not directly observable. In order to find proxies for these latent states, past research often focused on their physiological manifestations, such as heart rate or skin conductance. Although this research greatly improved our general understanding and theories of emotion, it is typically limited to the arousal dimension of emotions only. In general emotion theories (Cannon 1927; Schachter and Singer 1962; Scherer 2005), arousal refers to the general level of (physiological) activation due to an emotional state. Arousal is distinguished from the valence of emotions. In overarching emotion theories, valence is one dimension from positive to negative valence, whereas in theories of basic emotions valence dimensions are distinguished between emotions such as Anger, Happiness, Sadness, Surprise, Fear, and Disgust (Ekman 1992). In the latter system, emotional valence is categorized as six basic emotions that are universally recognized and that differ in evolutionary function. The communicative aspect of emotions provides the opportunity to differentiate both emotional arousal and valence and is therefore a vital pathway for the scientific study of emotions. Competing contemporary emotion theories disagree with the basic emotions approach of Ekman (1992). However, the vast majority of computerized facial emotion expression recognition tools are based on the basic emotions theory (Corneanu et al. 2016). Therefore, this chapter is restricted to basic emotion theory. We discuss limitations emerging from this restriction in the conclusion of this chapter.

Research on the communicative function of emotion requires a system to measure the nature and intensity of emotional communication signals. The most popular systematic approach in this domain is the Facial Action Coding System (FACS; Ekman and Friesen 1978). In the FACS, distinct facial movements, such as moving the corner of the lips upwards, for example as part of a facial expression evaluated as a smile, are categorized as Action Units (AUs). Furthermore, AUs are anatomically valid: for every AU, a single muscle or a group of facial muscles that initiate AU-movement is/are identified. For example, when frowning, which is coded as AU4, the musculus corrugator is active. FACS raters use these AUs to evaluate facial expressions of persons. Typically, raters are presented with a picture and rate the displayed expression for every AU on a 5pt-Likert scale. Basic emotions are characterized by a distinct pattern of activated AUs (Ekman 1992). These patterns are listed in the FACS Affect Interpretation Database (FACSAID; Ekman et al. 1998). For example, a happy expression consists of AU6 (cheek raiser) and AU12 (lip corner pull). If, among all AUs, only these were rated as active, then the FACS rater would conclude that the facial expression is happiness.

Although the FACS greatly advanced research on facial emotion expressions, much research in this field still relies on self-reports from participants experiencing or expressing the emotion. The main reason for this is probably that gathering FACS ratings is extremely effortful, time-consuming, and costly—to the point where conducting research with large samples of participants and/or many time-points becomes unfeasible. The predominant use of human raters also leads to a neglect of the dynamic nature of facially-expressed emotions. Clearly, videos should be preferred over pictures (Tcherkassof et al. 2007), but having raters evaluate each frame from a lengthy video is usually unrealistic due to the resources required to do so. In addition, due to common biases in human raters, multiple raters are required for achieving reliable and objective ratings, which multiplies the required rating effort. Consequently, less effortful, more objective, much quicker, and cheaper procedures for rating facial emotion expressions are essential for advancing research and applications in this field.

3.3 Computerized Face Perception

The evaluation of facial emotion expressions might meet the requirements of effort, prize, speed, and objectivity if accomplished through machines. Typically, this evaluation is 'taught' to computers akin to how these skills function in humans. In one widely acknowledged approach, Bruce and Young (1986) distinguish two very basic levels of information codes in face recognition that also apply to the functionality of face detection in computers: pictorial and structural code. Pictorial code refers to any information that constitutes a picture. For a digital file, that would be the pixels and related information, such as the color or brightness of every pixel in a picture.

Pictorial code is specific to an individual picture and not to an individual identity. Structural code refers to more abstract information that is derived from the pictorial code, such as the black circle shape of the pupil, which might appear in varying positions in different pictures, but still represent the same structure. In other words, structural code is about the invariant features of a face and it is built upon internal features of a face. For complete face detection, several structural codes are important, for example the eyes, the jawline, the lips, etc.

To detect structural code via software, texture analysis, such as Gabor filters, is used. This method mimics human visual perception via photoreceptor cells (Marĉelja 1980): it detects regularities (lines, curves, circles, complex shapes, etc.) in images by analyzing luminosity and color of pixels. Faces consist of many facial features that are combinations of such regularities. For example, the eyes consist of an outer roughly oval-shaped (as seen in neutral, frontal photographs of faces with open eyes) white structure (i.e. visible parts of the eyeball) and two inner circle shaped structures (i.e. the colored iris and the black pupil). Texture analysis tools—just like humans— are trained on large datasets to detect such structures and to thus separate structural code from pictorial code. Finally, humans and software classify the configuration of structural codes; for example, an eyeball, iris, and pupil "create" an eye, while two eyes, a nose, and a mouth "create" a face. This information is essential and indispensable to derive decisions about whether a picture is showing a (part of a) face or not. Furthermore, the organization of the structural code is essential for setting facial landmarks. Such landmarks are important for the analysis of changeable aspects of faces, for example when known identities express different emotions.

There is large variability in available algorithms for face detection (Zhang and Zhang 2010). Modern algorithms achieve high detection rates even in low quality recordings, such as with extreme lighting conditions or low resolutions. However, face detection algorithms are still developing. One common problem is that detection rates decrease when faces are not presented frontally. The cause for this deficit is 2D-based detection in such algorithms. Without the depth dimension, landmark position estimates are biased and therefore differ systematically from the real position in three-dimensional space. Consequently, recent developments focus on three dimensional face models generated from 2D video recordings (e.g. Jeni et al. 2015). This is a very important step to further approach human face detection abilities. However, real 3D recordings or 3D estimations from videos with sufficient information to estimate a 3D model are still rare (Zinkernagel et al. 2018). This is mostly due to the facts that (a) the possibility to do 3D recordings is still uncommon in regular commercial devices, such as smartphones, and (b) the computational power of common devices does not yet allow for real time 3D model estimation from 2D videos. Nevertheless, the field of computerized face detection is highly advanced and available tools continue to improve. The sophistication of contemporary face detection tools provides a strong basis for computerized facial emotion recognition.

3.4 Computerized Facial Emotion Recognition

Employing facial landmark detection established by face detection algorithms in the last decade provided the basis for the development of several proprietary and open-source tools for computerized facial emotion recognition (e.g. Affectiva Affdex, Affectiva 2018; OpenFace, Baltrusaitis et al. 2016; IntraFace, De la Torre et al. 2015; Facet, Emotient 2016a; FaceReader, Noldus Information Technology 2018; just to name a few). Although details of the algorithms are usually proprietary in commercial software, the general functionality is similar across tools. The steps required for this are outlined below and summarized in Table 3.1.

First, a face is detected from a picture and facial landmarks are located and set. In the next step, emotional expressions are recognized by comparing landmark positions relative to a neutral or unexpressive standard face. In available software, all these steps happen in 2D space. The choice of the standard face is crucial for the precision of

Table 3.1 Summary of typical steps and descriptions in computerized facial emotion expression recognition

Step #	Step name	Description
1	Data input	Data type (two- or three-dimensional) determines whether the following steps are conducted in 2D or 3D. Most current data are 2D, but with additional data (sufficient head movement and true physical length of landmark distances) 3D data can be inferred from 2D input
2	Face detection	Employing texture analysis (e.g. Gabor filter) structural code of faces is identified. If sufficient structural code to form a full face is identified, the face area is marked with a square (in 2D) or cube (3D). Subsequent steps only focus on the marked face area
3	Landmark detection	Facial landmarks (e.g. mouth corners, eye lids, eye brows, etc.) are identified and marked within the face area via texture analysis
4	Landmark movement/AU evaluation	Facial landmark positions relative to the face area are compared to a standard face or the previous image to estimate landmark movement/AU activity
5	Emotion evaluation	The combination of all landmark movement/AU activity scores is compared to prototypical emotional expressions and based on their overlap expression likelihood or intensity are evaluated. For example, if AU1, AU2, AU5, and AU26 are strongly active and all others are not, an expression is rated as a strongly surprised expression

the landmark movement estimation. Usually, this standard face is derived from large datasets of faces on which algorithms were trained via machine learning or similar methods (Corneanu et al. 2016). Reference to this standard face can be problematic, because it ignores individual differences in facial morphology and plasticity. For example, morphology can introduce a bias if a person has considerably broader lips than the standard face. Then, the landmark movement algorithm would score a movement of the lip corner landmarks even if the person shows a completely neutral expression. In other words, the neutral or baseline point (0) of landmark movement varies across persons and can therefore be biased to an unknown degree if a standard face is employed.

The intensity of a landmark movement (or AU) is also evaluated in comparison to a large dataset with maximally strong movement of the respective landmark. Again, machine learning is typically employed to retrieve the necessary parameters (Corneanu et al. 2016). However, similar to the individual baseline, the individual facial plasticity determines the maximal point of landmark movement. Consider for example the fact that, for reasons of individual differences in facial anatomy some individuals can maximally pull their lip corners wider than others. Persons with lower than average plasticity cannot reach the maximal landmark movement scores, even if the individual maximum movement is reached. Consequently, we recommend using an individual baseline and controlling for face plasticity if landmark movement is evaluated. This requires a short calibration, i.e. recordings of neutral faces and of maximal activation of a series of AUs.

For many applied and scientific purposes, it is important to provide a score that also expresses the intensity of an emotion expression. This scoring is usually based on landmark movements (or AUs) and scores are usually better if the full pattern of all landmark positions is considered. If the landmark movement pattern of an evaluated face closely resembles the typical pattern of an emotional face, which is derived based on large datasets of persons showing this expression, then the evaluated face receives a high score on the respective emotion. Importantly, the full pattern of all landmark movements should be evaluated and not only those that are part of the emotion evaluated. Consider, for example, the emotions surprise and fear. Following FACSAID, the AUs active in surprise (AU1, AU2, AU5 and AU26) are a complete subset of the AUs active in fear (AU1, AU2, AU4, AU5, AU7, AU20 and AU26). If only AUs active in surprise expressions were used for evaluation, then a maximal fear expression would always result in a maximal surprise score, too. Therefore, the evaluation of surprise expression needs to also consider inactive AUs, i.e. a high score is achieved if (among others) AU4, AU7, and AU20 are inactive. For a detailed depiction, see Fig. 3.1. Similar scenarios apply between the facial expression of states such as pain and orgasm (Chen et al. 2018).

Most programs offer scores for (a subset of) Ekman's (1992) basic emotions and the emotion contempt. Additionally, many programs also deliver AU scores or land-mark movement scores which indicate AU activity. These data provide the opportunity to generate alternative emotion scores or develop scores for other expressions. For example, just as when experiencing strong emotion, experiencing strong pain automatically elicits specific facial expressions (Werner et al. 2017). Research on

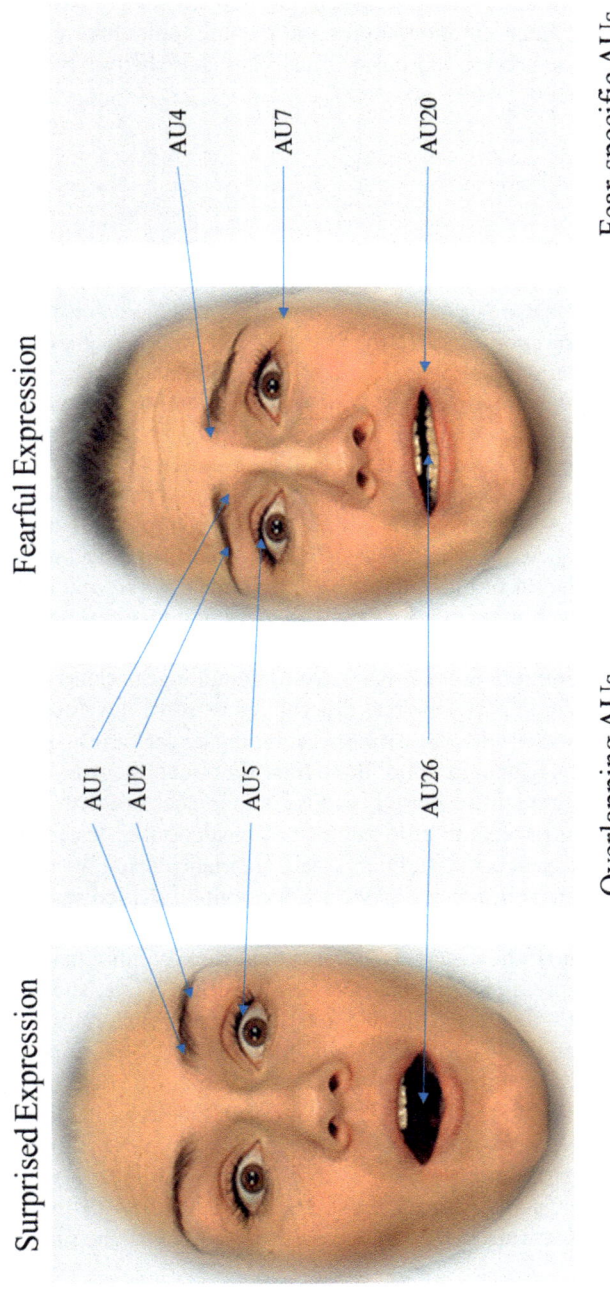

Fig. 3.1 An example of the Action Unit (AU) overlap of fearful and surprised expression, as described in the FACSAID (Ekman et al. 1998). AU1: inner brow raiser; AU2: outer brow raiser; AU4: brow lowerer; AU5: upper lid raiser; AU7: lid tightener; AU20: lip stretcher; AU26: jaw drop

pain expressions converges to a universal facial pain expression (Williams 2002; see also Chen et al. 2018), just like the universal emotion expressions of basic emotions. With this information, facial emotion expression recognition software that reports AU scores could also be used to score how much pain somebody experiences, for example in hospitals.

3.5   Limitations

Currently available software is not without limitations. The quality of decisions derived from programs that are based on machine learning methods (i.e. the precision of classification) hinges upon the validity of the underlying theory and employed training datasets (Zinkernagel et al. 2018). Obviously, the underlying theory, such as the comprehensive theory of basic emotions might be wrong or over simplistic. For example, the universality of basic emotions has been challenged by cultural differences in perceiving fear and disgust between Western and Eastern cultures (Jack et al. 2009). Other theories categorize emotions in largely different valence types (Scherer 2005). Assuming the facial expressions of fear and disgust were not universal, we would require culturally- or individually-adaptive software that judges an expression based on the cultural or individual background of a person. Such an endeavor is much more challenging than assuming that basic emotions are real, that they exist in every person, and that they have six prototypic manifestations representing distinct configurations of AUs—but this challenge is clearly an exciting new opportunity. In order to partly address this challenge, some developers began to train their algorithms on very diverse datasets, including a vast variety of ethnicities and achieved high performance for all of them (e.g. Emotient 2016b).

Yet, the software might still be biased because, so far, such software has usually been trained via supervised learning. In contrast to unsupervised learning, presuppositions about certain aspects of the training data are made during the training step during supervised learning (Hinton et al. 1999). For example, all pictures in a training dataset are pre-categorized according to western human FACS ratings in terms of the displayed basic emotion. If, however, there were culturally- or individually-different patterns of emotion expressions (as reported by Jack et al. 2009), such presuppositions would be wrong. Consequently, the software would have only learned the western configuration and would misjudge humans with the eastern configuration to express fear and disgust. Unsupervised machine learning might be deemed a solution for this problem. However, to be unbiased, this approach requires a facial expression training database that is perfectly representative of the whole world's population and would need constant training to account for changes over time, i.e. evolutionary conditioned changes in behavior. Acquiring such a dataset is currently inconceivable and consequently any software trained unsupervised would carry the sample bias of its training dataset.

Therefore, development and application of facial emotion expression recognition software must be cautious with respect to the cultural and individual background of their sample of training pictures and must consider measures such as improvement of data-basis, score adjustment, and constant training. In our example of eastern fear and disgust configuration, it might be sufficient to develop an alternative score. Jack and colleagues (2009) report a focus on the eye and nose region for eastern participants and a more holistic perception for western participants. Employing AU scores, it might be sufficient to only consider eye and nose AUs for eastern participants (e.g. fear AUs 1, 2, 4, 5 and 7) and all respective AUs (or the original fear emotion score) for western participants.

Another important limitation is that for proprietary software, relevant functional details are hidden in a black box. For proprietary products, it is usually not reported which databases are used to train the machine learning algorithms, how the databases are validated, and which machine learning algorithms were used for the different steps. Yet, for a sound assessment of such products information concerning the nature and use of the data bank are essential. For example: does it make a difference whether expressions stem from actual emotional states or were faked? Are the machine learning algorithms, such as support vector machines, neural networks and random forests relevant concerning categorization? This and further technical information are important for assessing the quality of a program and additional data are required to assess further aspects of classification accuracy.

Software developers report classification accuracy rates and often enough conclude that the software is valid and reliable (e.g. Emotient 2016a, c). Furthermore, results indicating that the software is not prone to systematic biases (such as lighting conditions) and recommendations for configuration (such as settings for video recording) are provided. However, evaluations independent of the software developers are required, too. Recent evaluations provide promising results. For example, scores of computerized facial emotion recognition software are more precise than untrained human raters and such scores provide reliable estimates of emotional intensity in dynamic expressions (Calvo et al. 2018). Software scores correlate strongly with corresponding intensities from electromyography measurement (Kulke et al. 2018). Yet, data show, too, that established software still has its limitations in judging blended emotions (Del Líbano et al. 2018).

In conclusion, although there is a variety of challenges for facial emotion recognition software, such software seems to work surprisingly well. Based on recent evaluations (Calvo et al. 2018; Kulke et al. 2018; Del Líbano et al. 2018) such software can be used in applied and scientific settings to reliably score facial emotion expression quicker, cheaper, and more objectively than with human FACS raters. And although there are still unresolved limitations, like the sole focus on 2D images and machine learning, recent approaches present ways to resolve these issues (Zinkernagel et al. 2018). Thus, presumably within the next few years, these limitations will have been overcome.

3.6   Current and Future Research

Recently, several studies (i.e. Calvo et al. 2018; Del Líbano et al. 2018; Hildebrandt et al. 2015; Kulke et al. 2018; Olderbak et al. 2014) have successfully introduced such software to answer basic and applied research questions. We will exemplify and highlight recent success with research on socio-emotional abilities. Socio-emotional abilities refer to human abilities related to interpersonal communication and action, including emotional abilities, such as emotion perception or regulation. Emotional abilities have also been categorized in the model of emotional intelligence (Mayer and Salovey 1997), which includes emotion perception, emotion expression, emotion understanding, and emotion management. Research mostly focused on receptive abilities, such as emotion perception and memory (Wilhelm et al. 2014). In order to measure such abilities, test takers are often asked to evaluate an item and choose amongst responses from a list of response options. Such receptive tasks ought to be distinguished from measures in which test takers create a response and the nature of that creation is the focus of evaluation. For example, productive measures have a long tradition in the measurement of written or spoken language proficiency (Reznick 1997). Often, performance evaluations for productive tasks are more difficult than assessing behavior in receptive tests, because defining a veridical answer to an item is more complex when the universe of possible responses is infinite. Nevertheless, if the goal is to evaluate aptitude and skill in emotion expression, we need productive tasks.

Novel technological developments allow assessing aptitude and skill in emotion expression by providing objective evaluations of facial expressions. We can now study how humans respond to specific forms of emotion elicitation and whether there are differences between subjects in these responses. We can now ask subjects to control their emotion expression and see how good they manage to do this. We can also ask participants to pose or imitate specific facial emotion expressions to study how good they do in deliberately using face expressions to communicate states (Hildebrandt et al. 2015). Beyond that, we can even study the dimensionality of typical facial expressions within persons to see what an individual's default expression looks like and to explore whether we find the same dimensions of typical emotion expression within and between subjects. We can even develop applications in which users respond with a landmark movement or an emotion expression rather than touching a device or responding verbally.

A prototypical procedure for studying such skills, aptitudes, or preferences could be that we ask test takers to react to emotion words (e.g. "happy") by producing corresponding facial emotion expressions. Likewise, we could ask them to react to a face by imitating its shown facial expression. A computerized evaluation of such expressions should proceed as follows. Presume a test taker is asked to express a happy face. The test-taker should receive a high score for an intense and prototypical happy expression, i.e. an expression that has a maximal probability to be recognized correctly by a random receiver. The evaluation is best accomplished if test takers are videotaped for a few seconds while they react to an item. Olderbak et al. 2014 discussed, how this data should be scored to represent expressive ability. Specifically, they recommend to first smooth the time series that could be based on AUs movements or represent emotion scores from readily available software and then extract maximum scores, as these represent this expressive ability (producing posed emotional expressions) with the least bias. This system is a sound way of scoring an emotion expression task and the procedures recommended here arguably offer a valid way of measuring abilities in the emotion expression domain.

Many of these procedures can be transferred to related fields. For instance, there is little research on emotion regulation ability employing actual ability tests, because evaluating the veridicality of a response to an emotion regulation item is complex. Just like with the posing of emotion expressions, we can ask test takers to regulate their facial expressions in different ways while experiencing emotional sensations. There are different ways to approach facial emotion regulation. For instance, we could provide strong stimuli to test takers and ask them to neutralize provoked expressions or to enhance the emotional response while they experience a congruent emotional sensation. In both cases, the evaluation of emotion expression should attempt to capture how successful test takers were in fulfilling the task (i.e. expressions under neutralizing instructions show no emotional expression beyond the baseline expression and enhanced expressions are strong in intensity and cannot be distinguished from authentic expressions).

3.7 Emotion Recognition Software in Mobile Devices

Apparently, emotion recognition software can easily be used in mobile devices, too. With constant improvements in the hardware of smartphones, alongside with increasing camera quality, the combination of mobile devices and computerized facial emotion expression recognition is an obvious and promising idea. An intriguing example for such applications was recently introduced by the artist Ruben van de Ven. Employing the Affectiva Software Developer Kit (Affectiva 2018), he developed the app Emotion Hero (van de Ven 2016a). In this gaming app, players practice and improve their facial expression skills by following instructions in a Guitar Hero-style fashion to facially express different emotions at varying time points and complex switches between emotions. The difficulty increases as the user progresses through levels and the app gives detailed individual feedback on how to improve facial expression in future rounds. Furthermore, players engage in a worldwide competition to become the 'Emotion Hero', motivating them to improve constantly.

Although the app is primarily meant as an art project and "a playful invitation to open up the box of expression analysis to reveal the assumptions that underlie this technology" (van de Ven 2016b), we also consider it a serious option for intervention studies on emotion expression abilities. For example, a much discussed key symptom of autism is reduced emotion expression ability (Jaswal and Akhtar 2018; Olderbak et al. 2019). Combining gamification and extremely detailed feedback, 'Emotion Hero' or similar tools could be just the right intervention for people with autism. Via smartphones such an intervention is easily embedded in everyday life, data indicating change in the key symptom can easily be recorded, stored, and automatically analyzed.

Detailed and frequent observation via smartphones also lead to a rise of experience sampling studies in the last few years. Some prominent examples are studies using emotion adjective lists to track mood fluctuations within a day or week (Wilt et al. 2011; Trampe et al. 2015). Although this is an intriguing approach to explore time dynamics of emotion, employing self-report questionnaires always has the cost of biased data caused by a plethora of response biases. With computerized facial emotion expression recognition software, less biased approaches might be possible. Instead of asking participants in an experience sampling study to provide ratings for adjective lists, asking for a selfie or for permission to record audio or video from time to time might be the better way to gather ecologically valid data in a noninvasive way. Just as in 'Emotion Hero', these recordings could directly be analyzed in the smartphone and only emotion scores stored and sent to the researchers to adhere to privacy considerations.

3.8 Conclusion

In sum, we see great—and so far, untapped—potential in facial emotion expression recognition software. Clearly, evaluation of facial expressions is an incredibly useful tool for studying behavior, skills, abilities, and preferences that were not easy to assess before. The application with mobile devices additionally removes constraints of location, time, and context in which face identity and facial expressions are studied. Although there are still many options for improvement, such software has already proven to surpass average human performance in emotion recognition and is comparable to trained FACS raters. At the same time, such software offers a quicker and more economical way of obtaining insight into human emotional states. In addition, as the software is increasing in precision and as existing biases are reduced or eliminated, in a few years the software will likely surpass FACS raters' accuracy rates. Research that seemed impossible only a few years ago now gives the opportunity to test hypotheses about the very core of emotion and emotional abilities. As a result, new interventions facilitated by such software will hopefully improve the lives of many people.

References

Affectiva (2018) Affectiva Affdex. https://www.affectiva.com/

Baltrusaitis T, Robinson P, Morency L-P (2016) OpenFace: an open source facial behavior analysis toolkit. https://doi.org/10.1109/wacv.2016.7477553

Bruce V, Young A (1986) Understanding face recognition. Br J Psychol 77:305–327. https://doi.org/10.1111/j.2044-8295.1986.tb02199.x

Calvo MG, Fernández-Martín A, Recio G, Lundqvist D (2018) Human observers and automated assessment of dynamic emotional facial expressions: KDEF-dyn database validation. Front Psych 9:2052

Cannon WB (1927) The James-Lange theory of emotions: a critical examination and an alternative theory. Am J Psychol 39:106–124. https://doi.org/10.2307/1415404

Chen C, Crivelli C, Garrod OGB et al (2018) Distinct facial expressions represent pain and pleasure across cultures. Proc Natl Acad Sci U S A 115:E10013–E10021. https://doi.org/10.1073/pnas.1807862115

Corneanu C, Simón MO, Cohn JF, Guerrero SE (2016) Survey on RGB, 3D, thermal, and multimodal approaches for facial expression recognition: history, trends, and affect-related applications. IEEE Trans Pattern Anal Mach Intell 38(8):1548–1568. https://doi.org/10.1109/TPAMI.2016.2515606

Darwin C (1987) Expression of the emotions in man and animal. Appleton and Company, New York

De la Torre F, Chu WS, Xiong X et al (2015) IntraFace. Presented at 11th IEEE international conference and workshops on automatic face and gesture recognition (FG), Ljubljana, Slovenia, 4–8 May 2015

Del Líbano M, Calvo MG, Fernández-Martín A, Recio G (2018) Discrimination between smiling faces: human observers vs. automated face analysis. Acta Psychol 187:19–29. https://doi.org/10.1016/j.actpsy.2018.04.019

Ekman P (1992) An argument for basic emotions. Cogn Emot 6:169–200. https://doi.org/10.1080/02699939208411068

Ekman P, Friesen WV (1978) Facial action coding system: a technique for the measurement of facial action. Manual for the Facial Action Coding System

Ekman P, Rosenberg E, Hager J (1998) Facial action coding system affect interpretation database (FACSAID). http://face-and-emotion.com/dataface/facsaid/description.jsp

Emotient (2016a) Facet Emotient Inc. www.emotient.com

Emotient (2016b) FACET 2.0 performance evaluation. https://imotions.com

Emotient (2016c) Emotient SDK 4.1 performance evaluation. https://imotions.com

Hildebrandt A, Olderbak S, Wilhelm O (2015) Facial emotion expression, individual differences. In: Wright JD (ed) International encyclopedia of the social & behavioral sciences, 2nd edn. Elsevier, Oxford, pp 667–675

Hinton GE, Sejnowski TJ, Poggio TA (1999) Unsupervised learning: foundations of neural computation. MIT Press, Cambridge

Jack RE, Blais C, Scheepers C et al (2009) Cultural confusions show that facial expressions are not universal. Curr Biol 19:1543–1548. https://doi.org/10.1016/j.cub.2009.07.051

Jaswal VK, Akhtar N (2018) Being vs. appearing socially uninterested: challenging assumptions about social motivation in autism. Behav Brain Sci 1–84. https://doi.org/10.1017/s0140525x18001826

Jeni LA, Cohn JF, Kanade T (2015) Dense 3D face alignment from 2D videos in real-time. Presented at 2015 11th IEEE international conference and workshops on automatic face and gesture recognition (FG), Ljubljana, Slovenia, 4–8 May 2015

Kulke L, Feyerabend D, Schacht A (2018) Comparing the Affectiva iMotions facial expression analysis software with EMG. PsyArXiv. https://doi.org/10.31234/osf.io/6c58y

Marĉelja S (1980) Mathematical description of the responses of simple cortical cells*. J Opt Soc Am, JOSA 70:1297–1300. https://doi.org/10.1364/JOSA.70.001297

Mayer JD, Salovey P (1997) What is emotional intelligence? In: Salovey P, Sluyter D (eds) Emotional development and emotional intelligence: implications for educators. Basic Books, New York, pp 3–31

Noldus Information Technology (2018) FaceReader. www.noldus.com

Olderbak S, Geiger M, Wilhelm O (2019) A call for revamping socio-emotional ability research in autism. Behav Brain Sci 42

Olderbak S, Hildebrandt A, Pinkpank T et al (2014) Psychometric challenges and proposed solutions when scoring facial emotion expression codes. Behav Res Methods 46:992–1006. https://doi.org/10.3758/s13428-013-0421-3

Reznick JS (1997) Intelligence, language, nature, and nurture in young twins. In: Sternberg RJ, Grigorenko E (eds) Intelligence, heredity, and environment. Cambridge University Press, New York

Schachter S, Singer J (1962) Cognitive, social, and physiological determinants of emotional state. Psychol Rev 69:379–399

Scherer KR (2005) What are emotions? And how can they be measured? Soc Sci Inf 44:695–729. https://doi.org/10.1177/0539018405058216

Tcherkassof A, Bollon T, Dubois M et al (2007) Facial expressions of emotions: a methodological contribution to the study of spontaneous and dynamic emotional faces. Eur J Soc Psychol 37:1325–1345. https://doi.org/10.1002/ejsp.427

Trampe D, Quoidbach J, Taquet M (2015) Emotions in everyday life. PLoS ONE 10:e0145450. https://doi.org/10.1371/journal.pone.0145450

van de Ven R (2016a) Emotion Hero. https://emotionhero.com/

van de Ven R (2016b) Emotion Hero. https://rubenvandeven.com/. Accessed 4 Oct 2018

Werner P, Al-Hamadi A, Walter S (2017) Analysis of facial expressiveness during experimentally induced heat pain. Presented at 2017 seventh IEEE international conference on affective computing and intelligent interaction workshops and demos (ACIIW). San Antonio, TX, USA, 23–26 October 2017

Wilhelm O, Hildebrandt A, Manske K et al (2014) Test battery for measuring the perception and recognition of facial expressions of emotion. Front Psychol 5. https://doi.org/10.3389/fpsyg.2014.00404

Williams AC de C (2002) Facial expression of pain: an evolutionary account. Behav Brain Sci 25. https://doi.org/10.1017/s0140525x02000080

Wilt J, Funkhouser K, Revelle W (2011) The dynamic relationships of affective synchrony to perceptions of situations. J Res Pers 45:309–321. https://doi.org/10.1016/j.jrp.2011.03.005

Zhang C, Zhang Z (2010) A survey of recent advances in face detection

Zinkernagel A, Alexandrowicz RW, Lischetzke T, Schmitt M (2018) The blenderFace method: video-based measurement of raw movement data during facial expressions of emotion using open-source software. Behav Res Methods. https://doi.org/10.3758/s13428-018-1085-9

Chapter 4
An Overview on Doing Psychodiagnostics in Personality Psychology and Tracking Physical Activity via Smartphones

Rayna Sariyska and Christian Montag

Abstract The aim of this chapter is to introduce and describe how digital technologies, in particular smartphones, can be used in research in two areas, namely (i) to conduct personality assessment and (ii) to assess and promote physical activity. This area of research is very timely, because it demonstrates how the ubiquitously available smartphone technology—next to its known advantages in day-to-day life—can provide insights into many variables, relevant for psycho-social research, beyond what is possible within the classic spectrum of self-report inventories and laboratory experiments. The present chapter gives a brief overview on first empirical studies and discusses both opportunities and challenges in this rapidly developing research area.

4.1 Introduction

With the rise of the digital era, human life is changing at a rapid pace (Scholz et al. 2018; Montag and Diefenbach 2018). Among others, this is mirrored in the growing use of portable devices such as laptops, smartphones, tablets and smartwatches. At the moment of writing this work, it is only 12 years after the introduction of the prominent iPhone, and in the meanwhile 2.71 billion smartphone users have been estimated worldwide for 2019 (Statista 2019a). Mobile devices such as the smartphone do not only help us in dealing with manifold everyday life issues, but they also determine to a large extent how people communicate, work, and navigate themselves to new places.

With the development of the Internet of Things (IoT), different devices can be easily connected and help us to communicate or exchange data. For example, the

R. Sariyska (✉) · C. Montag
Institute of Psychology and Education, Ulm University, Ulm, Germany
e-mail: rayna.sariyska@uni-ulm.de

C. Montag
MOE Key Lab for Neuroinformation, The Clinical Hospital of Chengdu Brain Science Institute, University of Electronic Science and Technology of China, Chengdu, China

© Springer Nature Switzerland AG 2019
H. Baumeister and C. Montag (eds.), *Digital Phenotyping and Mobile Sensing*,
Studies in Neuroscience, Psychology and Behavioral Economics,
https://doi.org/10.1007/978-3-030-31620-4_4

companies Amazon and Apple introduced products, which enable their users to play music or prepare the grocery list via voice commands (e.g. Apple's HomePod with Siri and Amazon Echo with Alexa). In this context also the smart home needs to be mentioned, enabling humans to operate, e.g. the heating system in an apartment via the Internet. There are many other examples of concepts and devices which have partly been developed or are developing as we speak such as smart cars, smart factories, and smart cities (Forbes 2016). Without doubt, this development towards a totally interconnected world brings many advantages, but it also has its drawbacks such as ethical and privacy issues and high developmental costs. Moreover, the massive amount of data stemming from the interaction with the IoT needs to be saved on servers and only powerful computers can process the available data. Challenges with respect to privacy and ethical issues will be only shortly addressed in this chapter since they will be described in more detail in Chap. 1 by Kargl et al. and Chap. 2 by Dagum and Montag of this book.

To illustrate how the study of traces from the IoT can be useful in the field of psychodiagnostics, and, thus, in many research fields such as health psychology, clinical psychology, behavioural medicine, sociology, cognitive and sport sciences, this chapter will provide the reader with some examples from the literature. We will start with an overview of a relatively new branch of personality psychology trying to link smartphone use, including its diverse applications, to personality. We will not refer to the many existing studies investigating smartphone use and addiction via self-report (Lachmann et al. 2017; Carvalho et al. 2018), but will exclusively concentrate on works tracking the daily smartphone interaction directly via the smartphone and link this to self-reported personality measures. A second area to be reviewed will deal with physical activity, a variable being of more and more relevance in different areas of research such as health psychology and sport sciences (Mammen and Faulkner 2013; Verburgh et al. 2014; Blondell et al. 2014; Schuch et al. 2016).

4.2   Definition of Personality and the Big Five of Personality

Personality definitions vary depending on the large number of available theories. Many of those theoretical frameworks agree upon personality representing relatively stable individual characteristics in cognition, emotion and motivation resulting in behavioural outcomes, also known as *traits* (e.g. McCrae and Costa 1997). In contrast, *states* are defined as the momentary mood of a person (Montag and Reuter 2017; for a new debate on personality definitions see the work by Baumert et al. 2017). Montag and Panksepp (2017) emphasized the relative stability of personality over time and to a lesser extent across situations. Referring to the time component in healthy personality development, it has been widely accepted that personality stabilizes in early adulthood with slight changes towards "better" persons in terms of higher agreeableness and conscientiousness across the life span (McCrae and Costa 1994; also see studies by Helson et al. 2002 and Edmonds et al. 2013, reporting differences regarding the temporal stability of different personality traits and changes in personality throughout adulthood). For new insights into personality change we refer to a recent review-work by Bleidorn et al. (2018). With respect to discussions on personality stability across diverse situations, please see the important works by Mischel and Shoda (1995) and Mischel (2004), but in general also the important latent trait-state theory (Steyer et al. 1999).

Among the many personality theories, the Big Five of Personality without doubt represents one of the most well-established theoretical frameworks, which has been found to describe personality in many different countries and cultures and has been applied to different research topics (for an overview see Davis and Panksepp 2018). The Big Five of Personality have been derived from a lexical approach with its early origins in the works of Cattell (1933), Allport and Odbert (1936), Fiske (1949) and Tupes and Christal (1992). The Five Factor Model of Personality formulating item sentences instead of relying "only" on adjectives to describe a person has been established by Costa and McCrae (1992); see also McCrae and John (1992). As a result of factorial analysis of human language with a focus on words describing persons, five factors appeared to describe personality: openness to experience, conscientiousness, extraversion, agreeableness and neuroticism (acronym OCEAN). High openness to experience is linked to curiosity and wide imagination, while conscientious people can be characterised by discipline, efficiency and reliability. Extraverts are energetic, outgoing and talkative. Agreeable people are generous and forgiving. Last, but not least, high neuroticism is linked to being anxious and tense (McCrae and Costa 1997). Please note that other taxonomies have been proposed going beyond the Big Five resulting in lower or higher number of factors to globally describe human personality. Perhaps of most relevance is the HEXACO model adding the dimension Honesty/Humility to the aforementioned Big Five factors (Ashton and Lee 2007). Personality traits have mostly been assessed via self-report inventories in the past. Therefore, we want to build an argument for the importance of also including directly observed behaviour to improve personality assessments in the next section.

4.3 On the Importance to Consider Directly Observed Behaviour in the Psychological Sciences

As mentioned, psychologists usually rely on self-report such as data collected via questionnaires or interviews, and comparably seldom on direct observation and implicit assessment methods (Paulhus and Vazire 2007; Baumeister et al. 2007). The examination of personality by means of self-report is very important and, in our opinion, will probably never be completely replaced by other research methods, since some information about a person can only be accessed via introspection (Montag and Elhai 2019). This is also the case in other research areas such as clinical psychology, where it is of special relevance to gain insights into a person's subjective experience on his/her well-being by directly asking the patient. This said, there are a lot of complicated issues with self-report, which will be discussed in more detail later in this work. Before doing this, we mention that we will not discuss the advantages and disadvantages of assessing personality via questionnaires on smartphones. Such research including studies comparing the psychometric quality of data derived via online questionnaires compared with paper-pencil questionnaires have been conducted abundantly (see Riva et al. 2003; or Weigold et al. 2013).

The need to study and directly observe behaviour to get insights into the personality structure of a person has been put forward many times. Among others, Baumeister et al. (2007) highlighted the importance to assess behaviour in the field of personality and social psychology. The authors reported in their work the progression of observational studies published throughout the years: while at the end of the 70s around 80% of studies included behavioural variables, in 2006 less than 20% of the studies published in the *Journal of Personality and Social Psychology* considered behavioural variables. Baumeister et al. (2007) stressed that next to self-report measures, behaviour needs to be assessed too, since actual and hypothetical behaviour are known to differ to some extent (e.g. estimating one's own smartphone use compared to one's actual use results in different numbers, Montag et al. 2015a). One solution to reverse this trend in the literature would be the inclusion of variables tracked during human-technology interaction. Here, the smartphone possesses high potential to enable researchers to "participate" in the life of a study's participant by tracking the interaction with a smartphone on a daily basis. This will give insights into variables such as movement, interaction with others via calls or WhatsApp activity, and the frequency and duration of smartphone use in general. These variables are discussed to serve as markers of personality as will become evident in the reviewed works in one of the next sections.

4.4   Limitations of Self-Report Assessment or Traditional Observation Methods

A large problem in research relying on self-report represents the tendency to answer in a socially desirable way, hence, to present oneself according to perceived social norms. Recall bias, or the difficulty to recall past events correctly or completely, is another problem often mentioned in this context. Moreover, self-report is often biased by impression management, self-deception and lack of conscious awareness (e.g. Paulhus and Vazire 2007).

With regard to traditional observation methods, standardised coding systems are often missing and this negatively impacts on inter-rater-reliabilities. Moreover, observer bias can negatively influence the quality of a study. Here, the expectations of the observer affect the way he/she perceives and evaluates the observed situation. A further disadvantage of many observation studies is that they usually take place in a controlled environment, thus, it is seldom possible to observe participants' behaviour in their natural environment. This limits the generalizability and ecological validity of the results in such a study. Of note, observation is a very time-consuming method where often considerable resources need to be invested including the training of the observers. Last, but not least, both self-report assessment and traditional observations are difficult to conduct as a part of a longitudinal study given limited resources.

4.5 Overcoming Limitations of Self-Report and Traditional Observation by Means of Smartphone App-Tracking

Compared to self-report or observation methods, the inclusion of digital technology in modern research facilitates real-time monitoring of a person's behaviour in a naturalistic (real-world) setting, thus, fostering ecological validity. Bidargaddi et al. (2017) defined digital footprints as "… data traces arising as a by-product of individuals' day-to-day interactions with mobile and/or Internet-connected technologies …" (p. 165). These digital footprints can be utilized to get insights in many variables via digital phenotyping or mobile sensing (e.g. Lane et al. 2010; Bidargaddi et al. 2017; Insel 2018). Such research activities could be seen as part of a new interdisciplinary research area called Psycho(neuro)informatics (Yarkoni 2012; Montag et al. 2016; Montag 2019). Examples for digital footprints are online traces left by (credit) card payments, data collected with wearable technology devices such as smartwatches, rings and glasses, with smartphones and tablets, and technology used as a part of a smart home or Facebook "likes" (Bidargaddi et al. 2017; see also Chap. 7 by Marengo and Settanni exclusively dealing with psychodiagnostics using data from social media platforms). For the present chapter, we mention the smartphone as an important source, because it is usually connected to the Internet and has powerful integrated sensors and features including Bluetooth, GPS, microphone, accelerometer, WiFi, light sensor, and proximity sensor, all of high relevance to study a myriad of variables (Miller 2012; Harari et al. 2016).

Due to the opportunity to record behaviour via a smartphone on repetitive occasions and to track many different behaviours at the same time (e.g. tracking calls and social network activity), human behaviour and related personality traits can be described in a more fine-grained way. This not only enhances the precision of results, but also improves research dealing with stability issues of behaviour (Allemand and Mehl 2017). Further advantages can be named: assessments via smartphones are usually passive and unobtrusive, since many smartphone applications record the interaction with the smartphone in the background. Digital tracking methods might also in particular be well suited to capture the behaviour of young children and ageing adults who cannot provide reliable self-report (Allemand and Mehl 2017). Due to the wide availability and acceptance of smartphones, it is easier to recruit large and representative samples compared to the pre-smartphone-era. Moreover, due to the affordability of this technology for masses, such digital assessments can take place at relatively low cost. However, the development of a smartphone application to track behaviour is still costly and brings ongoing efforts to always smoothly operate on the newest platform versions. The recent years have also seen a strong trend to track biological variables, using wearables, such as electrodermal activity, pulse, blood pressure, even headbands for recording EEG parameters. These biological variables can be brought together with the behaviour of a person recorded via a smartphone or another mobile device. Additionally, some applications are able to track users' subjective experiences through integrated questionnaires on the smartphone, next to the interactions with the smartphone (Montag et al. 2019). In so far, also the more classic method of "experience sampling" can be added to research in Psychoinformatics.

4.6   Examples for the Use of Smartphone Applications and Sensors to Assess Personality

In the following, results from studies showing associations between self-reported personality and (passively) recorded smartphone data will be presented. This also will provide first insights into the feasibility of personality assessment using smartphones via digital traces left as a side product of the daily smartphone interaction. Ultimately, researchers wonder if robust links between personality and certain patterns of smartphone interaction exist. Does knowledge about such patterns alone provide insights into the personality of a person? When interpreting the results from the following studies it is important to bear in mind that in most of the empirical works objectively measured smartphone use was correlated with self-reported personality, which might be biased, and, thus, not mirror the "true" personality of a person, as described in the previous sections.

De Montjoye et al. (2013) tested 69 participants with an open sensing framework on Android smartphones. Among others, calls and messages were recorded (please refer to Table 1 in the article by de Montjoye et al. (2013) for an overview of all examined parameters) and personality was assessed with the Big Five Inventory BFI-44. By applying a classification method, the authors were able to predict personality variables (high vs. average vs. low scores) with accuracies varying between 49 and 61%. Even though these prediction rates are not very high, the results still suggest a link between smartphone logs and self-reported personality.

Chittaranjan et al. (2013) presented data from 117 participants tracked for the duration of 17 months. Regarding the data collection, among others SMS logs, call logs and app logs were examined. Positive associations between extraversion and the duration of incoming calls, and negative correlations between conscientiousness and the usage of video/music apps were reported (those are only a few examples from this study, many more associations were reported). Moreover, the authors demonstrated that a set of smartphone variables might be able to predict different personality traits (in particular, based on the smartphone data it was possible to classify users into groups of high vs. low on a personality trait). However, it needs to be mentioned, that the associations found were rather weak.

The link between extraversion and call variables, reported by Chittaranjan et al. (2013) was supported in two independent studies by Montag et al. (2014, 2019). In those two studies participants from Germany were tested using an application called *Menthal* for the duration of four weeks and the NEO-Five-Factor Inventory (NEO-FFI) in Montag et al. (2014), and another smartphone application called *Insights* for the duration of 12 days and the Trait Self-Description Inventory (TSDI, see Olaru et al. 2015) to assess personality in Montag et al. (2019). Nearly all investigated call variables (e.g. number of calls, duration of calls) were positively linked to extraversion. Conscientiousness was positively linked to the duration of calls in Montag et al. (2014) and neuroticism was negatively associated with the number of incoming calls in Montag et al. (2019). Moreover, Montag et al. (2014) reported excellent reliabilities of the tracked variables (calls) for the period of four weeks, which indicates that call behaviour is a stable variable on the phones and that a few weeks/perhaps even few days of tracking might provide already sufficient insights into call behaviour.

Stachl et al. (2017) also demonstrated a positive association between the frequency of calls, recorded on a smartphone for 60 days, and extraversion. However, extraversion was negatively linked to the duration of calls. Furthermore, positive associations between extraversion and application use related to photography and communication were reported. Regarding agreeableness, a positive relation with the use of transportation apps was reported, while conscientiousness was linked to lower use of gaming apps.

In another study, Montag et al. (2015b) addressed the role of social networking channels used on the smartphone. Here, 2418 participants with an average age of 25 years were tracked using the same *Menthal* application, introduced earlier in this section for the duration of four weeks (for more information on the actual data analysis see the original work). Participants received feedback regarding their smartphone use (e.g. duration of smartphone use per day or most frequently used applications) through the *Menthal* application as an incentive for their participation. Self-reported personality data was collected with the ten-item version of the Big Five Inventory (BFI), integrated in the app. Results showed that participants spent 162 min per day on average on their smartphones, with about 20% of this time dedicated to WhatsApp use (ca. 32 min per day) and with less than 10% (ca. 15 min) dedicated to Facebook use. Additionally, results showed that younger age and the female gender were linked to longer WhatsApp use. With respect to personality, extraversion was positively and conscientiousness negatively associated with the duration of WhatsApp use. Similar, but weaker patterns could be observed for the duration of Facebook use. Again, the observed associations were rather weak.

Xu et al. (2016) investigated the installed apps on the smartphones of 2043 users and their relation to personality traits. Personality was assessed using the Big Five 44 questionnaire. Among others, it was demonstrated that high extraversion was negatively linked to installed apps in the gaming category and high neuroticism was positively associated with the adoption of photography and personalization apps. Moreover, high conscientiousness was negatively linked to the installed apps in the categories music and video, photography and personalization. While agreeableness was negatively linked to the adoption of personalization apps, openness to experience was not associated with any of the installed app categories.

Sariyska et al. (2018) demonstrated the use of the *Insights* smartphone application, mentioned earlier in this section, for personality-genetic research. In total, 117 participants were tested and smartphone data was recorded for 12 days. Among others, the association between a functional single nucleotide polymorphism (SNP) on the OXTR gene (rs2268498) and diverse smartphone-recorded variables was examined. The results demonstrated that the TT genotype (linked to higher empathy and better abilities in face recognition in previous studies, see Melchers et al. 2013; Christ et al. 2016), was associated with higher number of active contacts (contacts that one is in touch with as opposed to the total number of contacts in the phone book). Please note that this association got weaker after age was controlled for. Moreover, the variable active contacts was positively linked to extraversion. This study further demonstrated, how personality variables recorded using the methods of Psychoinformatics might be further used in the field of molecular psychology, the latter being a field aiming to understand the molecular (genetic) basis of individual differences in psychological variables (Montag 2018a, b). The feasibility to use smartphone recorded data in neuropsychological studies was also demonstrated in a study by Montag et al. (2017), where Facebook use (e.g. higher frequency of use per day) was associated with smaller grey matter volume of the nucleus accumbens, a brain region which is part of the reward system. In sum, both molecular-genetic and brain imaging variables have been successfully investigated in the context of smartphone-trackedreal-world behaviour. Indeed, the fusion of bio and med-tech might become an important research avenue in the near future (Montag and Dagum 2019).

In sum, the results of these studies indicate a link between objectively measured smartphone use and self-reported personality. Among others, the link between extraversion and call variables, recorded on the smartphone, was reported in a couple of independent studies, all using different self-report measures of personality. Thus, even after taking into account that self-reports might be possibly biased, the reported results strongly suggest an association between being extraverted and calling behavior. Moreover, app usage (e.g. social networking apps such as WhatsApp) seem to be linked to personality, too. It is important to bear in mind that most of the associations cited were rather weak. However, future studies need to address the question if a larger number of smartphone parameters taken together along with the use of machine learning approaches will be able to predict personality with a higher accuracy. This would not mean that digital traces can completely replace self-report measures of personality, since the subjective experiences of a person can only be assessed using self-report. However, depending on the study design and research question, objective measures derived with smartphones might complement self-report measures or even be more suitable in longitudinal designs where behavior is easily passively recorded and at a lower cost.

4.7 Examples of Studies Assessing Physical Activity

In the following, the feasibility to use smartphone applications to assess physical activity and the effect of interventions to promote physical activity using smartphone applications will be shortly discussed in order to give a second perspective on smartphone app usage in assessing behaviour. Please note that our target was not to provide an exhaustive literature review (a few very recent systematic reviews and meta-analyses are reported below), but to give an overview of the recent literature and, thus, provide examples for the viability to use smartphones for physical activity tracking and promotion.

The study of physical activity is of high importance, because it is linked to many different health benefits such as the prevention of heart disease. Moreover, physical activity increases muscular and bone health, and reduces the risk for diabetes and obesity (World Health Organization 2018a). Ischaemic heart disease and diabetes mellitus are among the top ten global causes of death for 2016 according to the WHO (World Health Organization 2018b). Before we provide an overview of relevant studies investigating physical activity with app technologies, we shortly introduce the concept of physical activity and how it is currently assessed.

Physical activity is defined as "any bodily movement produced by skeletal muscles that results in energy expenditure" and includes any activity as a part of leisure time (e.g. sports, exercise, playing, dancing), work (e.g. household chores, gardening) and transport (e.g. walking, cycling) (World Health Organization 2018a; Caspersen et al. 1985, p. 126). For definitions on different kinds of physical activity we also refer to an older work of Caspersen et al. (1985). According to the World Health Organization (World Health Organization 2018a) globally 81% of adolescents aged 11–17 and 23% of adults at 18 years of age or older showed insufficient physical activity in 2010. For adolescents the WHO recommended 60 min of moderate to vigorous physical activity per day, whereas for adults a minimum of 150 min of moderate-intensity or 75 min of vigorous-intensity physical activity per week are recommended. Guthold et al. (2018) reported levels of insufficient physical activity between 2001 and 2016, in a study including 1.9 million adult participants from 168 countries. The percentage of insufficient physical activity in 2016 was estimated 27.5% globally, with women (compared to men) and high-income countries (compared to low-income countries) demonstrating higher levels of insufficient activity according to the WHO guidelines. Althoff et al. (2017) conducted another large-scale study, including more than 700.000 participants from 111 countries (some of the analyses were conducted with only 46 countries). However, in this study physical activity was assessed using smartphone accelerometers as compared to the above study, based on self-report measures (accelerometers are sensors that measure the acceleration of a moving body, which can be applied to assess physical activity variables such as the number of steps per day; Evenson and Terry 2009). Some of the findings with regard to cross-cultural and gender differences were similar to those, reported by Guthold et al. (2018), thus, demonstrating similar results across different methods.

The WHO is giving advice on how to increase physical activity as presented in *The Global Action Plan for the Prevention and Control of Noncommunicable Diseases 2013–2020* and *The Global Action Plan for Physical Activity 2018–2030*, where a road map and guidelines have been developed, including policy recommendations for increasing physical activity on a global and national level. Similar to the problems reported regarding self-report measures of personality assessment, over- and under-estimation of physical activity in self-report has been demonstrated as compared to objective measures such as accelerometers, pedometers or heart rate monitors (e.g. Prince et al. 2008; Adamo et al. 2009).

Due to the high penetration rate of smartphones in many societies around the globe, it is meaningful to assess their potential in promoting physical activity in this context. Seifert et al. (2017) demonstrated in a representative sample from Switzerland (n = 1013) that 20.5% of participants older than 50 years used mobile devices to track physical activity, thus showing that such devices are even used by a substantial amount of people in the higher age range. Among those participants, 55.1% used tablets or smartphones to track physical activity, while 6.1% used smartwatches and 38.8% used activity trackers. However, when comparing a group of participants tracking their physical activity with those who did not, it was demonstrated that the first group of participants was younger, was more involved in physical exercise and was more often male than female. Thus, such potential differences need also be considered as confounding factors in future research.

Regarding physical activity research Bort-Roig et al. (2014) conducted a systematic review, including articles published between 2007 and 2013, and reported a varying measurement accuracy of physical activity assessment via smartphones. In this study accuracy varied between 52 and 100% (please note that physical activity data here included findings from studies using accelerometers built in the smartphones, as well as external measurement devices, connected to the smartphone). Lu et al. (2017) demonstrated the feasibility to use smartphone accelerometers for tracking different types of physical activity such as daily living activities (e.g. walking, jogging, sitting and standing) and sports activities (e.g. race walking and basketball playing). Höchsmann et al. (2018) tested the accuracy of step counting for different smartphones and activity trackers in 20 participants in a laboratory setting (on a treadmill) and in a "free-walking" condition, and compared the results with video recordings. It was demonstrated that smartphone accelerometers (in particular the iPhone SE) accurately recorded the number of steps independent of their body position (in the pocket or bag) or walking speed in the laboratory condition, while they tended to slightly underestimate the number of steps in the "free-walking" condition. However, different levels of accuracy were reported for different smartphone brands (e.g. iPhone SE with Apple Health vs. Samsung Galaxy S6 Edge with Samsung S Health). Different accelerometer applications (not built in, but installed on the smartphone in this case on the iPhone SE; e.g. Runtastic Pedometer, Accupedo and Pacer), were also evaluated and all showed high accuracy for step counting. Hekler et al. (2015) focussed on the evaluation of Android smartphones (HTC MyTouch, Google Nexus One and Motorola Cliq) for physical activity tracking as compared to an Actigraph (an accelerometer-based activity/sleep monitor by the company Actigraph) in a laboratory setting (15 participants) and a "free-living" setting (23 participants). Different activities were assessed such as sitting, standing, walking with different speeds and bicycling. For the laboratory setting correlations between 0.77 and 0.82 between the Actigraph and the smartphones were demonstrated (these correlations got stronger when bicycling and standing were excluded). The correlations in the "free-living" setting were weaker, and varied depending on the intensity of the activity (0.38–0.67). In sum, these results suggest that smartphones offer a promising way to assess physical activity as their measurement accuracy is comparable with the accuracy of devices used by researchers in the laboratory. Last but not least, due to the opportunity to connect to the Internet via the smartphone, the smartphones' reasonably big displays, and features such as microphone and a speaker, smartphones can be used to provide feedback and motivate humans to actively engage in physical activity. Thus, smartphones have a large potential also as intervention tools.

It is noteworthy that first scientific evidence supports the idea that smartphone-based interventions are effective in increasing physical activity. Coughlin et al. (2016) conducted a review to evaluate the efficacy of different smartphone apps to promote physical activity and to lose weight. The review included six qualitative studies as well as eight randomized control trials. Here, mostly positive effects of the use of smartphone applications in promoting physical activity and reducing weight were reported (for effect sizes please refer to the individual studies reviewed in the article). In a recent meta-analysis Romeo et al. (2019) examined six randomized control studies on the influence of physical activity apps on objectively measured steps per day. Even though the intervention groups demonstrated increased activity in steps, the difference to the control groups was not significant. However, it was demonstrated that shorter interventions (< 3 months) and the sole focus on physical activity without considering further health benefits were more effective than longer interventions and additionally targeting diet. Overall, more research is needed on the effectiveness of physical activity interventions using smartphones. Moreover, the implementation of psychological theories which incorporate techniques for behavioural change, such as the Social Cognitive Theory, at the basis of functioning of the app interventions might help to enhance the effectiveness of such interventions (see Chap. 17 by Baumeister et al.). Some examples of such techniques include features allowing to monitor one's own activity or compare it to others, or to gain social support by sharing one's own achievements via online social networks (for an overview of theories on behavioural change and their current application in the realm of activity tracking please refer to Sullivan and Lachman 2017). Such features might not only support the behavioural intervention, but also seem to be facilitating factors for the use of smartphone accelerometers in the first place as reported in the review by Bort-Roig et al. (2014).

Last but not least, the link between personality and physical activity has been of interest for researchers to predict levels of physical activity, participation in different kinds of sports, and to answer the question of what distinguishes competitive sportsmen from amateur athletes, to name a few (e.g. Malinauskas et al. 2014; Wilson and Dishman 2015; Monasterio et al. 2016). By collecting further evidence on the link between those variables, smartphone interventions can be organized in a tailor-made manner, meaning that different strategies for promoting physical activity will be applied, e.g. with highly extraverted individuals as compared to highly neurotic individuals.

The availability and affordability of smartphones as well as the positive user attitude towards smartphones offers a large potential for smartphones as measurement tools and their role in intervention studies. However, there are also some drawbacks, which will be discussed in the next section.

4.8   Challenges and Further Implications

In the following, the challenges of smartphone tracking devices will be discussed and further implications will be drawn (see Table 4.1). One important issue to tackle in the near future to successfully implement smartphone technologies in research might be low compliance on the side of participants. Moreover, through the minute-to-minute tracking of different activities on the smartphone, participants might feel observed and might also try and adapt their behaviour (e.g. spent less time on Facebook; see also Montag et al. 2016). This falls in the context of social desirability, as discussed at the beginning of this chapter. A second concern is related to privacy issues and ethical considerations. Since tracking one's interaction with a smartphone can be described as an invasion of privacy, it is very important to ensure high levels of security when transferring the data from the phones of participants and storing them on a server (for an infrastructure of such a set up see Markowetz et al. 2014). Here, it should be also considered not to track everything what is possible, but ask (a) what variables do I really need to record to answer the research question at hand and (b) on what granularity level needs this behaviour be recorded? This all is of tremendous relevance to increase trust in a research project and elevate participation rates. Again, we will not go into detail on this topic because privacy issues and ethical concerns are addressed in Chap. 1 by Kargl et al. and Chap. 2 by Dagum and Montag.

There are also challenges to be met concerning a researcher's skill set. In psychological sciences, researchers so far seldom have the know-how to develop a smartphone application or use smartphone sensors to track behaviour. Thus, technical advice from a computer scientist is usually needed regarding hardware and software requirements, because large amounts of data are saved and processed. Moreover, the researchers might benefit from training in screening, organising (e.g. defining variables) and analysing big data, so that it can be interpreted in a meaningful way. Classical statistical approaches such as descriptive or inferential statistics might only superficially grasp what actually could be found in a data set. New analysis techniques from the computer science including machine learning clearly might be of advantage in Psychoinformatics. For example, by using deep learning approaches on large data sets one can achieve astonishingly accurate predictions/classifications comparable to those achieved by humans (e.g. Ronao and Cho 2016). These sophisticated methods have a tremendous potential to be used in research in general, but

Table 4.1 Overview of the challenges and future tasks of app tracking in research

Challenges of app tracking	Future tasks
Sensitive data and low compliance	Exhaustive information about the study design, to be collected data, where data is saved and how it is processed, who has access to data and what happens after data collection is finished
Privacy issues	See Chap. 1 of this book for a detailed overview on privacy issues
Ethical considerations	See Chap. 2 of this book
No official standards of use and data collection	Develop standard operating procedures (SOPs) in an interdisciplinary setting
Missing information on the reliability and validity of variables assessed with a smartphone application	Execution of "proof of concept" studies and development of SOPs based on the results of those studies
(Naturally) limited researcher's skill set regarding data analyses	Interdisciplinary cooperation and knowledge exchange
Precision of assessing a certain variable	Use of multiple parameters and combination of smartphone assessment with physiological data collection using wearable technology
High costs of developing such tracking applications	Consider the advantages of mobile tracking together with a reduction of costs in other areas such as for the administration of paper-pencil questionnaires or for staff to test participants on multiple occasions on site; moreover app tracking technologies are getting more affordable (see also Montag et al. 2019)
Reachability of participants who do not own a smartphone	Growing number of smartphone owners worldwide. Consideration of statistics for the particular country when representative studies are to be conducted using smartphone applications
Restrictions to a specific operating system such as Android	Development of applications that support all or most used operating systems; in addition a new work suggests that differences in certain variables across used operation systems are rather low. (Götz et al. 2017)

especially in fields such as health and clinical psychology as well as medicine, where accurate prediction or classification of e.g. symptoms is the foundation for successful prevention and intervention (Obermeyer and Emanuel 2016).

Additionally, the costs for the development of app tracking technologies are considerably high. However, in the long run the costs in other related research areas are tending to get lower, e.g. one has no costs for the administration of paper-pencil questionnaires or achieving a high sample size when conducting longitudinal studies.

In a still young research field such as Psychoinformatics, researchers need to ensure that their developed applications are reliable and valid measures of concepts such as emotion, personality or physical activity. In so far, the next years will surely see much research with "proof of concept studies". Moreover, standard operating procedures (SOPs) for app tracking need to be developed (e.g. How many days of recordings are necessary to guarantee a reliable measure?). An additional drawback for researchers is that not everyone owns a smartphone yet and that there are considerable differences in the number of smartphone owners in different countries (Sullivan and Lachman 2017). Thus, this might hamper the recruitment of representative samples. However, this will likely change very soon, because penetration rates of smartphones are growing at a rapid pace. In Germany alone more than 61 million users (ca. 74% of the population) were estimated to use a smartphone in 2019 (Statista 2019b).

An important issue to be dealt with also concerns the precision in assessing a certain variable. As can be seen from the aforementioned reviewed personality studies, associations between personality and smartphone variables are currently moderate at best, and certain characteristics can be predicted better than others. In the literature reviewed earlier in this chapter, it is the link between extraversion and different smartphone variables that stands out, while the association with other personality dimensions was demonstrated less often. This inherently makes sense, because the smartphone or social media applications are often used to communicate with other persons, hence such variables tap in the social aspects of personality most strongly anchored in extraversion (see also the Chap. 7 by Marengo and Settanni). In sum, it is important to deal with the question what variables of the rapid developing IoT are of largest importance to best capture the manifold existing (psychological) variables (see also Montag and Elhai 2019)? Again, we stress the importance to think about how smartphone-based assessments can be accompanied by wearable devices collecting physiological data (biosensors for neural activity, heart rate variability and skin conductance such as smartwatches, shirts/shoes/socks with sensors or a headband to record EEG; see Malhi et al. 2017). Ultimately, it would be best, if one device could capture most of what is needed for the research question at hand. For a summary of propositions about implementation of psychological theories in smartphone-based research, study design and types of smartphone data necessary to test a particular hypothesis, we refer to the review by Sullivan and Lachman (2017) and to the article by Harari et al. (2016).

4.9 Summary and Conclusions

In sum, smartphones can be considered a valuable addition to the toolbox of personality psychologists, but also in other research areas such as health psychology or sport sciences to track physical activity. Their potential in research is yet to further unfold and for this to happen multidisciplinary collaborations are needed.

References

Adamo KB, Prince SA, Tricco AC et al (2009) A comparison of indirect versus direct measures for assessing physical activity in the pediatric population: a systematic review. Int J Pediatr Obes 4(1):2–27. https://doi.org/10.1080/17477160802315010

Allemand M, Mehl MR (2017) Personality assessment in daily life: a roadmap for future personality development research. In: Personality development across the lifespan. Elsevier, pp 437–454

Allport GW, Odbert HS (1936) Trait-names: a psycho-lexical study. Psychol Monogr 47(1):i–171. https://doi.org/10.1037/h0093360

Althoff T, Sosič R, Hicks JL et al (2017) Large-scale physical activity data reveal worldwide activity inequality. Nature 547(7663):336–339. https://doi.org/10.1038/nature23018

Ashton MC, Lee K (2007) Empirical, theoretical, and practical advantages of the HEXACO model of personality structure. Pers Soc Psychol Rev 11(2):150–166. https://doi.org/10.1177/1088868306294907

Baumeister RF, Vohs KD, Funder DC (2007) Psychology as the science of self-reports and finger movements: whatever happened to actual behavior? Perspect Psychol Sci 2(4):396–403. https://doi.org/10.1111/j.1745-6916.2007.00051.x

Baumert A, Schmitt M, Perugini M et al (2017) Integrating personality structure, personality process, and personality development. Eur J Pers 31(5):503–528. https://doi.org/10.1002/per.2115

Bidargaddi N, Musiat P, Makinen V-P et al (2017) Digital footprints: facilitating large-scale environmental psychiatric research in naturalistic settings through data from everyday technologies. Mol Psychiatr 22(2):164–169. https://doi.org/10.1038/mp.2016.224

Bleidorn W, Hopwood CJ, Lucas RE (2018) Life events and personality trait change. J Pers 86(1):83–96. https://doi.org/10.1111/jopy.12286

Blondell SJ, Hammersley-Mather R, Veerman JL (2014) Does physical activity prevent cognitive decline and dementia?: a systematic review and meta-analysis of longitudinal studies. BMC Publ Health 14(1):510. https://doi.org/10.1186/1471-2458-14-510

Bort-Roig J, Gilson ND, Puig-Ribera A et al (2014) Measuring and influencing physical activity with smartphone technology: a systematic review. Sports Med 44(5):671–686. https://doi.org/10.1007/s40279-014-0142-5

Carvalho LF, Sette CP, Ferrari BL (2018) Problematic smartphone use relationship with pathological personality traits: systematic review and meta-analysis. Cyberpsychology 12(3). https://doi.org/10.5817/cp2018-3-5

Caspersen CJ, Powell KE, Christenson GM (1985) Physical activity, exercise, and physical fitness: definitions and distinctions for health-related research. Publ Health Rep 100(2):126–131

Cattell RB (1933) Temperament tests. I. Temperament. Br J Psychol 23(3):308–329

Chittaranjan G, Blom J, Gatica-Perez D (2013) Mining large-scale smartphone data for personality studies. Pers Ubiquit Comput 17(3):433–450. https://doi.org/10.1007/s00779-011-0490-1

Christ CC, Carlo G, Stoltenberg SF (2016) Oxytocin receptor (OXTR) single nucleotide polymorphisms indirectly predict prosocial behavior through perspective taking and empathic concern. J Pers 84(2):204–213. https://doi.org/10.1111/jopy.12152

Costa PT, McCrae RR (1992) NEO PI-R professional manual. Psychological Assessment Resources, Odessa, FL

Coughlin SS, Whitehead M, Sheats JQ et al (2016) A review of smartphone applications for promoting physical activity. Jacobs J Community Med 2(1)

Davis KL, Panksepp J (2018) The emotional foundations of personality: a neurobiological and evolutionary approach. W. W. Norton & Company

de Montjoye YA, Quoidbach J, Robic F, Pentland AS (2013) Predicting personality using novel mobile phone-based metrics. In: International conference on social computing, behavioral-cultural modeling, and prediction. Springer: Berlin, Heidelberg, pp 48–55

Edmonds GW, Goldberg LR, Hampson SE, Barckley M (2013) Personality stability from childhood to midlife: relating teachers' assessments in elementary school to observer- and self-ratings 40 years later. J Res Pers 47(5):505–513. https://doi.org/10.1016/j.jrp.2013.05.003

Evenson KR, Terry JW (2009) Assessment of differing definitions of accelerometer nonwear time. Res Q Exerc Sport 80(2):355–362. https://doi.org/10.1080/02701367.2009.10599570

Fiske DW (1949) Consistency of the factorial structures of personality ratings from different sources. J Abnorm Soc Psychol 44(3):329–344. https://doi.org/10.1037/h0057198

Forbes (2016) The future is now: smart cars and IoT in cities. https://www.forbes.com/sites/pikeresearch/2016/06/13/the-future-is-now-smart-cars/#3000ff19509c

Götz FM, Stieger S, Reips UD (2017) Users of the main smartphone operating systems (iOS, Android) differ only little in personality. PLoS One 12(5):e0176921

Guthold R, Stevens GA, Riley LM, Bull FC (2018) Worldwide trends in insufficient physical activity from 2001 to 2016: a pooled analysis of 358 population-based surveys with 1·9 million participants. Lancet Glob Health 6(10):e1077–e1086. https://doi.org/10.1016/S2214-109X(18)30357-7

Harari GM, Lane ND, Wang R et al (2016) Using smartphones to collect behavioral data in psychological science: opportunities, practical considerations, and challenges. Perspect Psychol Sci 11(6):838–854. https://doi.org/10.1177/1745691616650285

Hekler EB, Buman MP, Grieco L et al (2015) Validation of physical activity tracking via android smartphones compared to actigraph accelerometer: laboratory-based and free-living validation studies. JMIR mHealth uHealth 3(2):e36. https://doi.org/10.2196/mhealth.3505

Helson R, Kwan VSY, John OP, Jones C (2002) The growing evidence for personality change in adulthood: findings from research with personality inventories. J Res Pers 36(4):287–306. https://doi.org/10.1016/S0092-6566(02)00010-7

Höchsmann C, Knaier R, Eymann J et al (2018) Validity of activity trackers, smartphones, and phone applications to measure steps in various walking conditions. Scand J Med Sci Sports 28(7):1818–1827. https://doi.org/10.1111/sms.13074

Insel TR (2018) Digital phenotyping: a global tool for psychiatry. World Psychiatr 17(3):276–277. https://doi.org/10.1002/wps.20550

Lachmann B, Duke É, Sariyska R, Montag C (2017) Who's addicted to the smartphone and/or the internet? Psychol Pop Media Cult 8(3):182–189. https://doi.org/10.1037/ppm0000172

Lane ND, Miluzzo E, Lu H et al (2010) A survey of mobile phone sensing. IEEE Commun Mag 48(9)

Lu Y, Wei Y, Liu L et al (2017) Towards unsupervised physical activity recognition using smartphone accelerometers. Multimed Tools Appl 76(8):10701–10719. https://doi.org/10.1007/s11042-015-3188-y

Malhi GS, Hamilton A, Morris G et al (2017) The promise of digital mood tracking technologies: are we heading on the right track? Evid Based Mental Health 20(4):102–107. https://doi.org/10.1136/eb-2017-102757

Malinauskas R, Dumciene A, Mamkus G, Venckunas T (2014) Personality traits and exercise capacity in male athletes and non-athletes. Percept Mot Skills 118(1):145–161. https://doi.org/10.2466/29.25.PMS.118k13w1

Mammen G, Faulkner G (2013) Physical activity and the prevention of depression: a systematic review of prospective studies. Am J Prev Med 45(5):649–657. https://doi.org/10.1016/j.amepre.2013.08.001

Markowetz A, Błaszkiewicz K, Montag C et al (2014) Psycho-informatics: big data shaping modern psychometrics. Med Hypotheses 82(4):405–411. https://doi.org/10.1016/j.mehy.2013.11.030

McCrae RR, Costa PT Jr (1997) Personality trait structure as a human universal. Am Psychol 52(5):509–516. https://doi.org/10.1037/0003-066X.52.5.509

McCrae RR, Costa PT Jr (1994) The stability of personality: observations and evaluations. Curr Dir Psychol Sci 3(6):173–175

McCrae RR, John OP (1992) An introduction to the five-factor model and its applications. J Personality 60:175–215. https://doi.org/10.1111/j.1467-6494.1992.tb00970.x

Melchers M, Montag C, Markett S, Reuter M (2013) Relationship between oxytocin receptor genotype and recognition of facial emotion. Behav Neurosci 127(5):780–787. https://doi.org/10.1037/a0033748

Miller G (2012) The smartphone psychology manifesto. Perspect Psychol Sci 7(3):221–237. https://doi.org/10.1177/1745691612441215

Mischel W (2004) Toward an integrative science of the person. Annu Rev Psychol 55(1):1–22. https://doi.org/10.1146/annurev.psych.55.042902.130709

Mischel W, Shoda Y (1995) A cognitive-affective system theory of personality: reconceptualizing situations, dispositions, dynamics, and invariance in personality structure. Psychol Rev 102(2):246–268. https://doi.org/10.1037/0033-295X.102.2.246

Monasterio E, Mei-Dan O, Hackney AC et al (2016) Stress reactivity and personality in extreme sport athletes: the psychobiology of BASE jumpers. Physiol Behav 167:289–297. https://doi.org/10.1016/j.physbeh.2016.09.025

Montag C (2018a) Eine kurze Einführung in die Molekulare Psychologie: Band I: Definition und molekulargenetische Grundbegriffe. Springer

Montag C (2018b) Eine kurze Einführung in die Molekulare Psychologie: Band II: Von Kandidatengenen bis zur Epigenetik. Springer

Montag C (2019) The neuroscience of smartphone/social media usage and the growing need to include methods from 'Psychoinformatics'. In: Information Systems and Neuroscience. Springer, Cham, pp 275–283

Montag C, Baumeister H, Kannen C et al (2019) Concept, possibilities and pilot-testing of a new smartphone application for the social and life sciences to study human behavior including validation data from personality psychology. J 2(2):102–115. https://doi.org/10.3390/j2020008

Montag C, Błaszkiewicz K, Lachmann B et al (2014) Correlating personality and actual phone usage: evidence from psychoinformatics. J Individ Differ 35(3):158–165. https://doi.org/10.1027/1614-0001/a000139

Montag C, Błaszkiewicz K, Lachmann B et al (2015a) Recorded behavior as a valuable resource for diagnostics in mobile phone addiction: evidence from psychoinformatics. Behav Sci 5(4):434–442. https://doi.org/10.3390/bs5040434

Montag C, Błaszkiewicz K, Sariyska R et al (2015b) Smartphone usage in the 21st century: who is active on WhatsApp? BMC Res Notes 8(1):331. https://doi.org/10.1186/s13104-015-1280-z

Montag C, Dagum P (2019) Molecular psychology: a modern research endeavour. http://edition.pagesuite-professional.co.uk/Launch.aspx?EID=14e30abb-c333-43f5-b63d-31e069aee049

Montag C, Diefenbach S (2018) Towards homo digitalis: important research issues for psychology and the neurosciences at the dawn of the internet of things and the digital society. Sustainability 10(2):415. https://doi.org/10.3390/su10020415

Montag C, Duke É, Markowetz A (2016) Toward psychoinformatics: computer science meets psychology. Comput Math Methods Med 2016:1–10. https://doi.org/10.1155/2016/2983685

Montag C, Elhai JD (2019) A new agenda for personality psychology in the digital age? Pers Individ Differ 147:128–134. https://doi.org/10.1016/j.paid.2019.03.045

Montag C, Markowetz A, Blaszkiewicz K et al (2017) Facebook usage on smartphones and gray matter volume of the nucleus accumbens. Behav Brain Res 329:221–228. https://doi.org/10.1016/j.bbr.2017.04.035

Montag C, Panksepp J (2017) Primary emotional systems and personality: an evolutionary perspective. Front Psychol 8. https://doi.org/10.3389/fpsyg.2017.00464

Montag C, Reuter M (2017) Molecular genetics, personality, and internet addiction revisited. In: Internet addiction. Springer, pp 141–160

Obermeyer Z, Emanuel EJ (2016) Predicting the future—big data, machine learning, and clinical medicine. N Engl J Med 375(13):1216–1219. https://doi.org/10.1056/NEJMp1606181

Olaru G, Witthöft M, Wilhelm O (2015) Methods matter: testing competing models for designing short-scale big-five assessments. J Res Pers 59:56–68. https://doi.org/10.1016/j.jrp.2015.09.001

Paulhus DL, Vazire S (2007) The self-report method. In: Robins RW, Fraley RC, Krueger RF (eds) Handbook of research methods in personality psychology, 1st edn. The Guilford Press, New York, pp 224–239

Prince SA, Adamo KB, Hamel M et al (2008) A comparison of direct versus self-report measures for assessing physical activity in adults: a systematic review. Int J Behav Nutr Phys Act 5(1):56. https://doi.org/10.1186/1479-5868-5-56

Riva G, Teruzzi T, Anolli L (2003) The use of the internet in psychological research: comparison of online and offline questionnaires. Cyberpsychol Behav 6(1):73–80. https://doi.org/10.1089/109493103321167983

Romeo A, Edney S, Plotnikoff R et al (2019) Can smartphone apps increase physical activity? systematic review and meta-analysis. J Med Internet Res 21(3):e12053. https://doi.org/10.2196/12053

Ronao CA, Cho S-B (2016) Human activity recognition with smartphone sensors using deep learning neural networks. Expert Syst Appl 59:235–244. https://doi.org/10.1016/j.eswa.2016.04.032

Sariyska R, Rathner E-M, Baumeister H, Montag C (2018) Feasibility of linking molecular genetic markers to real-world social network size tracked on smartphones. Front Neurosci 12:945. https://doi.org/10.3389/fnins.2018.00945

Scholz R, Bartelsman E, Diefenbach S et al (2018) Unintended side effects of the digital transition: European scientists' messages from a proposition-based expert round table. Sustainability 10(6):2001. https://doi.org/10.3390/su10062001

Schuch FB, Vancampfort D, Richards J et al (2016) Exercise as a treatment for depression: a meta-analysis adjusting for publication bias. J Psychiatr Res 77:42–51. https://doi.org/10.1016/j.jpsychires.2016.02.023

Seifert A, Schlomann A, Rietz C, Schelling HR (2017) The use of mobile devices for physical activity tracking in older adults' everyday life. Digit Health 3:205520761774008. https://doi.org/10.1177/2055207617740088

Stachl C, Hilbert S, Au J-Q et al (2017) Personality traits predict smartphone usage. Eur J Pers 31(6):701–722. https://doi.org/10.1002/per.2113

Statista (2019a) Number of smartphone users worldwide from 2014 to 2020 (in billions). https://www.statista.com/statistics/330695/number-of-smartphone-users-worldwide/

Statista (2019b) Number of smartphone users in Germany from 2015 to 2022 (in millions). https://www.statista.com/statistics/467170/forecast-of-smartphone-users-in-germany/

Steyer R, Schmitt M, Eid M (1999) Latent state–trait theory and research in personality and individual differences. Eur J Pers 13(5):389–408. https://doi.org/10.1002/(SICI)1099-0984(199909/10)13:5%3c389:AID-PER361%3e3.0.CO;2-A

Sullivan AN, Lachman ME (2017) Behavior change with fitness technology in sedentary adults: a review of the evidence for increasing physical activity. Front Publ Health 4. https://doi.org/10.3389/fpubh.2016.00289

Tupes EC, Christal RE (1992) Recurrent personality factors based on trait ratings. J Personality 60(2):225–251. https://doi.org/10.1111/j.1467-6494.1992.tb00973.x

Verburgh L, Königs M, Scherder EJA, Oosterlaan J (2014) Physical exercise and executive functions in preadolescent children, adolescents and young adults: a meta-analysis. Br J Sports Med 48(12):973–979. https://doi.org/10.1136/bjsports-2012-091441

Weigold A, Weigold IK, Russell EJ (2013) Examination of the equivalence of self-report survey-based paper-and-pencil and internet data collection methods. Psychol Methods 18(1):53–70. https://doi.org/10.1037/a0031607

Wilson KE, Dishman RK (2015) Personality and physical activity: a systematic review and meta-analysis. Pers Individ Differ 72:230–242. https://doi.org/10.1016/j.paid.2014.08.023

World Health Organization (2018a) Physical activity. http://www.who.int/news-room/fact-sheets/detail/physical-activity

World Health Organization (2018b) The top 10 causes of death. https://www.who.int/news-room/fact-sheets/detail/the-top-10-causes-of-death

Xu R, Frey RM, Fleisch E, Ilic A (2016) Understanding the impact of personality traits on mobile app adoption—insights from a large-scale field study. Comput Human Behav 62:244–256. https://doi.org/10.1016/j.chb.2016.04.011

Yarkoni T (2012) Psychoinformatics: new horizons at the interface of the psychological and computing sciences. Curr Dir Psychol Sci 21(6):391–397. https://doi.org/10.1177/0963721412457362

Chapter 5
Smartphones in Personal Informatics: A Framework for Self-Tracking Research with Mobile Sensing

Sumer S. Vaid and Gabriella M. Harari

Abstract Recent years have seen a growth in the spread of digital technologies for self-tracking and personal informatics. Smartphones, in particular, stand out as being an ideal self-tracking technology that permits both active logging (via self-reports) and passive tracking of information (via phone logs and mobile sensors). In this chapter, we present the results of a literature review of smartphone-based personal informatics studies across three different disciplinary databases (computer science, psychology, and communication). In doing so, we propose a conceptual framework for organizing the smartphone-based personal informatics literature. Our framework situates self-tracking studies based on their substantive focus across two domains: (1) the measurement domain (whether the study uses subjective or objective data) and (2) the outcome of interest domain (whether the study aims to promote insight or change in physical and/or mental characteristics). We use this framework to identify and discuss research trends and gaps in the literature. For example, most research has been concentrated on tracking of objective measurements to change either physical or mental characteristics, while less research used subjective measures to study a physical outcome of interest. We conclude by pointing to promising future directions for research on self-tracking and personal informatics and emphasize the need for a greater appreciation of individual differences in future self-tracking research.

Keywords Self-tracking · Smartphones · Mobile sensing · Personal informatics

The tracking of physical (e.g., weight, physical activity) and mental (e.g., mood, stress) characteristics has long fascinated individuals and scientists alike (e.g., Li et al. 2010; Wolf 2010). The practice of self-tracking involves a process of "…collecting data about oneself on a regular basis and then recording and analyzing the data to produce statistics and other data (such as images) relating to regular habits, behaviors, and feelings" (Lupton 2014, p. 1). According to the Pew Research Center, a recent national survey of adults in the United States found that nearly 69% of American adults track at least one health-related physical characteristic or behavior (e.g.,

S. S. Vaid (✉) · G. M. Harari
Department of Communication, Stanford University, 94305 Stanford, CA, USA
e-mail: sumer@stanford.edu

© Springer Nature Switzerland AG 2019   65
H. Baumeister and C. Montag (eds.), *Digital Phenotyping and Mobile Sensing,*
Studies in Neuroscience, Psychology and Behavioral Economics,
https://doi.org/10.1007/978-3-030-31620-4_5

weight, diet, sickness symptoms, exercise routine), and that individuals with chronic health conditions were more likely to track such behaviors compared to healthy individuals. Among the self-trackers, approximately half reported recording their histories "in their head" (49%) and reported using notebooks or digital technology to record their physical health behaviors (55%; Fox and Duggan 2013).

The practice of self-tracking is typically associated with the goal of inducing behavior change through self-insight and self-monitoring (Kersten-van Dijk et al. 2017). That is, most individuals track their behaviors with the intention of changing unhealthy patterns and improving upon their general well-being. Today, the advent of ubiquitous sensor-driven technologies (e.g., smartphones, wearable devices) has revolutionized the way individuals self-track their physical and mental characteristics, and how they interact with personal informatics in general (e.g., Swan 2012).

5.1 Personal Informatics and Self-Tracking Technologies

Personal informatics (PI) broadly defines a set of self-tracking technologies that help individuals collect and reflect on personal information (Li et al. 2010). Self-tracking technologies include diverse forms of digital technology, such as web-based applications (i.e. Mint Financial planner), wearables (i.e. Apple Watch, Nike + Band), mobile phone applications (i.e. WeRun; Li et al. 2010). Personal informatics systems operate under the pretext of three interrelated goals (Kersten-van Dijk et al. 2017): (a) to accurately measure the target domain (e.g., physical behaviors, mental states) using data produced from the use of digital technology, (b) to produce meaningful analysis of this data and (c) to communicate this analysis to the user in a comprehensible manner.

Past research has compared two working models of personal informatics (Kersten-van Dijk et al. 2017). The first model is a stage-based model consisting of distinctive consecutive states: preparation, collection, integration, reflection, and action (Li et al. 2010). The reflection stage constitutes periods of self-reflection resulting from the use of PI systems, which leads users to change their behavioral trajectories after self-reflection. The second competing model maintains that the use of PI systems is too continuous to be discretely modeled in a stage-wise manner because participants using personal informatics systems often simultaneously engage in the activities described in the discrete stage model (Epstein et al. 2015). For example, the collection of self-tracking data usually occurs in conjunction with processes of self-reflection, as participants' experience of self-tracking induces them to reflect on their behavioral patterns. Despite their differences, both models of personal informatics (Li et al. 2010; Epstein et al. 2015) converge on the idea that behavior change is the ultimate outcome of an engagement with personal informatics systems (see Kersten-van Dijk et al. 2017 for complete review and comparison of both models).

The use of self-tracking technologies in daily life is likely to continue increasing at a rapid rate in the near future, as digital self-tracking is already becoming a pervasive and ubiquitous phenomenon (Paré et al. 2018). Thus far, the majority of commercial digital technologies for self-tracking target health and fitness as areas of application (e.g., Samsung Gear Fit, Apple Watch). Yet, digital self-tracking applications are accessible for none to marginal costs, and portable fitness hardware such as pedometers are relatively affordable (Rooksby et al. 2014). Moreover, scholars have collectively recognized the growing popularity of smartphone applications, wearable sensing technology, and other digital self-tracking platforms (Lupton 2013; Rooksby et al. 2014; Sanders 2017). And movements such as Quantified Self (2015) have developed a variety of digital technologies that facilitate the tracking of diverse behaviors (e.g., mobility patterns from GPS data; Parecki 2018).

Smartphones, in particular, stand out as a digital technology with much promise for self-tracking and personal informatics because they permit both active logging (via surveys) and passive tracking (via mobile sensing; Harari et al. 2017). Mobile sensing technologies permit the unobtrusive collection of data from mobile sensors and system logs embedded in the smartphone (microphones, accelerometers, app usage logs) to recognize human activity (e.g., sociability, physical activity, digital media use; Choudhury et al. 2008; Lane et al. 2010). By automating the continuous detection of a person's behavioral patterns and surrounding context (Harari et al. 2017b; Harari et al. 2018), mobile sensing is poised to play an important role in the development of effective personal informatics systems that induce positive behavior change. To examine the effects of self-tracking with smartphones in personal informatics, here we review and provide an organizing framework for existing and future scholarship on self-tracking research with mobile sensing.

5.2 A Framework for Self-Tracking Research

We have two interrelated aims in writing this chapter. First, we aim to provide a review of the existing trends, gaps, and directions in the research literature on smartphones in personal informatics. We focus specifically on reviewing the existing literature that uses smartphone-based self-tracking technologies to collect user data (e.g., via self-report questions or mobile sensing), analyze user data, and/or that use the smartphone to communicate results aggregated from a variety of sources to the user. Given the interdisciplinary nature of self-tracking research and the variety of application domains into which smartphones have been deployed in personal informatics systems, we conducted our literature review across three databases selected to represent the primary disciplines engaging in such research: PsychINFO (representing psychology), ACM (representing computer science), and Communication and Mass Media Complete (representing communication). We note that nearly all of the articles that met our inclusion criteria were indexed in the ACM database, while our inclusion criteria yielded few articles from the PsychINFO and Communication and Mass Media Complete databases. We provide a detailed description of our literature review procedure in Table 5.1 (keywords used), Table 5.2 (derivation of keywords and their search arrangement), and Table 5.3 (filtering process). Additionally, Fig. 5.1

Table 5.1 Methodology of literature review

Mobile sensing keywords	Mobile sensing; mobile-sensing; mobile sense; smartphone sensing; smartphone sensing; smartphone sense
Personal informatics keywords	Self-monitoring; self monitoring; self-tracking; self-track; self tracking; self track; quantified self; life-logging; lifelogging; life logging; personal informatics
Behavior change keyword	Behavior change
Categorical keywords	*Physical activities:* Physical health; Activity, walking, steps, sedentary; Running; Exercise, fitness, workouts; Illness, symptoms
	Physiological: Physiological; Heart rate; Blood pressure; Nutrition; Food diet; Calories; Sugar intake; Water intake; Alcohol; Vitamins; Medications
	Sleeping patterns: Sleeping patterns; Duration, Quality, Schedule; Rest
	Productivity: Productivity; Study habits; Laziness; Focus; Time management, time spent working; School-life balance taking breaks; Academics; Class Schedule
	Mood: Mood, Emotions; Depression; Sadness; Anxiety; Stress; Happy; Content; Relaxed; Relaxation; Anger; Curiosity; Kindness; Consideration; Positive Thoughts; Negative thoughts
	Socializing: Socializing; Conversation Quality; Conversation Duration; People Spending Time With; Time Spent Alone; Dating, Romance; Social Media Communications
	Digital Media Use: Digital media use; TV Time; Screen Time; Computer Use; Internet use; Phone use; Gaming; Social Media Use
	Daily Activities: Daily activities; Hobbies; Reading; Learning; Location; Hygiene

shows the number of search results returned at and filtered during the various stages of the filtration process .

Second, we aim to provide a conceptual organizing framework to situate the substantive contributions of past and future smartphone-based personal informatics research. We believe a framework is needed to help situate the contributions of a given self-tracking study within the broader literature with regard to its measurement and outcome of interest domains.

Generally, existing research on smartphones in personal informatics has largely focused on describing the development, deployment, and effectiveness of individual personal informatics systems that measure different behaviors in a variety of contexts.

Less is known about the substantive focus of these different studies, across different measurement domains (tracking of physical and/or mental characteristics) and outcome of interest domains (aiming to promote insight or change in physical and/or mental characteristics). To provide an organizing conceptual framework, we present a two-dimensional space that resembles the Cartesian coordinate plane (as shown in Fig. 5.2) that maps out four different quadrant areas representing the substantive focus of smartphone-based self-tracking research: (1) objective measurement—physical outcome of interest (e.g., measuring accelerometer to infer physical activity levels), (2) objective measurement- mental outcome of interest (e.g., measuring mobility data using GPS to infer depressed mood), (3) subjective measurement—mental outcome of interest (e.g., measuring self-reported experience sampling surveys to assess psychological states), and (4) subjective measurement—physical outcome of interest (e.g., measuring self-reported experience sampling surveys to assess subsequent changes in sensed physical activity).

To illustrate our framework's potential for conceptually organizing domains of interest in self-tracking research, we coded each reviewed article into one of the four quadrants based on its substantive research contribution with regard to its measurement-outcome of interest domains. To verify the accuracy of our framework classifications, a research assistant independently coded the articles using the

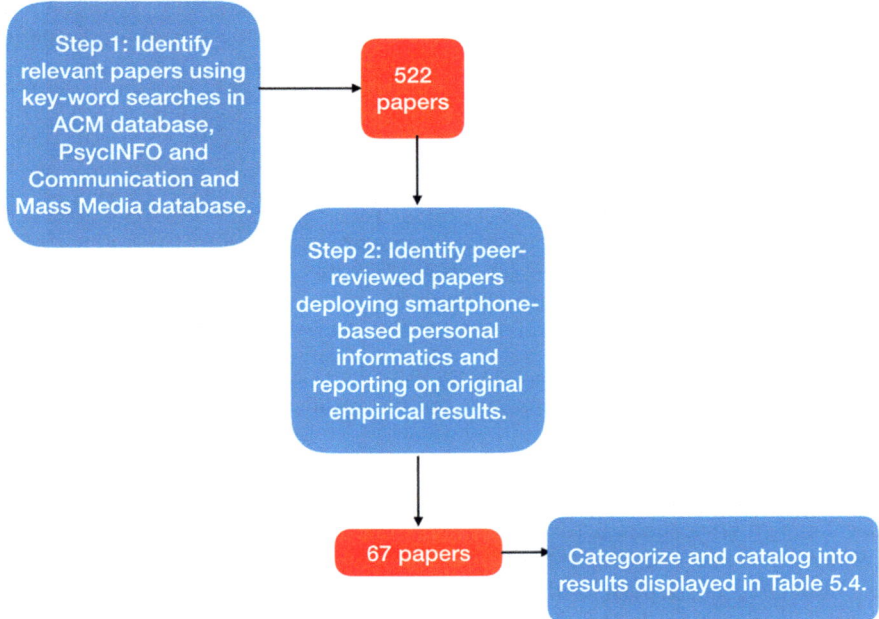

Fig. 5.1 Procedural flowchart and number of search results returned

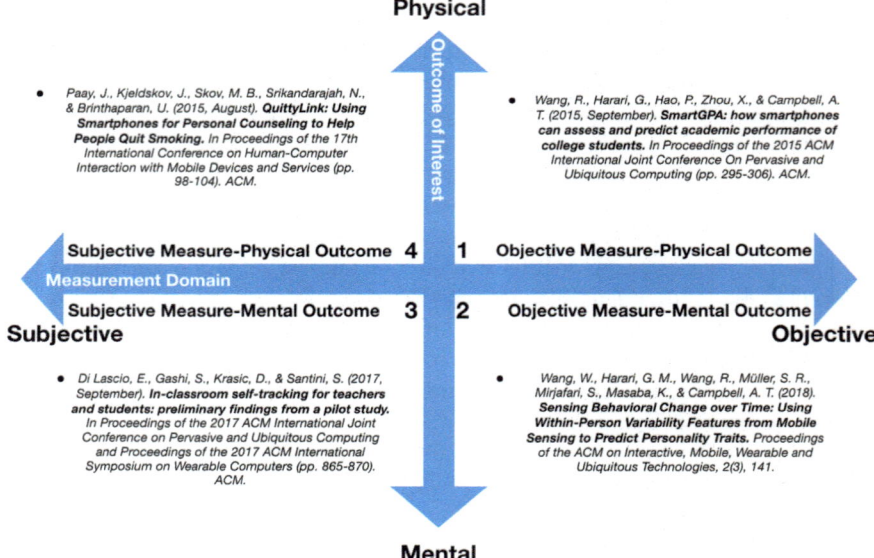

Fig. 5.2 Conceptual framework for organizing the smartphone-based PI literature. To organize the surveyed mobile sensing literature, we present a two dimensional conceptual framework. The conceptual framework resembles the Cartesian coordinate plane, consisting of two axis that encode magnitude relative to space. As shown in Fig. 5.1, the x-axis represents the measurement domain of the surveyed literature—specifically, it identifies the extent to a given article was focused on collecting either subjectively measured data (i.e. experience sampling surveys) or objectively measured data (e.g., mobile sensors). For instance, some papers described collecting physical activity data using the accelerometer (Wang et al. 2015), whereas others were focused on collecting mood or depression related information using phone-based ecological momentary assessments (Di Lascio et al. 2017). The y-axis represents the extent to which the researchers used their measurements to assess either physical or mental outcomes of interests. For instance, while some researchers were focused on using collected physical activity data to categorize various kinds of physical movements and social interactions (Harari et al. 2017b), others were focused on using physical activity data to infer mood or depression scores (Mehrotra et al. 2016)

four quadrants as well. The classifications were then compared, and any discrepancies were resolved through discussion. The full results of the literature review are presented in Table 5.4.

Much of the existing research has focused on the different components or "stages" that are characteristic of personal informatics systems. Some studies were focused on developing high-accuracy activity classifiers from sensors embedded in smartphones and wearables (e.g., Madan et al. 2010), whereas others were anchored around creating optimal feedback systems that were effective at inducing behavior change (e.g., Bentley et al. 2013). Below we discuss the literature on smartphones in personal informatics, focusing our discussion of previous research based on the type of physical and mental characteristics being tracked (Table 5.4).

Table 5.2 Details of literature review

Procedure	Categorical keyword source
In order to determine the extant literature linking the themes of smartphone-based personal informatics and behavior change, we used the following formatting of keywords to return searches in each database: [mobile-sensing keywords separated by OR] AND [self-tracking keywords separated by OR) AND ["behavior change"] AND [coding category keywords]	The coding category keywords were extracted from another study in which student's responded to questions asking them about motivations to self-track different aspects of their lifestyle To conduct our literature review, we utilized categorical keywords that were extracted from qualitative responses that 1706 young adults provided to the following question: "What would motivate you personally to self-track, and which behaviors would you track?" The qualitative responses were content analyzed to obtain an exhaustive list of 75 individual self-tracking categories, which could be described by 8 broader categories (See Table 5.1 for the full list of categories): physical health, physiological, sleeping patterns, productivity, mood, socializing, digital media use, and daily activities Hence, we conducted the full literature review in eight stages, operationalizing one self-tracking category (with all of its individual sub-category topics) through our keyword patterns in each stage

Table 5.3 Steps for filtering literature review results

Step No.	Description of filter
Step 1	The result must be a peer-reviewed paper and report on original empirical work
Step 2	The result must discuss at least one smartphone-based technology that supports a collection of human characteristics and/or acts a mediator of relevant feedback information on behavioral patterns

Note The smartphone or mobile may be involved either in the data collection stage or the feedback generation and communication stage of the personal informatics ecosystem deployed

Physical Activity. Physical activity was the substantive focus of many of the papers we reviewed. The majority of studies on physical activity fell under either the objective measurement-physical outcome of interest or objective measurement-mental outcome of interest quadrants of our conceptual framework. For instance, Harari et al. (2017a) deployed a smartphone sensing application, StudentLife, which measured daily durations of physical activity using data collected from the accelerometer sensor of the smartphone in a college student population. Notably, the authors found that individual differences in ethnicity and academic class were predictive of changes in physical activity. While Harari et al. (2017a) were not focused on the feedback component of personal informatics systems, other researchers were especially interested in developing an effective feedback system developed from the collected physical activity data in order to engage users in self-reflection. Kocielnik et al. (2018) developed Reflect Companion, a mobile conversational software that facilitated immersion in the reflection of activity levels as aggregated from fitness trackers. Their results indicated that mini-dialogues were successful in inducing reflection from the users, on their physical activity levels. Some studies were focused on using the personal informatics system to induce systematic behavior change that increases physical activity levels in individuals. To incentivize higher levels of physical activity, these studies gamified the objective of increasing physical activity by either individual into motivated competing teams (e.g., Ciman et al. 2016; Zuckerman and Gal-Oz 2014) or by tapping into their existing social networks (Gui et al. 2017).

A large majority of studies in this category were single deployment studies that examined the efficacy of personal informatics tools deployed to monitor physical activity. While some studies attempted to situate their work under a theoretical model of behavior change (e.g., Theory of Planned Behavior; Ajzen 1985; Du et al. 2014) virtually no research attempted to integrate with extant personal informatics models of behavior change such as those proposed by Li et al. (2010). Instead, different researchers tended to make use of different theoretical models of behavior change originating from a range of behavioral disciplines. For example, the Transtheoretical Model of Behavior Change (Glanz et al. 2008) and the Social Cognitive Theory of Behavior Change (Bandura 2004) have been employed in physical activity intervention studies (e.g., in the design of applications; Marcu et al. 2018). However, there is a general lack of integration between these theoretical models and personal informatics models of behavior change. Such an approach has led to some studies reporting conflicting behavior change outcomes and high participant attrition, which may be a result of using models of behavior change that are not specifically adapted to the use of personal informatics systems.

Physiological. There were only four studies that we categorized as pertaining to physiology in our literature search. Hwang and Pushp (2018) deployed a system called "StressWatch", which was aimed towards assisting users in triangulating the sources of their stress in their daily life. Using a concert of smartphone and smartwatch systems, StressWatch monitored the context of users in concert with their heart rate variability. Subsequently, the StressWatch extracted stress levels from the heart rate variability data and "matched" these patterns with the changing contexts of the user in order to suggest possible origins of stress in daily life. In a single subject pilot deployment, the authors refined the design of StressWatch by deciding that stress levels could be accurately detected when eating and working, but not when walking.

In a similar line of work, Bickmore et al. (2018) developed a virtual conversational agent that counseled patients suffering from chronic heart condition atrial fibrillation using data collected from a heart rhythm monitor attached to a smartphone. In a randomized trial with 120 patients, the authors found that the conversational agents led to a significant positive change in the self-reported quality of life scores as compared to a control group who did not receive the agent-based counseling. Cumulatively,

Table 5.4 Overview of literature review results organized according to our framework

References	Framework classification	Physical and mental characteristics								
		Physical activity	Physio-logical	Nutrition	Sleeping patterns	Productivity	Mood	Socializing	Digital media use	Daily activities
1. Objective Measurement—Physical Outcome of Interest										
Athukorala et al. (2014)	1								X	X
Bentley et al. (2013)	1	X		X						X
Bexheti et al. (2015)	1	X								X
Brewer et al. (2015)	1								X	X
Chen et al. (2016)	1	X		X				X		
Ciman et al. (2016)	1	X								X
Du et al. (2017)	1	X								X
Fang et al. (2016)	1	X								X
Fujiki et al. (2007)	1	X		X				X		
Gouveia et al. (2015)	1	X								X

(continued)

Table 5.4 (continued)

References	Framework classification	Physical and mental characteristics								
		Physical activity	Physio-logical	Nutrition	Sleeping patterns	Productivity	Mood	Socializing	Digital media use	Daily activities
Grimes et al. (2010)	1	X		X						X
Gweon et al. (2018)	1	X							X	X
Harari et al. (2017b)	1	X						X		
Hirano et al. (2013)	1	X								X
Johansen et al. (2017)	1	X								X
Jylhä et al. (2013)	1	X								X
Kadomura et al. (2014)	1			X						
Kamphorst et al. (2014)	1	X								X
Ko et al. (2015)	1			X					X	X
Kocielnik et al. (2018b)	1	X								X

(continued)

Table 5.4 (continued)

References	Framework classification	Physical and mental characteristics								
		Physical activity	Physiological	Nutrition	Sleeping patterns	Productivity	Mood	Socializing	Digital media use	Daily activities
Lacroix et al. (2008)	1	X					X			X
Lee et al. (2014)	1	X	X	X						X
Lee et al. (2017)	1	X		X	X					X
Li et al. (2017)	1								X	X
Madan et al. (2010)	1	X		X				X		
Mollee et al. (2017)	1	X								X
Muaremi et al. (2013)	1	X		X						X
Pipke et al. (2013)	1									X
Rabbi et al. (2015)	1	X		X						X
Simon et al. (2012)	1								X	X

(continued)

Table 5.4 (continued)

References	Framework classification	Physical and mental characteristics								
		Physical activity	Physio-logical	Nutrition	Sleeping patterns	Productivity	Mood	Socializing	Digital media use	Daily activities
Tang et al. (2013)	1								X	X
Tulusan et al. (2012)	1					X				X
Van Bruggen et al. (2013)	1								X	X
Wang et al. (2015)	1	X								X
Weiss et al. (2012)	1					X			X	X
Zheng et al. (2008)	1									X
Zuckerman and Gal-Oz (2014)	1	X	X	X						
2. *Objective Measurement—Mental Outcome of Interest*										
Abney et al. (2014)	2								X	X
Bai et al. (2013)	2				X			X		X

(continued)

Table 5.4 (continued)

References	Framework classification	Physical and mental characteristics								
		Physical activity	Physiological	Nutrition	Sleeping patterns	Productivity	Mood	Socializing	Digital media use	Daily activities
Bickmore et al. (2018)	2	X	X							X
Canzian and Musolesi (2015)	2						X			X
Chaudhry et al. (2016)	2	X		X						
Cuttone and Larsen (2014)	2									X
Doryab et al. (2015)	2	X								X
Greis et al. (2017)	2								X	X
Huang et al. (2016)	2						X	X		X
Hwang and Pushp (2018)	2		X				X			X
Mehrotra et al. (2016)	2						X		X	X

(continued)

Table 5.4 (continued)

References	Framework classification	Physical and mental characteristics								
		Physical activity	Physio-logical	Nutrition	Sleeping patterns	Productivity	Mood	Socializing	Digital media use	Daily activities
Meyer et al. (2016)	2	X								X
Wang et al. (2016)	2	X			X		X			
Wang et al. (2018a)	2	X					X			
Wang et al. (2018b)	2	X			X	X		X		X
3. Subjective Measurement—Mental Outcome of Interest										
Barbarin et al. (2018)	3	X		X						
Bentley and Tollmar (2013)	3	X			X		X			X
Di Lascio et al. (2017)	3					X	X	X		
Kuo et al. (2018)	3	X							X	X
Paredes et al. (2014)	3	X		X				X		.

(continued)

Table 5.4 (continued)

References	Framework classification	Physical and mental characteristics								
		Physical activity	Physio-logical	Nutrition	Sleeping patterns	Productivity	Mood	Socializing	Digital media use	Daily activities
Sasaki et al. (2018)	3						X			X
Springer et al. (2018)	3						X			
4. Subjective Measurement—Physical Outcome of Interest										
Du et al. (2014)	4	X		X						X
Gui et al. (2017)	4	X						X		X
Hsu et al. (2014)	4	X		X						X
Kocielnik et al. (2018a)	4								X	X
Li et al. (2015)	4	X						X		X
Luhanga (2015)	4	X		X						
Marcu et al. (2018)	4	X								X
Möller et al. (2013)	4								X	X
Paay et al. (2015)	4	X								X

Table 5.4 shows the following information: (1) the quadrant of our self-tracking framework that references were categorized under and (2) the categories that were a focus of the specified references, as denoted by an "X" in the relevant columns

studies in the Physiology category provided promising avenues for detecting and modeling feedback based on real-time heart-rate data.

Nutrition. The large majority of studies on nutrition were categorized as objective measurement-physical outcome of interest quadrants of our conceptual framework. For example, Rabbi et al. (2015) developed and tested the efficacy of the MyBehavior application using the Theory of Planned Behavior (Ajzen 1985). MyBehavior was a smartphone application that integrated inferences of physical activity levels and dietary behaviors to produce personalized recommended changes to these patterns in order to promote a healthier lifestyle. The authors found that their personal informatics system led to an increase in physical activity and a decrease in food calorie intake, as compared to a control condition of participants not using the MyBehavior application. Other studies were focused especially on target populations—such as women suffering from obesity (Barbarin et al. 2018). Instead of implementing behavior change interventions directly, these studies were focused on identifying the unique needs of clinical populations.

Sleeping Patterns. The large majority of studies on sleeping patterns were categorized in either the objective measurement-physical outcome of interest or objective measurement-mental outcome of interest quadrants of our conceptual framework. Studies categorizing this theme are focused on assessing sleeping patterns from smartphone or wearable sensors and also on the digitized manual tracking of sleeping patterns, to delineate resulting changes in behavior. The emphasis individual studies place on different components of personal informatics varies. For instance, Bai et al. (2013) assessed changes in sleeping patterns using data collected from a mobile phone and through self-report surveys. They did not provide a feedback mechanism for participants, but instead used parts of the collected data to train their model on sleep-related habit formation patterns.

In contrast to this feedback-agnostic approach, Bentley et al. (2013) were exclusively focused on creating a tool to help individuals derive meaningful feedback from smartphone sensing data. The researchers constructed Health Mashups, a system designed to detect meaningful connections that are stable over time between a variety of behaviors and sensed data. One of these behaviors was sleep—the researchers collected sleeping pattern data from Fitbit. The researchers performed statistical analyses on the sleep data each night and then displayed natural language statements to individuals about observed associations (i.e. "On days when you sleep more, you get more exercise") on their smartphones (Bentley et al. 2013). A comprehensive PI deployment, incorporating elements from both of the previously cited sleep-related studies, was performed by Lee et al. (2017), who developed a wearable and smartphone-based system to manage unconscious itching behaviors that occur while individuals slept. Developed over the duration of two experiments and deployed in a full pilot study, the Itchtector was deemed helpful by many of the participants in the study, as revealed through qualitative interviews.

Productivity. Research examining productivity in personal informatics systems is relatively sparse and focused on the objective measurement-physical outcome of interest, objective measurement-mental outcome of interest, and subjective measurement-mental outcome of interest quadrants of our conceptual framework. Di Lascio et al. (2017) refocused the attention of personal informatics from fitness and personal health onto the "work environment". The researchers identified broad aims that a Quantified Workplace personal informatics system would need to be designed to address: choosing valuable data sources, deriving insights relevant to the workplace from this data, and driving change from these insights. The researchers then implemented a personal informatics system in a university setting to explore potential answers to their three questions. A metric called the Emotional Shift was developed in order to assess changes in affective states over the course of a university lecture. The authors reported tracking this metric using the PI system to show how emotional trajectories manifest in a real-life productivity-based environment. Since these researchers relied on surveys to collect data to assess mood changes in a work environment, this paper fit into the subjective measurement-mental variable of interest quadrant of our conceptual framework.

Mood. The large majority of studies on mood were distributed over the objective measurement-mental outcome of interest and mental measurement-mental outcome of interest quadrants. Studies categorizing this theme typically relied on self-reported mood information at pre-specified daily frequencies to track individual trajectories of mood over time. For instance, the EmotiCal personal informatics system tracked mood and provided predictive emotional analytics to individuals with the intention of facilitating participant understanding of mood and "trigger events" (Hollis et al. 2017; Springer et al. 2018). The EmotiCal system also implemented a feature to generate remedial plans by recommending new behaviors with the aim of increasing positive emotion. The researchers found that mood forecasting improved mood and emotional self-awareness in comparison to control condition participants, implying that positive behavior change had occurred as a result of using the PI system. In another example, mood-driven PI systems deployed amongst targeted populations— such as bipolar patients—did not result in systematic behavior changes (Doryab et al. 2015). The researchers deployed the MONARCA system, which patients used to report their daily mood scores. Additionally, the MONARCA system was able to sense behavioral traces, and it aimed to identify the effect of specific behaviors on the daily mood scores. While the authors identified sleeping patterns and physical activity as the main drivers of mood, none of the 78 participants in the pilot deployment reported any mood improvements as a result of using MONARCA. The authors used their insights to develop a "mood inference" engine for the existing MONARCA app.

The majority of the studies categorized by this theme did not set out to induce and measure the behavior change resulting from the use of their platforms. This was a concern, as without such an approach, the efficacy of different mood-targeted personal informatics systems cannot be assessed. Moreover, there was a distinct absence of passive mood detection technologies— all the surveyed papers relied on participant input to collect mood-related information.

Socializing. The large majority of studies on socializing were distributed over the objective measurement-physical outcome of interest and objective measurement-mental outcome of interest quadrants. Studies categorizing this theme typically assessed sociability from existing online social networks or from the microphone contained in smartphones and attempted to relate variations in behaviors and traits to the observed variances in sensed sociability. For instance, Harari et al. (2017a) examined behavior change in sociability patterns amongst a cohort of 48 students that participated in a 10-week smartphone-sensing study. The results suggested that sociability was typically high during the initial weeks of a semester but then decreased during the first half of the semester as the midterm examination period approached. In the second half, sociability increased and individual differences in sociodemographic characteristics (ethnicity and academic class) predicted sociability trajectories during the semester.

In a similar line of work across the objective measurement-physical outcome of interest quadrant, Madan et al. (2010) investigated how health-related behaviors spread as a result of face-to-face interactions with peers, by deploying a mobile sensing study that collected relevant data about participant location and ambient conversation using the microphone. The researchers found that the health behaviors exhibited by participants were correlated with the behaviors of peers that they interacted with over sustained periods of time, and this type of sensing could be implemented using just the sensor technology already embedded in a smartphone. This work suggests that future behavior change interventions might benefit from relying on the sensing of social interactions, given the strengths of these technologies for passively and non-intrusively collecting data about interpersonal interactions as they unfold in the course of daily life.

Hence, studies categorizing this self-tracking theme were typically focused on detecting sociability from online social networks or through smartphone sensors, in order to assess how social trajectories manifest in an ecologically valid manner. Some studies were focused on examining the impact of this sociability on real-world habits and behaviors, including fitness and diet. Studies under this theme occasionally attempted to assess behavioral change quantitatively (for instance, see Chen et al. 2016) but a large number of studies did not aim to operationalize behavior change or failed to do so in a quantitative manner.

Digital Media Use. Studies on digital media use were distributed over the objective measurement-physical outcome of interest, and the objective measurement-mental outcome of interest quadrants. For instance, FamiLync is a self-tracking app designed to measure digital media use and abuse in collaboration with one's family (Ko et al. 2015). The app promoted the non-use of smartphones in certain settings, contained a 'virtual public space' which facilitated social awareness on the use of the smartphone and contained tools that discouraged or prevented the use of digital media technologies (i.e., locking the screen and snoozing social media notifications). A three-week user study spanning 12 weeks indicated that the app improved the understanding of smartphone usage behavior and therefore allowed for more efficient parental mediation of excessive digital media use. Such positive changes in digital media use are likely to boost the mental health of participants, as excessive or very frequent smart-phone usage is shown to negatively influence mental illness (Choi et al. 2012). In a related approach that went one further step by operationalizing participant pre-dictions of future behaviors, Greis et al. (2017) deployed a PI system that tracked the number of times individuals unlocked their phone on any given day. Participants were required to indicate how many times they predicted to unlock their phone on any given day during each morning of the study. The results indicated that the exer-cise of self-predicting future behavioral patterns led to the automatic discovery of new insights into patterns of smartphone use (Greis et al. 2017). Hence, while the literature on digital media use trackers is sparse, this is a particularly important area for PI deployment with past research showing that it is possible to obtain behavioral assessments of "smartphone addiction" tendencies from smartphone data (Montag et al. 2015). This seems like a promising direction for behavior change interven-tions given the adverse risks that excessive digital media use (e.g., smartphone or social media addiction) poses to developing adolescents and adults. The research reviewed here indicates that digital media use tracking PI systems may influence positive behavioral change through increased self-tracking habits that reduced usage through increased awareness of use (Greis et al. 2017; Ko et al. 2015).

Daily Activities. The large majority of studies on daily activities were distributed over the objective measurement -physical outcome of interest, objective measurement-mental outcome of interest, and subjective measurement-physical outcome of inter-est quadrants. For instance, Wang et al. (2018) investigated how patterns of behavior change in daily activities, as sensed through the smartphone, could be used to pre-dict personality traits of young adults. Specifically, the researchers examined how trends in within-person ambient audio amplitude, exposure to human voice, phys-ical activity, phone usage, and location data predicted self-reported Big Five per-sonality traits (Extraversion, Agreeableness, Conscientiousness, Neuroticism, and Openness). Hence, this study fits into the objective measurement -mental variable of interest quadrant because it used data generated from physical measure (i.e., accelerometer data, microphone data) to make mental inferences (i.e., personality traits). The results indicated that personality traits could be modeled with high accu-racy from these within-person variations in daily activities.

While Wang et al. (2018) did not overtly focus on administering feedback to indi-viduals, Hirano et al. (2013) deployed a PI system with a focus on providing effective feedback to participants, specifically about their walking behavior. The researchers designed an app that detected participant motion, contained a manual digital logging feature of daily step count, and regularly notified participants to engage in physical activity. The researchers found that participants reported becoming more self-aware of their bodies and were wary of the time they spent sitting.

Some studies examined how daily activities are indicative of personality traits and mental states. For instance, one mobility study, focused on target populations of depressed individuals, succeeded in predicting depression states from mobility data of participants (Canzian and Musolesi 2015). Other studies developed and tested the efficacy of developing feedback techniques that nudge users towards engaging in healthier physical routines during their daily activities (i.e., Hirano et al. 2013). While the reviewed work suggests that daily activity PI systems can induce self-insight and self-awareness (e.g., Hirano et al. 2013), the effects of these systems on behavioral change outcomes remain relatively under-investigated and ambiguous.

5.3 Discussion

In this chapter, we surveyed the existing literature to identify an organizing framework for situating past and future scholarship using smartphones in personal informatics research. Our review findings suggest that a two-dimensional conceptual framework can be used to organize the smartphone-based self-tracking literature. Our review suggests that most of the smartphone-based self-tracking literature is concentrated in the first two quadrants of the conceptual framework: they tend to use objectively collected measurements (e.g., using mobile sensing) to deploy interventions aimed towards influencing mental or physical outcomes of interest. However, research in the other two quadrants was relatively sparse: fewer studies attempted to collect subjective measurements of physical and mental states to assess or influence physical outcomes of interest. Thus, future research should focus on filling in this gap in the literature by evaluating personal informatics systems that collect mental state information using subjective measures and quantify behavior change in relation to changing physical and mental states resulting from the intervention. There is especially a need to develop passive mobile sensing systems that can collect mood-related information unobtrusively from users (e.g., LiKamWa et al. 2013). The growth of mobile sensing systems in the physical domain has been rapid, and there is immense potential for this growth to percolate into the surrounding conceptual quadrants—namely to the subjective measurement-mental outcome of interest and subjective measurement—physical outcome of interest quadrants

Generally, research has prioritized certain components of personal informatics systems over others. For instance, some researchers focused on developing accurate sensing technologies in favor of administering user feedback, while other researchers were entirely focused on exploring optimal ways of generating actionable feedback for the user. Our results indicated that personal informatics work is currently dominated by computer science researchers, indicating a timely opportunity for behavioral researchers to get into the fray. Furthermore, we found limited theoretical integration in most of the extant literature, with findings indicating a shift in behavioral trajectories typically being considered in relation to one or two behavior change theories disciplinary-specific theories. While the use of classical theories in designing personal informatics is valuable, future work needs to further deploy theories developed specifically for the use of personal informatics systems (e.g., Li et al. 2010) in order to directly address the needs and habits of personal informatics application users. Such a consistency in theoretical integration will also ease cross-domain comparisons of the effectiveness of different personal informatics apps in inducing self-awareness and causing behavior change. Future theoretical work should integrate theories of personal informatics (e.g., Kersten-van Dijk et al. 2017) with behavior change theories (e.g., Prochaska and Velicer 1997), in order to develop a cross-disciplinary theoretical framework for designing optimal personal informatics applications.

5.4 Future Directions

The vast majority of reviewed papers in this literature review originated from the ACM database, suggesting that our results are skewed towards research produced by computer science and technically-oriented researchers. Furthermore, we found that an appreciation for individual differences in demographic and personality traits was generally absent from the reviewed empirical work, presumably because these were not variables of interest to technical researchers (e.g., Götz et al. 2017). In order to sustain behavior change interventions using digital self-tracking data, future work should identify how variations in individual differences are related to patterns of behavior change resulting from digitally engineered interventions. Indeed, there is a need for more work in the domain of understanding how an individual's personality and demographic traits relate to their motivations to self-track, and how these then influence the sustainability of the resulting behavior change.

An increase in cross-disciplinary dialogue between technologists and social scientists may facilitate an appreciation for individual differences in psychosocial characteristics (e.g., demographics, personality traits) during the design, implementation, and evaluation of smartphone-based self-tracking systems. Such interdisciplinary efforts are underway in the form of workshops (e.g., Campbell and Lane 2013), conferences (e.g., Rentfrow and Gosling 2012), and research initiatives (e.g., Life Sensing Consortium; lifesensingconsortium.org) that bring mobile sensing researchers from diverse disciplines into conversation with one other. An increase in interdisciplinary collaboration and widespread adoption of personal informatics models are likely to engender the next generation of personal informatics tools that customize their feedback to an individuals' psychological characteristics. We believe that customizing interventions according to individual differences will play an important role in facilitating self-reflection and sustained behavior change for the next generation of personal informatics systems.

It is encouraging to see that there are abundant deployment studies of different types of personal informatics studies in the literature. The diversity of applications of personal informatics system is particularly impressive, as is the commitment of researchers to pilot their proposed systems with real participants. While results pertaining to induced behavioral change vary across studies, we generally find that participants respond favorably to interventions and report increased feelings of self-awareness as a result. This work suggests that the future for personal informatics is bright, as more and more individuals are likely to adopt self-tracking methods as wearables and sensor-laden smartphones penetrate further into the human population. Future work should especially build upon two domains that contained sparse sensing literature: productivity and digital media use. Developing personal informatics systems for productivity tracking can assist organizations in monitoring employee productivity by displaying the times and locations at which an employee is at their most productive, and can further assist employees in maintaining adequate work-life balance by allowing employees to set time-based goals for work and recreational activities (see Mashhadi et al. 2016 for review). Similarly, sensing tools to examine detailed patterns of engagement with digital media can help curb concerns of social media abuse and its resulting detriments on mental health by prompting users to limit their time on the internet if it exceeds some predetermined threshold (i.e. Pardes 2019). More sophisticated applications could sense different kinds of social media use (i.e., active vs. passive use; Gerson et al. 2017) and alert users when they are engaged in types of social media usage that are typically associated with declines in mental well-being.

Moreover, future research should focus on deploying sensing work in non-Western settings. Virtually every cited study in this paper sampled from predominantly Western populations, as has been the tradition in behavioral science (Henrich et al. 2010). However, there are several Eastern social media platforms that rival the size and reach of their Western counterparts, such as WeChat (Lien and Cao 2014, Montag et al. 2018) and Weibo (Sullivan 2014). Similarly, over the last few years, the Quantified Self movement has diffused from the western hemisphere to the eastern hemisphere, especially to China (Yangjingjing 2012). The potential for personal informatics to succeed in the developing world is also bolstered by increasing smartphone penetration rates in fast-growing democratic nations such as India (Singh 2012). Hence, the spread of personal informatics around the world, coupled by the growth of non-Western social media platforms makes it essential for future research to focus on non-Western, multicultural samples in designing and deploying smartphone-based personal informatics systems. By deploying digital self-tracking platforms in developing countries around the world, we can accomplish two interrelated goals: (a) we can make digitized self-tracking a tool used at large by diverse individuals and (b) we can use the generated data to examine how richness in individual differences guides human-computer interaction. We look forward to social scientists and technologists collectively embracing the potential of self-tracking technologies in conducting interdisciplinary research.

Acknowledgements We thank Leela Srinivasan for assistance with the literature review and helpful feedback on earlier versions of the work presented in this manuscript.

References

Abney A, White B, Glick J, Bermudez A, Breckow P, Yow J, Heath P et al (2014) Evaluation of recording methods for user test sessions on mobile devices. In: Proceedings of the first ACM SIGCHI annual symposium on Computer-human interaction in play. ACM, pp 1–8

Ajzen I (1985) From intentions to actions: a theory of planned behavior. In: Action control. Springer, Berlin, pp 11–39

Athukorala K, Lagerspetz E, Von Kügelgen M, Jylhä A, Oliner AJ, Tarkoma S, Jacucci G (2014) How carat affects user behavior: implications for mobile battery awareness applications. In: Proceedings of the SIGCHI conference on human factors in computing systems. ACM, pp 1029–1038

Bai Y, Xu B, Jiang S, Yang H, Cui J (2013) Can you form healthy habit?: predicting habit forming states through mobile phone. In: Proceedings of the 8th international conference on body area networks. ICST (Institute for Computer Sciences, Social-Informatics and Telecommunications Engineering), pp 144–147

Bandura A (2004) Health promotion by social cognitive means. Health Educ Behav 31(2):143–164

Barbarin AM, Saslow LR, Ackerman MS, Veinot TC (2018) Toward health information technology that supports overweight/obese women in addressing emotion-and stress-related eating. In: Proceedings of the 2018 CHI conference on human factors in computing systems. ACM, p 321

Bentley F, Tollmar K (2013) The power of mobile notifications to increase wellbeing logging behavior. In: Proceedings of the SIGCHI conference on human factors in computing systems. ACM, pp 1095–1098

Bentley F, Tollmar K, Stephenson P, Levy L, Jones B, Robertson S, Wilson J et al (2013) Health Mashups: presenting statistical patterns between wellbeing data and context in natural language to promote behavior change. ACM Trans Comput-Human Interact (TOCHI) 20(5):30

Bexheti A, Fedosov A, Findahl J, Langheinrich M, Niforatos E (2015) Re-live the moment: visualizing run experiences to motivate future exercises. In: Proceedings of the 17th international conference on human-computer interaction with mobile devices and services adjunct. ACM, pp 986–993

Bickmore TW, Kimani E, Trinh H, Pusateri A, Paasche-Orlow MK, Magnani JW (2018) Managing chronic conditions with a smartphone-based conversational virtual agent. In: Proceedings of the 18th international conference on intelligent virtual agents. ACM, pp 119–124

Brewer RS, Verdezoto N, Holst T, Rasmussen MK (2015) Tough shift: exploring the complexities of shifting residential electricity use through a casual mobile game. In: Proceedings of the 2015 annual symposium on computer-human interaction in play. ACM, pp 307–317

Campbell AT, Lane ND (2013) Smartphone sensing: a game changer for behavioral science. Workshop held at the summer institute for social and personality psychology. The University of California, Davis

Canzian L, Musolesi M (2015) Trajectories of depression: unobtrusive monitoring of depressive states by means of smartphone mobility traces analysis. In: Proceedings of the 2015 ACM international joint conference on pervasive and ubiquitous computing. ACM, pp 1293–1304

Chaudhry BM, Schaefbauer C, Jelen B, Siek KA, Connelly K (2016) Evaluation of a food portion size estimation interface for a varying literacy population. In: Proceedings of the 2016 CHI conference on human factors in computing systems. ACM, pp 5645–5657

Chen Y, Randriambelonoro M, Geissbuhler A, Pu P (2016) Social Incentives in pervasive fitness apps for obese and diabetic patients. In: Proceedings of the 19th ACM conference on computer supported cooperative work and social computing companion. ACM, pp 245–248

Choi HS, Lee HK, Ha JC (2012) The influence of smartphone addiction on mental health, campus life and personal relations-focusing on K university students. J Korean Data Inf Sci Soc 23(5):1005–1015

Choudhury T, Borriello G, Consolvo S, Haehnel D, Harrison B, Hemingway B, LeGrand L et al (2008) The mobile sensing platform: an embedded activity recognition system. IEEE Pervasive Comput 7(2):32–41

Ciman M, Donini M, Gaggi O, Aiolli F (2016) Stairstep recognition and counting in a serious game for increasing users' physical activity. Pers Ubiquit Comput 20(6):1015–1033

Cuttone A, Larsen JE (2014) The long tail issue in large scale deployment of personal informatics. In: Proceedings of the 2014 ACM international joint conference on pervasive and ubiquitous computing: adjunct publication. ACM, pp 691–694

Di Lascio E, Gashi S, Krasic D, Santini S (2017) In-classroom self-tracking for teachers and students: preliminary findings from a pilot study. In: Proceedings of the 2017 ACM international joint conference on pervasive and ubiquitous computing and proceedings of the 2017 ACM international symposium on wearable computers. ACM, pp 865–870

Doryab A, Frost M, Faurholt-Jepsen M, Kessing LV, Bardram JE (2015) Impact factor analysis: combining prediction with parameter ranking to reveal the impact of behavior on health outcome. Pers Ubiquit Comput 19(2):355–365

Du H, Youngblood GM, Pirolli P (2014) Efficacy of a smartphone system to support groups in behavior change programs. In: Proceedings of the wireless health 2014 on national institutes of health. ACM, pp 1–8

Du J, Wang Q, de Baets L, Markopoulos P (2017) Supporting shoulder pain prevention and treatment with wearable technology. In: Proceedings of the 11th EAI international conference on pervasive computing technologies for healthcare. ACM, pp 235–243

Epstein DA, Ping A, Fogarty J, Munson SA (2015) A lived informatics model of personal informatics. In: Proceedings of the UbiComp 2015 international joint conference on pervasive and ubiquitous computing. ACM, New York

Fang B, Xu Q, Park T, Zhang M (2016) AirSense: an intelligent home-based sensing system for indoor air quality analytics. In: Proceedings of the 2016 ACM international joint conference on pervasive and ubiquitous computing. ACM, pp 109–119

Fox S, Duggan M (2013) Tracking for health. Available from http://www.pewinternet.org/2013/01/28/tracking-for-health

Fujiki Y, Kazakos K, Puri C, Pavlidis I, Starren J, Levine J (2007) NEAT-o-games: ubiquitous activity-based gaming. In: CHI2007 extended abstracts on Human factors in computing systems. ACM, pp 2369–2374

Gerson J, Plagnol AC, Corr PJ (2017) Passive and active facebook use measure (PAUM): validation and relationship to the reinforcement sensitivity theory. Personality Individ Differ 117:81–90

Glanz K, Rimer BK, Viswanath K (eds) (2008) Health behavior and health education: theory, research, and practice. Wiley, New York

Götz FM, Stieger S, Reips UD (2017) Users of the main smartphone operating systems (iOS, Android) differ only little in personality. PLoS ONE 12(5):e0176921

Gouveia R, Karapanos E, Hassenzahl M (2015) How do we engage with activity trackers?: a longitudinal study of habito. In: Proceedings of the 2015 ACM international joint conference on pervasive and ubiquitous computing. ACM, pp 1305–1316

Greis M, Dingler T, Schmidt A, Schmandt C (2017) Leveraging user-made predictions to help understand personal behavior patterns. In: Proceedings of the 19th international conference on human-computer interaction with mobile devices and services. ACM, p 104

Grimes A, Kantroo V, Grinter RE (2010) Let's play!: mobile health games for adults. In: Proceedings of the 12th ACM international conference on Ubiquitous computing. ACM, pp 241–250

Gui X, Chen Y, Caldeira C, Xiao D, Chen Y (2017) When fitness meets social networks: investigating fitness tracking and social practices on werun. In: Proceedings of the 2017 CHI conference on human factors in computing systems. ACM, pp 1647–1659

Gweon G, Kim B, Kim J, Lee KJ, Rhim J, Choi J (2018) MABLE: mediating young children's smart media usage with augmented reality. In: Proceedings of the 2018 CHI conference on human factors in computing systems. ACM, p 13

Harari GM, Gosling SD, Wang R, Chen F, Chen Z, Campbell AT (2017a) Patterns of behavior change in students over an academic term: A preliminary study of activity and sociability behaviors using smartphone sensing methods. Comput Hum Behav 67:129–138

Harari GM, Müller SR, Aung MS, Rentfrow PJ (2017b) Smartphone sensing methods for studying behavior in everyday life. Curr Opin Behav Sci 18:83–90

Harari GM, Müller SR, Gosling SD (2018) Naturalistic assessment of situations using mobile sensing methods. In: The Oxford handbook of psychological situations

Henrich J, Heine SJ, Norenzayan A (2010) The weirdest people in the world? Behav Brain Sci 33(2–3):61–83

Hirano SH, Farrell RG, Danis CM, Kellogg WA (2013) WalkMinder: encouraging an active lifestyle using mobile phone interruptions. In: CHI2013 extended abstracts on human factors in computing systems. ACM, pp 1431–1436

Hollis V, Konrad A, Springer A, Antoun M, Antoun C, Martin R, Whittaker S (2017) What does all this data mean for my future mood? actionable analytics and targeted reflection for emotional well-being. Human–Computer Interact 32(5–6):208–267

Hsu A, Yang J, Yilmaz YH, Haque MS, Can C, Blandford AE (2014) Persuasive technology for overcoming food cravings and improving snack choices. In: Proceedings of the SIGCHI conference on human factors in computing systems. ACM, pp 3403–3412

Huang Y, Xiong H, Leach K, Zhang Y, Chow P, Fua K, Barnes LE et al (2016) Assessing social anxiety using GPS trajectories and point-of-interest data. In: Proceedings of the 2016 ACM international joint conference on pervasive and ubiquitous computing. ACM, pp 898–903

Hwang C, Pushp S (2018) A mobile system for investigating the user's stress causes in daily life. In: Proceedings of the 2018 ACM international joint conference and 2018 international symposium on pervasive and ubiquitous computing and wearable computers. ACM, pp 66–69

Johansen B, Petersen MK, Pontoppidan NH, Sandholm P, Larsen JE (2017) Rethinking hearing aid fitting by learning from behavioral patterns. In: Proceedings of the 2017 CHI conference extended abstracts on human factors in computing systems. ACM, pp 1733–1739

Jylhä A, Nurmi P, Sirén M, Hemminki S, Jacucci G (2013) Matkahupi: a persuasive mobile application for sustainable mobility. In: Proceedings of the 2013 ACM conference on pervasive and ubiquitous computing adjunct publication. ACM, pp 227–230

Kadomura A, Li CY, Tsukada K, Chu HH, Siio I (2014) Persuasive technology to improve eating behavior using a sensor-embedded fork. In: Proceedings of the 2014 ACM international joint conference on pervasive and ubiquitous computing. ACM, pp 319–329

Kamphorst BA, Klein MC, Van Wissen A (2014) Autonomous E-coaching in the wild: empirical validation of a model-based reasoning system. In: Proceedings of the 2014 international conference on autonomous agents and multi-agent systems. International Foundation for Autonomous Agents and Multiagent Systems, pp 725–732

Kersten-van Dijk ET, Westerink JH, Beute F, IJsselsteijn WA (2017) Personal informatics, self-insight, and behavior change: a critical review of current literature. Human–Computer Interact 32(5–6):268–296

Ko M, Choi S, Yang S, Lee J, Lee U (2015) FamiLync: facilitating participatory parental mediation of adolescents' smartphone use. In: Proceedings of the 2015 ACM international joint conference on pervasive and ubiquitous computing. ACM, pp 867–878

Kocielnik R, Avrahami D, Marlow J, Lu D, Hsieh G (2018) Designing for workplace reflection: a chat and voice-based conversational agent. In: Proceedings of the 2018 on designing interactive systems conference 2018. ACM, pp 881–894

Kocielnik R, Xiao L, Avrahami D, Hsieh G (2018b) Reflection companion: a conversational system for engaging users in reflection on physical activity. Proc ACM Interact Mob Wearable Ubiquitous Technol 2(2):70

Kuo PYP (2018) Design for self-experimentation: participant reactions to self-generated behavioral prompts for sustainable living. In Proceedings of the 2018 ACM international joint conference and 2018 international symposium on pervasive and ubiquitous computing and wearable computers. ACM, pp 802–808

Lacroix J, Saini P, Holmes R (2008) The relationship between goal difficulty and performance in the context of a physical activity intervention program. In: Proceedings of the 10th international conference on Human computer interaction with mobile devices and services. ACM, pp 415–418

Lane ND, Miluzzo E, Lu H, Peebles D, Choudhury T, Campbell AT (2010) A survey of mobile phone sensing. IEEE Commun Mag 48(9):140–150

Lee ML, Dey AK (2014) Real-time feedback for improving medication taking. In: Proceedings of the SIGCHI conference on human factors in computing systems. ACM, pp 2259–2268

Lee J, Cho D, Kim J, Im E, Bak J, Lee KH, Kim J (2017) Itchtector: a wearable-based mobile system for managing itching conditions. In: Proceedings of the 2017 CHI conference on human factors in computing systems. ACM, pp 893–905

Li I, Dey A, Forlizzi J (2010) A stage-based model of personal informatics systems. In: Proceedings of the SIGCHI conference on human factors in computing systems. ACM, pp 557–566

Li N, Zhao C, Choe EK, Ritter FE (2015) HHeal: a personalized health app for flu tracking and prevention. In: Proceedings of the 33rd annual ACM conference extended abstracts on human factors in computing systems. ACM, pp 1415–1420

Li Y, Cao Z, Wang J (2017) Gazture: design and implementation of a gaze based gesture control system on tablets. Proc ACM Interact, Mob, Wearable Ubiquitous Technol 1(3):74

Lien CH, Cao Y (2014) Examining WeChat users' motivations, trust, attitudes, and positive word-of-mouth: Evidence from China. Comput Hum Behav 41:104–111

LiKamWa R, Liu Y, Lane ND, Zhong L (2013) Moodscope: building a mood sensor from smartphone usage patterns. In: Proceeding of the 11th annual international conference on Mobile systems, applications, and services. ACM, pp 389–402

Luhanga ET (2015) Evaluating effectiveness of stimulus control, time management and self-reward for weight loss behavior change. In: Adjunct proceedings of the 2015 ACM international joint conference on pervasive and ubiquitous computing and Proceedings of the 2015 ACM international symposium on wearable computers. ACM, pp 441–446

Lupton D (2013) Quantifying the body: monitoring and measuring health in the age of mHealth technologies. Crit Public Health 23(4):393–403

Lupton D (2014) Self-tracking cultures: towards a sociology of personal informatics. In: Proceedings of the 26th Australian computer-human interaction conference on designing futures: the future of design. ACM, pp 77–86

Madan A, Moturu ST, Lazer D, Pentland AS (2010) Social sensing: obesity, unhealthy eating and exercise in face-to-face networks. In: Wireless Health 2010. ACM, pp 104–110

Marcu G, Misra A, Caro K, Plank M, Leader A, Barsevick A (2018) Bounce: designing a physical activity intervention for breast cancer survivors. In: Proceedings of the 12th EAI international conference on pervasive computing technologies for healthcare. ACM, pp 25–34

Mashhadi A, Kawsar F, Mathur A, Dugan C, Shami NS (2016) Let's talk about the quantified workplace. In: Proceedings of the 19th ACM conference on computer supported cooperative work and social computing companion. ACM, pp 522–528

Mehrotra A, Hendley R, Musolesi M (2016) Towards multi-modal anticipatory monitoring of depressive states through the analysis of human-smartphone interaction. In: Proceedings of the 2016 ACM international joint conference on pervasive and ubiquitous computing: adjunct. ACM, pp 1132–1138

Meyer J, Heuten W, Boll S (2016) No effects but useful? long term use of smart health devices. In: Proceedings of the 2016 ACM international joint conference on pervasive and ubiquitous computing: adjunct. ACM, pp 516–521

Mollee JS, Middelweerd A, Velde SJT, Klein MC (2017) Evaluation of a personalized coaching system for physical activity: user appreciation and adherence. In: Proceedings of the 11th EAI international conference on pervasive computing technologies for healthcare. ACM, pp 315–324

Möller A, Kranz M, Schmid B, Roalter L, Diewald S (2013) Investigating self-reporting behavior in long-term studies. In: Proceedings of the SIGCHI conference on human factors in computing systems. ACM, pp 2931–2940

Montag C, Błaszkiewicz K, Lachmann B, Sariyska R, Andone I, Trendafilov B, Markowetz A (2015) Recorded behavior as a valuable resource for diagnostics in mobile phone addiction: evidence from psychoinformatics. Behav Sci 5(4):434–442

Montag C, Becker B, Gan C (2018) The multi-purpose application WeChat: a review on recent research. Front Psychol 9:2247

Muaremi A, Seiter J, Tröster G, Bexheti A (2013) Monitor and understand pilgrims: data collection using smartphones and wearable devices. In: Proceedings of the 2013 ACM conference on pervasive and ubiquitous computing adjunct publication. ACM, pp 679–688

Paay J, Kjeldskov J, Skov MB, Srikandarajah N, Brinthaparan U (2015) QuittyLink: using smartphones for personal counseling to help people quit smoking. In: Proceedings of the 17th international conference on human-computer interaction with mobile devices and services. ACM, pp 98–104

Pardes A (2019) Curb your time wasted on the web with this browser extension. Retrieved from https://www.wired.com/story/habitlab-browser-extension/

Paré G, Leaver C, Bourget C (2018) Diffusion of the digital health self-tracking movement in canada: results of a national survey. J Med Internet Res 20(5)

Parecki A (2018) My GPS Logs. Retrieved from https://aaronparecki.com/gps/

Paredes P, Gilad-Bachrach R, Czerwinski M, Roseway A, Rowan K, Hernandez J (2014) PopTherapy: coping with stress through pop-culture. In: Proceedings of the 8th international conference on pervasive computing technologies for healthcare. ICST (Institute for Computer Sciences, Social-Informatics and Telecommunications Engineering), pp 109–117

Pipke RM, Wegerich SW, Saidi A, Stehlik J (2013) Feasibility of personalized nonparametric analytics for predictive monitoring of heart failure patients using continuous mobile telemetry. In: Proceedings of the 4th conference on wireless health. ACM, p 7

Prochaska JO, Velicer WF (1997) The transtheoretical model of health behavior change. Am J Health Promot 12(1):38–48

Quantified Self Labs (2015) Quantified self—self knowledge through numbers. Retrieved from http://www.quantifiedself.com

Rabbi M, Aung MH, Zhang M, Choudhury T (2015) MyBehavior: automatic personalized health feedback from user behaviors and preferences using smartphones. In: Proceedings of the 2015 ACM international joint conference on pervasive and ubiquitous computing. ACM, pp 707–718

Rentfrow PJ, Gosling SD (2012) Using smart-phones as mobile sensing devices: a practical guide for psychologists to current and potential capabilities. In: Preconference for the annual meeting of the Society for personality and social psychology. San Diego, CA

Rooksby J, Rost M, Morrison A, Chalmers MC (2014) Personal tracking as lived informatics. In: Proceedings of the 32nd annual ACM conference on human factors in computing systems. ACM, pp 1163–1172

Sanders R (2017) Self-tracking in the digital era: biopower, patriarchy, and the new biometric body projects. Body Soc 23(1):36–63

Sasaki W, Nakazawa J, Okoshi T (2018) Comparing ESM timings for emotional estimation model with fine temporal granularity. In: Proceedings of the 2018 ACM international joint conference and 2018 international symposium on pervasive and ubiquitous computing and wearable computers. ACM, pp 722–725

Simon J, Jahn M, Al-Akkad A (2012) Saving energy at work: the design of a pervasive game for office spaces. In: Proceedings of the 11th international conference on mobile and ubiquitous multimedia. ACM, p 9

Singh P (2012) Smartphone: the emerging gadget of choice for the urban Indian. The Nielsen Company Retrieved from http://www.nielsen.com/content/dam/corporate/india/reports/2012/Featured

Springer A, Hollis V, Whittaker S (2018) Mood modeling: accuracy depends on active logging and reflection. Pers Ubiquitous Comput 1–15

Sullivan J (2014) China's Weibo: Is faster different? New Media Soc 16(1):24–37

Swan M (2012) Sensor mania! the internet of things, wearable computing, objective metrics, and the quantified self 2.0. J Sens Actuator Netw 1(3):217–253

Tang LY, Hsiu PC, Huang JL, Chen MS (2013) iLauncher: an intelligent launcher for mobile apps based on individual usage patterns. In: Proceedings of the 28th annual ACM symposium on applied computing. ACM, pp 505–512

Tulusan J, Staake T, Fleisch E (2012) Providing eco-driving feedback to corporate car drivers: what impact does a smartphone application have on their fuel efficiency? In: Proceedings of the 2012 ACM conference on ubiquitous computing. ACM, pp 212–215

Van Bruggen D, Liu S, Kajzer M, Striegel A, Crowell CR, D'Arcy J (2013) Modifying smartphone user locking behavior. In: Proceedings of the ninth symposium on usable privacy and security. ACM, p 10

Wang R, Harari G, Hao P, Zhou X, Campbell AT (2015) SmartGPA: how smartphones can assess and predict academic performance of college students. In: Proceedings of the 2015 ACM international joint conference on pervasive and ubiquitous computing. ACM, pp 295–306

Wang R, Aung MS, Abdullah S, Brian R, Campbell AT, Choudhury T, Tseng VW et al (2016) CrossCheck: toward passive sensing and detection of mental health changes in people with schizophrenia. In: Proceedings of the 2016 ACM international joint conference on pervasive and ubiquitous computing. ACM, pp 886–897

Wang W, Harari GM, Wang R, Müller SR, Mirjafari S, Masaba K, Campbell AT (2018a) Sensing behavioral change over time: using within-person variability features from mobile sensing to predict personality traits. Proc ACM Interact Mob Wearable Ubiquitous Technol 2(3):141

Wang R, Wang W, daSilva A, Huckins JF, Kelley WM, Heatherton TF, Campbell AT (2018b) Tracking depression dynamics in college students using mobile phone and wearable sensing. Proc ACM Interact Mob Wearable Ubiquitous Technol 2(1):43

Weiss M, Staake T, Mattern F, Fleisch E (2012) PowerPedia: changing energy usage with the help of a community-based smartphone application. Pers Ubiquit Comput 16(6):655–664

Wolf G (2010) The data-driven life. N Y Times 28:2010

Yangjingjing X (2012) The science of the self. Global Times. Retrieved from http://www.globaltimes.cn/content/750476.shtml

Zheng Y, Li Q, Chen Y, Xie X, Ma WY (2008) Understanding mobility based on GPS data. In: Proceedings of the 10th international conference on Ubiquitous computing. ACM, pp 312–321

Zuckerman O, Gal-Oz A (2014) Deconstructing gamification: evaluating the effectiveness of continuous measurement, virtual rewards, and social comparison for promoting physical activity. Pers Ubiquitous Comput 18(7):1705–1719

Chapter 6
Digital Brain Biomarkers of Human Cognition and Mood

Paul Dagum

Abstract By comparison to the functional metrics available in other medical disciplines, conventional measures of neuropsychiatric and neurodegenerative disorders have several limitations. They are obtrusive, requiring a subject to break from their normal routine. They are episodic and provide sparse snapshots of a patient only at the time of the assessment. They require subjects to perform a task outside of the context of everyday behavior. And lastly, they are poorly scalable, taxing limited resources. We present validation studies that demonstrate the clinical efficacy of a new approach in reproducing gold-standard neuropsychological measures. We discuss the neuroscience constructs and mathematical underpinnings of cognition and mood measurement from human-computer interaction data. We conclude with a discussion on four areas that we predict will be impacted by these new clinical measurements: (i) understanding of the interdependency between cognition and mood; (ii) nosology of psychiatric illnesses; (iii) drug discovery; and (iv) delivery of healthcare services.

6.1 Introduction

The twentieth century introduced electroencephalography, magnetic resonance imaging, genome sequencing and other novel techniques that advanced our understanding of neuroscience. But despite these advances, our progress in treating brain illnesses has been slow. We have struggled to translate discoveries into insights of psychopathology that improve functional outcomes. This struggle is rooted in how we measure brain function and how we diagnose. These measures whether for disorders of cognition or mood share several limitations.

The laboratory nature of existing clinical instruments are not objective measures of functional performance in the real world. This limits their utility in correlating the effect of exogenous and endogenous factors such as duress, lifestyle, illness, and

P. Dagum (✉)
Mindstrong Health, 303 Bryant St, Mountain View, CA 94041, USA
e-mail: paul@mindstronghealth.com

© Springer Nature Switzerland AG 2019
H. Baumeister and C. Montag (eds.), *Digital Phenotyping and Mobile Sensing*,
Studies in Neuroscience, Psychology and Behavioral Economics,
https://doi.org/10.1007/978-3-030-31620-4_6

drugs with symptom burden. A similar limitation existed in other disciplines such as cardiology, where for example a normal resting electrocardiogram revealed nothing about a person's ability to climb a flight of stairs. Similar to a stress electrocardiogram, we need to measure a person's ecological resilience in their natural environment. But unlike a stress electrocardiogram, ecological brain measurements need to be unobtrusive and continuous.

Our measurements today are not ecological. They are obtrusive, requiring a subject to break from their normal routine. They are episodic and provide sparse snapshots of a patient only at the time of the assessment. They require subjects to perform a task outside of the context of everyday behavior. And lastly, they are poorly scalable, taxing limited resources.

Ecological measurement of brain function was unthinkable even a decade ago. But the emergence of ubiquitous mobile digital devices created the opportunity to passively and continuously collect digital signals from individuals. Early applications of these digital signals included cybersecurity where they were used to create digital fingerprints of cybercriminals (Dagum 2018a).

In 2013, following our early work in cybersecurity, we launched several clinical studies that were the first to demonstrate the feasibility of creating digital biomarkers that correlate with laboratory assessments of mood and cognition using passively acquired data from the daily use of a smartphone (Kerchner et al. 2015; Dagum 2018b). These digital biomarkers meet our ecological requirements of providing insight into day-to-day functional performance. They are unobtrusive, placing no burden on the subject beyond the normal use of a smartphone. They provide dense daily assessments with potential insight into hourly or daily variations in brain function. They scale globally to over two billion smartphone users today (Statistica 2019).

In the real-world, brain function is affected by illness and environment influences such as drugs and insomnia, and performance will vary from what we measure in a laboratory setting. We postulate that the daily variability in the digital biomarkers will provide rich temporal insight into state-dependent signatures of cognition and emotional health that may be predictive of disease and environmental effects. These biomarker signatures define digital phenotypes (Insel 2017) that similar to cardiac phenotypes may have distinct clinical profiles and outcomes. This has the potential to create the foundation of a new nosology of mental illness.

6.2   Human-Computer Interactions

The interface between humans and computers has evolved over the past seven decades and continues to evolve. Interfaces today range from ubiquitous touch-screens found in smartphones and automobile dashboards, to mixed-reality interfaces in smart-glasses, and to voice user interfaces (VUI). We have recently witnessed the emergence of direct brain-computer interfaces (BCI) that may someday replace human-computer interfaces (HCI).

Regardless of the interface, whether it is today's HCI or tomorrow's BCI, we can decompose the interaction model into three fundamental components. The first component is *information presentation* by the computer. This information presentation is broad and often domain and modality specific. Visual examples predominate and include a scrollable list of selectable items, keyboard input, navigation map, or performance indicators. Examples of auditory presentation are fewer, analog, carry lower information capacity and include voice responses in VUI systems.

The second component is *mental processing* by the human of the information presented. Mental processing relies on the integrity of neural circuits. The cognitive constructs needed for mental processing depend on the nature of the information presented and what is expected of the human. These constructs range from attention (selective, sustained), memory (working, visual, delayed), speed of information processing, to verbal fluency and executive function.

The third component is the *human response* that follows the mental processing step. Both the type of response and the time to respond to the presented information are measurable by the interface. The total response time from presentation includes the motor response time. By comparison to mental processing time, that exhibits significant variability, long duration and high correlation with total response time, the motor response time shows low variability, shorter duration and poor correlation (Botwinick and Thompson 1966). Figure 6.1 illustrates the three components of the human-computer interaction model.

The speed at which different types of information is processed by humans varies across a population and modality of presentation. For a specific modality, the population variation captures normal variation linked to a general factor of intelligence and differences in mental processing that include the effect of injuries and illnesses. This has been confirmed by a century of traditional neurocognitive testing from paper and pencil tests, to computer tests, to auditory tests (Strauss et al. 2006). These tests involve completing repeated simple primitive tasks that are scored on completion time. Like HCI models, these gold-standard tests can be decomposed into information presentation, mental processing and human response.

Ecological measurement moves us beyond a population distribution of functional capacity to individual longitudinal measures of functional performance. Moment-to-moment functional performance is affected by ecological causes that are measurable from human-computer interactions as delays or errors in the human response component of an individual. Common ecological causes include fatigue, mood disorders, and chronic medical illnesses that affect brain function whether from poorly regulated blood glucose (e.g., diabetes mellitus), acid-base imbalance (e.g., chronic-obstructive pulmonary disease), or inadequate cerebral perfusion (e.g., ischemic heart disease). Monitoring ecological biomarkers together with behavior observed through passive capture of phone behavioral data (Markowetz et al. 2014) or self-journaling (Kubiak and Smyth 2019) may help bridge our current behavioral nosology of mental illness with proximal digital measures of central nervous system function.

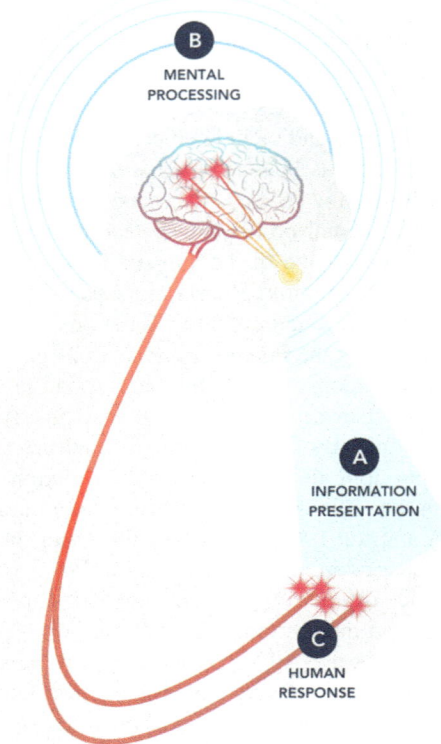

Fig. 6.1 An illustration of **a** information presentation, **b** mental processing and **c** human response that comprise the fundamental components of human-computer interaction. The type of information presentation, for example a choice selection, and the response time when measured repeatedly by the day-to-day use of a smartphone provide passive, ecological, in-the-moment insight into a user's cognitive and emotional state. Standard neuropsychological instruments of cognition and mood lie in low-dimensional submanifolds of the high-dimensional space formed by the different types of information presentation stimuli and features

6.3   Digital Biomarkers of Cognition

In January of 2014, we launched a study to demonstrate the feasibility of decoding human-computer interactions into clinical measures of neuropsychological function. We recruited 27 subjects (ages 18–34 years, education 14.1 ± 2.3 years, M:F 8:19) volunteered for neuropsychological assessment and installed an app on their smartphone that passively captured their HCI from touchscreen activity. All participants, recruited via social media, signed an informed consent form. Inclusion criteria required participants to be functional English speaking and active users of a smartphone. The protocol involved 3 h of psychometric assessment followed by installation of an app on their smartphone. The test battery is shown in the first column of Table 6.1. A single psychometrician performed all testing in a standard assessment clinic. The app on the phone ran passively in the background and captured patterns and timings of touch-screen user activity that included swipes, taps, and keystroke events, comprising the human-computer interactions.

From the HCI events we identified 45 interaction patterns using high-dimensionality reduction. Each pattern represents a task that is repeated up to several hundred times per day by a user during normal use of their phone. Most patterns consisted of two successive events, such as tapping on the space-bar followed by the first

Table 6.1 Fourteen neurocognitive assessments covering five cognitive domains and dexterity were performed by a neuropsychologist. Shown are the group mean and standard deviation, range of score, and the correlation between each test and the cross-validated prediction constructed from the digital biomarkers for that test

Cognitive predictions			
	Mean (SD)	Range	R (predicted), p-value
Working memory			
Digits forward	10.9 (2.7)	7–15	0.71 ± 0.10, 10^{-4}
Digits backward	8.3 (2.7)	4–14	0.75 ± 0.08, 10^{-5}
Executive function			
Trail A	23.0 (7.6)	12–39	0.70 ± 0.10, 10^{-4}
Trail B	53.3 (13.1)	37–88	0.82 ± 0.06, 10^{-6}
Symbol digit modality	55.8 (7.7)	43–67	0.70 ± 0.10, 10^{-4}
Language			
Animal fluency	22.5 (3.8)	15–30	0.67 ± 0.11, 10^{-4}
FAS phonemic fluency	42 (7.1)	27–52	0.63 ± 0.12, 10^{-3}
Dexterity			
Grooved pegboard test (dominant hand)	62.7 (6.7)	51–75	0.73 ± 0.09, 10^{-4}
Memory			
California verbal learning test (delayed free recall)	14.1 (1.9)	9–16	0.62 ± 0.12, 10^{-3}
WMS-III logical memory (delayed free recall)	29.4 (6.2)	18–42	0.81 ± 0.07, 10^{-6}
Brief visuospatial memory test (delayed free recall)	10.2 (1.8)	5–12	0.77 ± 0.08, 10^{-5}
Intelligence scale			
WAIS-IV block design	46.1 (12.8)	12–61	0.83 ± 0.06, 10^{-6}
WAIS-IV matrix reasoning	22.1 (3.3)	12–26	0.80 ± 0.07, 10^{-6}
WAIS-IV vocabulary	40.6 (4.0)	31–50	0.67 ± 0.11, 10^{-4}

character of a word, or tapping the "delete" key followed by another "delete" key tap. Some patterns were collected in a specific context of use. For example, tapping on a character followed by another character could be collected at the beginning of a word, middle of a word, or end of a word. Each pattern generated a time-series composed of the time interval between patterns. The time-series were segmented into daily time-series. To each daily time-series we applied 23 mathematical transforms to produce 1,035 distinct daily features that we term *digital biomarkers*.

For each participant we selected the first 7 days of data following their test date. A biomarker was considered a candidate for a neurocognitive test if over the 7 day window the 7 correlations between sorted biomarker values and the test scores were stable (meaning of the same sign). The 2-dimensional design matrix for the supervised kernel principal component analysis (PCA) was constructed by selecting the peak value of each candidate biomarker over the 7 days. For each test, we constructed a linear reproducing Hilbert space kernel from the biomarkers and used a supervised kernel principal component analysis (Dagum 2018b) with leave-one-out-cross-validation (LOOCV) as follows. To predict the 1st participant test result the model fitting algorithm was run on the remaining participants without access to the 1st participant's data, and so forth iterating 27 times to generate the 27 predictions. Cross-validation allows us to control the risk of over-fitting the data because each participant's data is never used in fitting the model used to predict that participant's results.

These preliminary results suggested that passive HCI measures from smartphone use could be a continuous ecological surrogate for laboratory-based neuropsychological assessment. Smartphone human-computer interaction data from the 7 days following the neuropsychological assessment showed a range of correlations from 0.62 to 0.83 with the neurocognitive test scores. Table 6.1 shows the correlation between each neurocognitive test and the cross-validated predictions of the supervised kernel PCA constructed from the biomarkers for that test. Figure 6.2 shows each participant test score and the digital biomarker prediction for (a) digits backward, (b) symbol digit modality, (c) animal fluency, (d) Wechsler Memory Scale-3rd Edition (WMS-III) logical memory (delayed free recall), (e) brief visuospatial memory test (delayed free recall), and (f) Wechsler Adult Intelligence Scale-4th Edition (WAIS-IV) block.

An obvious limitation of this study was the small size (n = 27) relative to the large number of potential biomarkers (n = 1,035). To counter the risk of over-fitting these results, predictions were made using LOOCV, stringent confidence level (p $< 10^{-4}$) and a simple linear kernel that was regularized. A further limitation was that the neuropsychological assessment occurred at one time point and the digital features were collected ecologically over the first 7 days following the assessment. For clinical assessments, we have found that the real-world, continuous assessment yields critical information relevant to function.

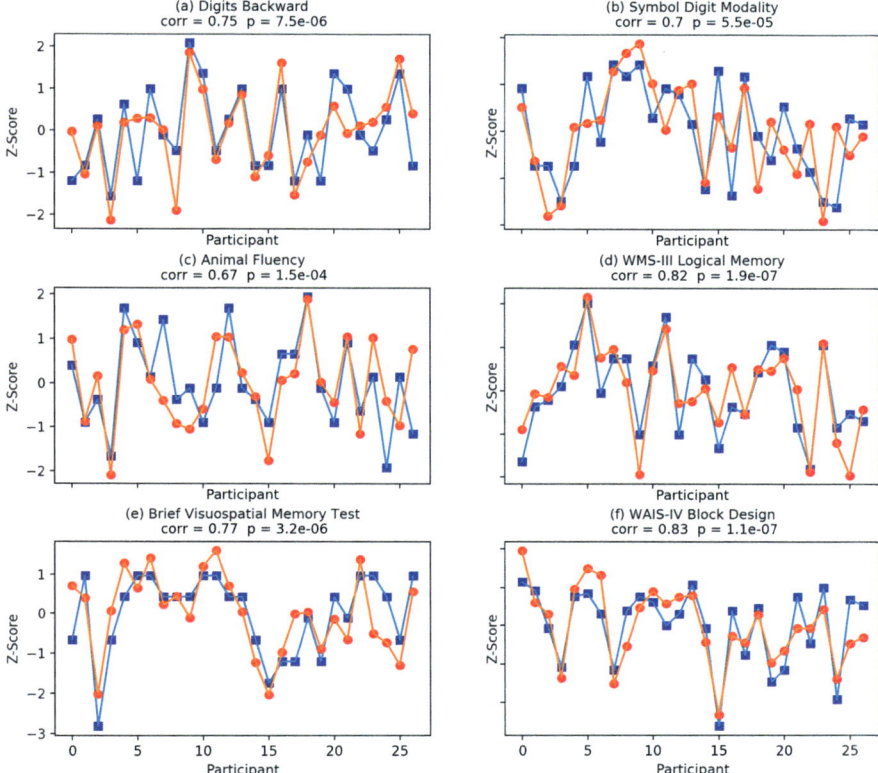

Fig. 6.2 A blue square represents a participant test Z-score normed to the 27 participant scores and a red circle represents the digital biomarker prediction Z-score normed to the 27 predictions. Test scores and predictions shown are **a** digits backward, **b** symbol digit modality, **c** animal fluency, **d** Wechsler memory Scale-3rd Edition (WMS-III) logical memory (delayed free recall), **e** brief visuospatial memory test (delayed free recall), and **f** Wechsler adult intelligence scale-4th Edition (WAIS-IV) block design

6.4 Digital Biomarkers of Mood

A core feature of major depressive disorder (MDD) is anhedonia defined as a diminished interest to previously rewarding stimuli. Anhedonia is present in approximate 40% of patients with MDD and may represent a physiologically distinct sub-group of patients with motivational and reward processing deficits (Keedwell et al. 2005). To identify digital biomarkers associated with depression and anhedonia we analyzed human-computer interaction from the smartphone of subjects enrolled in a multi-center, double blind-clinical trial for the treatment of MDD (Madrid et al. 2017; Smith et al. 2018). Of the 15 subjects randomized (ages 18–48 years, M:F 3:12) 10 were Caucasians, 4 were African-Americans and 1 was Asian, 2 were Hispanic and 13 were not Hispanic.

We used 7 days of smartphone data collected in the first week following baseline assessment. Baseline assessment included the Montgomery-Asberg Depression Rating Scale (MADRS) and Snaith Hamilton Pleasure Scale (SHAPS) for anhedonia. We used the same panel of 1,035 daily digital biomarkers generated from the touch-screen human-computer interactions of each subject. We again used a supervised kernel PCA with a 2-dimensional design matrix similarly constructed by selecting the peak value of each candidate biomarker over the 7 days. For each test, we constructed a linear reproducing Hilbert space kernel from the biomarkers and used a supervised kernel principal component analysis with LOOCV. The correlation between the MADRS and SHAPS assessment of depression and anhedonia and the cross-validated predictions of the supervised kernel PCA constructed from the biomarkers for that assessment where Spearman $r = 0.67$, $p = 0.024$ and Spearman $r = 0.82$, $p = 0.001$ respectively.

In another research study from 2017 to 2018 (ClinicalTrials.gov 2017), we demonstrated that digital brain biomarkers are sensitive and specific to temporal changes in mood and clinical severity. Ten participants (ages 44 ± 10 years, education 15 ± 1 years, M:F 5:5) were enrolled from a private outpatient psychiatric clinic with a diagnosis of either Major Depressive Disorder (MDD) or Bipolar Depression (BD) confirmed using the Structured Clinical Interview for DSM Disorders-Clinical Trials (SCID-CT). Participants consented to repeat weekly clinical assessment for eight months and to installing the smartphone app on their smartphone during this entire period. All clinical assessments were performed by the same person.

For each participant, we used the HCI data collected on the day of the clinical assessment to create the digital biomarkers. This differs from our prior studies where we used data collected over a 7-day period to select the digital biomarkers. We treated the repeated clinical assessments as independent measures and used an extension of supervised kernel PCA (Barshan et al. 2011) to longitudinal data (Staples et al. 2018). The 2-dimensional design matrix of the supervised longitudinal kernel PCA for a target clinical assessment was constructed using digital biomarkers selected from the full panel only if they satisfied a false-discovery rate (FDR) of 5% and 0.5% for mood and clinical severity, respectively, using the Benjamini-Hochberg procedure (Benjamini and Hochberg 1995). As an additional FDR control, we computed the permutation p-value of the results using 100 random permutations of the training targets (Winkler et al. 2014). Using an FDR of 5% to select the digital biomarkers used in the model for a neuropsychological test guarantees that at most 1 in 20 selected biomarkers is a false positive.

The output of the supervised longitudinal kernel PCA was evaluated using both leave-one-out cross-validation (LOOCV) and out-of-bag prediction error using bootstrap aggregation. The supervised longitudinal kernel PCA treated repeated samples as independent in the model, required for the validity of cross validation at the patient-test level in this longitudinal study (Bergmeir et al. 2018). Because of the small sample size, we further performed permutation testing to validate the results. These tests confirmed invariance of the LOOCV results following permutation of the

Table 6.2 Subjects were instructed to complete the Patient Health Questionaire-9 based on their assessment over the past 24-h only. Clinical severity was scored by a single clinician using the Clinical Global Impression for Severity scale. Shown are the group mean and standard deviation, range of score, total number of observations, the correlation and correlation p-value between the instrument and the cross-validated prediction constructed from the digital biomarkers

Mood predictions

	Patient Health Questionaire-9 (24 h)	Clinical Global Impression—Severity
Mean (SD)	13.3 (6.1)	4.0 (1.4)
Range	0–26	2–6
Observations	170	128
R (predicted), p-value	0.63, 2.8e–20	0.71, 7.0e–21

participant's repeated clinical assessments as would be expected if the repeated samples were truly independent in the model. In contrast, re-fitting the cross-validated models following permutation of the targets between participants, that is between participant's test scores, led to no statistically significant predictions. This last permutation test would yield statistically significant results if the model-fitting procedure was over-fitting the data.

To assess mood we used the Patient Health Questionnaire 9 (PHQ-9) scale modified by asking participants to respond from their experience over the past 24 h only. Clinical severity was assessed using the Clinical Global Impression—Severity (CGI-S). Table 6.2 shows the correlation between each of the two assessments and the cross-validated predictions from the supervised longitudinal kernel. Figure 6.3 shows each participant assessments and the digital biomarker predictions for (a) PHQ-9 24 h and (b) CGI-S.

The results from both these studies demonstrate that the panel of digital biomarkers constructed from passively acquired human-computer interactions could be a continuous ecological surrogate for laboratory-based assessments for mood disorders and of clinical severity. These studies demonstrated feasibility and early clinical validity of our approach. The risk of over-fitting the data and multiple comparisons was controlled using strict false-discovery rates and permutation testing.

We have shown through these studies that standard cognitive and mood assessments performed by a neuropsychologist lie in a high-dimensional space of 1,035 digital biomarkers generated from touchscreen HCI patterns. Reconstructing an assessment reduces to identifying the minimal set of biomarkers that spans a low-dimensional submanifold containing the assessment. To visualize this biomarker space against the many cognitive and mood assessments, Figs. 6.4 and 6.5 show a heatmap of the correlations between specific digital biomarkers and target measures. For all correlations, p-values were corrected to limit the false discovery rate to 1 in 100. Correlations with a p-value below this threshold are shown in gray, and any biomarkers that did not exceed this threshold on at least one target measure are not shown. Rows and columns are sorted using hierarchical clustering so rows and

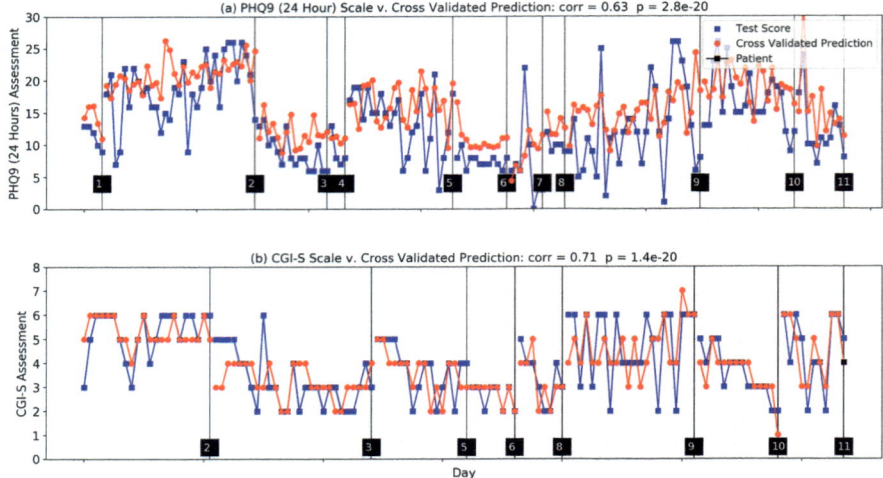

Fig. 6.3 A blue square represents a participant assessment score and a red circle represents the digital biomarker prediction. Assessment scores and predictions shown are **a** patient health questionnaire 9 using a 24-h recall (PHQ 9 24 h) and **b** clinical global impression scale for severity (CGI-S)

columns that are more similar are positioned closer in the plot and have a shorter dendrogram line.

6.5 Future Directions

Nobel laurate Sydney Brenner has said that "progress in science depends on new techniques, new discoveries and new ideas, probably in that order". We have introduced a new paradigm for measuring cognition and mood that is based on ubiquitous human-computer interactions from smartphone use. With the ubiquitous use of smartphones (Andone et al. 2016), we anticipate that this paradigm may have profound implications in our understanding of brain disorders, functional outcomes, and lead to new therapeutic discoveries. We specifically discuss four areas that could be impacted: (i) understanding of the interdependency between cognition and mood; (ii) nosology of psychiatric illnesses; (iii) drug discovery; and (iv) delivery of healthcare services.

Mental disorders that include mood and psychosis show remitting and relapsing behavior that can be exacerbated by triggers or improved with therapy. The potential for rapid clinical fluctuations in these disorders can have profound effects on brain function. These state-dependent changes in cognition are often more disabling than the affective symptoms. Our understanding of this interdependency has been hampered by episodic clinic measurements of cognition that seek to measure trait level

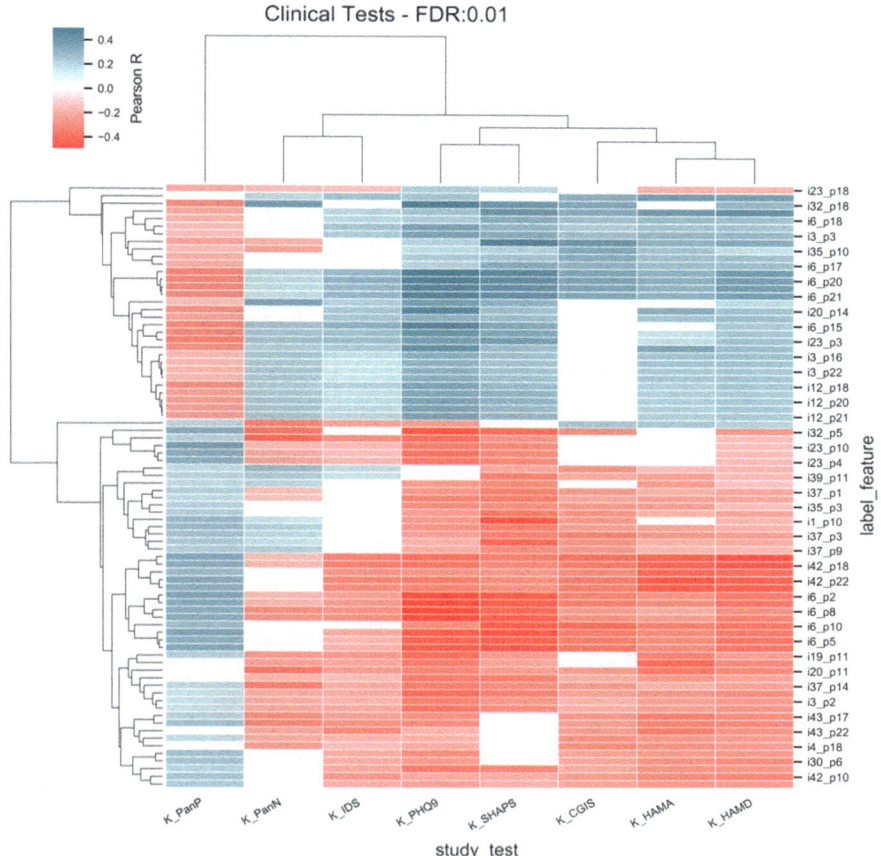

Clinical Tests		
Abbreviation	**Test**	**Construct**
PAN_P	PANAS Positive	Positive Affect
PAN_N	PANAS Negative	Negative Affect
IDS	Depression Scale	Depression
PHQ 9	Patient Health Questionnaire	Depression
SHAPS	Snaith-Hamilton Pleasure Scale	Anhedonia
CGIS	Clinician Global Impression- Severity	Global psychiatric functioning
HAMA	Hamilton Anxiety Scale	Anxiety
HAMD	Hamilton Depression Scale	Depression

Fig. 6.4 Correlations between the top Mindstrong Biomarkers and Clinical measures. The p-values from all correlations are corrected to limit the false discovery rate to 1%. Correlations with a p-value below this threshold are shown in gray

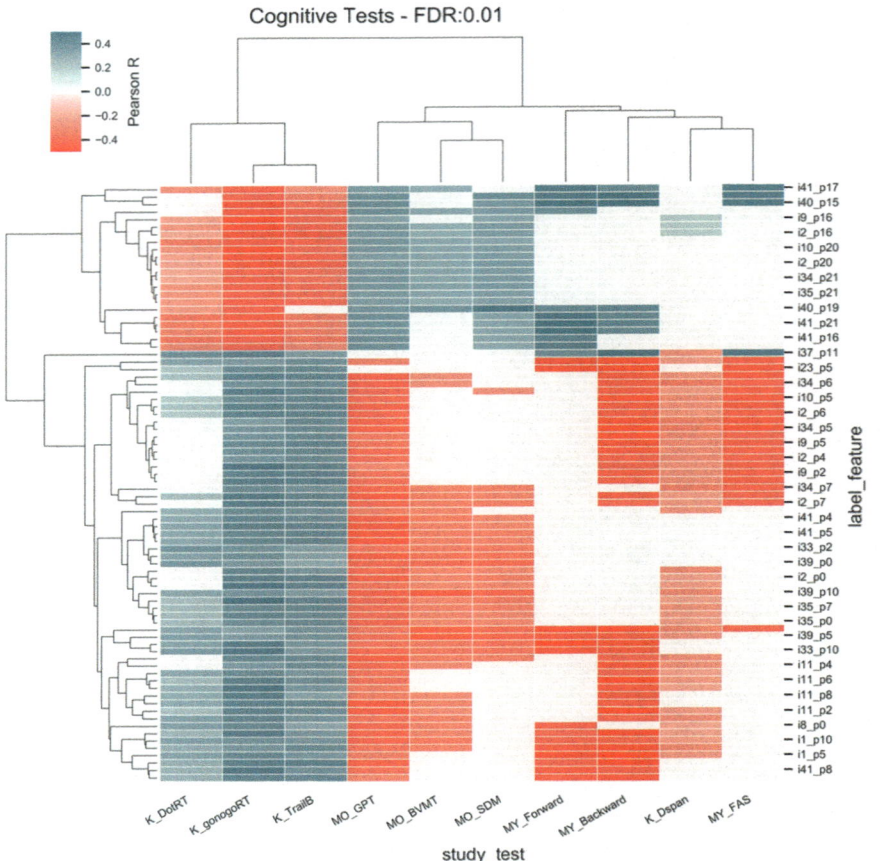

Cognitive Tasks		
Abbreviation	**Test**	**Construct**
DotRT	Choice reaction time	Attention and Motor Speed
gonogoRT	Go/no-go reaction time (RT)	Attention and Behavioral Inhibition
TrailB	Trail Making Task B	Executive Function
GPT	Grooved Pegboard Task	Visuomotor Coordination
BVMT	Brief Visuospatial Memory Test	Visuospatial Memory
SDM	Symbol Digit Modalities Test	Executive Functioning
Forward	Digit Span Forward	Working Memory
Backward	Digit Span Backward	Executive Functioning
Dspan	Digit Span Visual	Working Memory
FAS	F-A-S fluency task	Phonemic Verbal Fluency

Fig. 6.5 Correlations between the top Mindstrong Biomarkers and Cognitive tasks. (Details are the same as for Fig. 6.3)

functional capacity in the laboratory. The ability to measure both mood and cognition ecologically and continuously could improve our understanding of how mood regulates cognition and cognition regulates mood. This could lead to new therapies and improved function.

Diagnosis of mental disorders today involves a constellation of clinical symptoms that define a human behavioral phenotype. There has been significant effort over the past decade to create a new nosology of mental disease, one that is based on clinical measurements that are proximal to the psychopathology. Mental illness is syndromal and identification of biological substrates of disease risk and burden would enable significant progress. Digital biomarkers of mood and cognition, with their ecological and continuous measurement, may be the right clinical constructs to correlate with mental illness morbidity.

Drug development for mental and neurological diseases costs far more than other medications to develop and is least likely to receive U.S. Food and Drug Administration approval. Common strategies for reducing clinical trial duration and cost include patient stratification, study enrichment, and early endpoint detection. But all these strategies require sensitive and specific measures of disease pathology which have been hard to create for most brain disorders. This is compounded further by the recognition that our inadequate understanding of disease etiology, onset, progression, treatment and outcome has limited drug development to drugs that address symptoms, not pathology.

Pharmaceutical companies have been quick to embrace the promise of digital biomarkers of mood and cognition in drug development (Business Wire 2017; PR Newswire 2018). With completion of active phase II clinical trials, we will have a clearer picture of the extent to which these digital biomarkers can contribute to a new generation of drugs that are safer, more efficacious, phenotype-focused therapeutic agents. Digital biomarkers create the possibility of companion diagnostics and targeted therapeutics that are more closely tailored to specific patient populations.

The measurement challenge in psychiatry has contributed significantly to the cost of care and poor outcomes of mental illnesses. Even with existing therapies, medication adherence is poor and treatment effectiveness inadequate because of therapeutic inertia or lack of symptom reporting by patients. At Mindstrong we are using these ecological digital biomarkers in outpatient telehealth to quantify disease activity, trigger treatment review and determine the need for early intervention (Mindstrong 2019). This provides a novel opportunity to reshape the brick-and-mortar care model of chronic disease into one that is ecological to the patient, measurement based, and can preempt disease progression.

We conclude with our views on privacy. Healthcare delivery in the United States is a highly regulated industry. Patients must consent to receive care and providers must comply with the Health Insurance Portability and Accountability Act (HIPAA) of 1996 that legislates provisions for data privacy and security to safeguard medical information. Digital brain biomarkers and digital phenotypes are not an exception. These are clinical measures of cognitive and emotional state with the potential to predict a person's behavior and response to illness, therapy and societal challenges. They represent medical tests that should be taken exclusively under a patient-provider consent-to-care relationship, governed by HIPAA privacy and security policies and used in the context of early intervention and therapeutic decision support by providers.

References

Andone I, Błaszkiewicz K, Eibes M et al (2016) How age and gender affect smartphone usage. In: Proceedings of the 2016 ACM international joint conference on pervasive and ubiquitous computing adjunct—UbiComp '16. Heidelberg, Germany, 12–16 September 2016

Barshan E, Ghodsi A, Azimifar Z, Zolghadri Jahromi M (2011) Supervised principal component analysis: visualization, classification and regression on subspaces and submanifolds. Pattern Recognit 44(7):1357–1371. https://doi.org/10.1016/j.patcog.2010.12.015

Benjamini Y, Hochberg Y (1995) Controlling the false discovery rate: a practical and powerful approach to multiple testing. J R Stat Soc Series B Stat Methodol 57(1):289–300. https://doi.org/10.1111/j.2517-6161.1995.tb02031.x

Bergmeir C, Hyndman RJ, Koo B (2018) A note on the validity of cross-validation for evaluating autoregressive time series prediction. Comput Stat Data Anal 120:70–83. https://doi.org/10.1016/j.csda.2017.11.003

Botwinick J, Thompson LW (1966) Premotor and motor components of reaction time. J Exp Psychol 71(1):9–15. https://doi.org/10.1037/h0022634

Business Wire (2017) BlackThorn Therapeutics announces innovative clinical collaboration agreement with Mindstrong Health. https://www.businesswire.com/news/home/20170607005392/en/BlackThorn-Therapeutics-Announces-Innovative-Clinical-Collaboration-Agreement

ClinicalTrials.gov (2017) Testing the value of smartphone assessments of people with mood disorders. A pilot, exploratory, longitudinal study (Identifier No. NCT03429361). https://clinicaltrials.gov/ct2

Dagum P (2018a) The long journey to digital brain biomarkers. npj Digital Medicine. https://npjdigitalmedcommunity.nature.com/users/89405-paul-dagum/posts/31757-the-long-journey-to-digital-brain-health

Dagum P (2018b) Digital biomarkers of cognitive function. NPJ Digit Med 1(1):10. https://doi.org/10.1038/s41746-018-0018-4

Insel TR (2017) Digital phenotyping: technology for a new science of behavior. JAMA Netw 318(13):1215–1216. https://doi.org/10.1001/jama.2017.11295

Keedwell PA, Andrew C, Williams SCR et al (2005) The neural correlates of anhedonia in major depressive disorder. Biol Psychiatry 58(11):843–853. https://doi.org/10.1016/j.biopsych.2005.05.019

Kerchner GA, Dougherty RF, Dagum P (2015) Unobtrusive neuropsychological monitoring from smart phone use behavior. Alzheimers Dement 11(7):272–273. https://doi.org/10.1016/j.jalz.2015.07.358

Kubiak T, Smyth JM (2019) Connecting domains—ecological momentary assessment in a mobile sensing framework. Mobile sensing and psychoinformatics. Springer, Berlin, pp x–x

Madrid A, Smith D, Alvarez-Horine S et al (2017) Assessing anhedonia with quantitative tasks and digital and patient reported measures in a multi-center double-blind trial with BTRX-246040 for the treatment of major depressive disorder. Neuropsychopharmacology 43:372–372

Markowetz A, Błaszkiewicz K, Montag C et al (2014) Psycho-informatics: big data shaping modern psychometrics. Med Hypotheses 82(4):405–411. https://doi.org/10.1016/j.mehy.2013.11.030

Mindstrong (2019). Mindstrong Health. https://mindstronghealth.com/

PR Newswire (2018) Mindstrong Health and Takeda partner to explore development of digital biomarkers for mental health conditions. https://www.prnewswire.com/news-releases/mindstrong-health-and-takeda-partner-to-explore-development-of-digital-biomarkers-for-mental-health-conditions-300604553.html

Smith DG, Saljooqi K, Alvarez-Horine S et al (2018) Exploring novel behavioral tasks and digital phenotyping technologies as adjuncts to a clinical trial of BTRX-246040. International Society of CNS Clinical Trials and Methodology

Staples P, Ouyang M, Dougherty RF et al (2018) Supervised kernel PCA for longitudinal data. arXiv preprint arXiv:180806638

Statistica (2019) StatSoft EUROPE. www.statistica.com

Strauss E, Sherman E, Spreen O (2006) A compendium of neuropsychological tests. Administration, norms, and commentary, 3rd edn. Oxford University Press, New York

Winkler AM, Ridgway GR, Webster MA et al (2014) Permutation inference for the general linear model. Neuroimage 92:381–397. https://doi.org/10.1016/j.neuroimage.2014.01.060

Chapter 7
Mining Facebook Data for Personality Prediction: An Overview

Davide Marengo and Michele Settanni

Abstract Users' interaction with Facebook generates trails of digital footprints, consisting of activity logs, "Likes", and textual and visual data posted by users, which are extensively collected and mined for commercial purposes, and represent a precious data source for researchers. Recent studies have demonstrated that features obtained using these data show significant links with users' demographic, behavioral, and psychosocial characteristics. The existence of these links can be exploited for the development of predictive models allowing for the unobtrusive identification of online users' characteristics based on their recorded online behaviors. Here, we review the literature exploring use of different forms of digital footprints collected on Facebook, the most used social media platform, for the prediction of personality traits. Then, based on selected studies, meta-analytic calculations are performed to establish the overall accuracy of predictions based on the analyses of digital footprints collected on Facebook. Overall, the accuracy of personality predictions based on the mining of digital footprints extracted from Facebook appear to be moderate, and similar to that achievable using data collected on other social media platforms.

7.1 Introduction

Since its creation in 2004, Facebook has experienced a steady increase in active users, reaching a total of 2.41 billion monthly active users as of the second quarter of 2019 (Statista 2019). In spite of growing competition by other social media platforms—such as Instagram (which is also currently owned by Facebook), Twitter, and Snapchat—and mounting controversies concerning the handling of user privacy (e.g., see the recent Cambridge Analytica scandal, Cadwalladr and Graham-Harrison 2018), as of 2019, Facebook remains the most used social media platform worldwide. Every day, millions of Internet users from different cultural contexts express their thoughts, emotions, and beliefs by writing, posting, and sharing content on Facebook,

D. Marengo (✉) · M. Settanni
Department of Psychology, University of Turin, Turin, Italy
e-mail: davide.marengo@unito.it

© Springer Nature Switzerland AG 2019   109
H. Baumeister and C. Montag (eds.), *Digital Phenotyping and Mobile Sensing*,
Studies in Neuroscience, Psychology and Behavioral Economics,
https://doi.org/10.1007/978-3-030-31620-4_7

which can be commented and/or endorsed ("liked" in the Facebook jargon) by their network of Facebook friends, or the overall Facebook community if the user profile is public. This unceasing interactive process produces a massive dataset of user-generated data, also referred to as "digital footprints", "digital records", or "digital traces" (e.g., Settanni and Marengo 2015; Youyou et al. 2015; Farnadi et al. 2016) consisting of personal information, activity logs, texts, pictures, and videos, with potential connections to users' offline behavioral and psychosocial characteristics.

In the research area of psycho-informatics (Markowetz et al. 2014), an increasing number of studies have explored the feasibility of mining digital footprints from Facebook, as well as those collectable from other social media platforms, in efforts to infer individual psychosocial and behavioral characteristics, such as personality (e.g., Golbeck et al. 2011; Kosinski et al. 2013; Liu et al. 2016; Farnadi et al. 2016) psychological distress (e.g., Choudhury et al. 2013; Settanni and Marengo 2015); for an overview see Settanni et al. 2018) and engagement in offline risk behaviors (e.g., alcohol use, Curtis et al. 2018). One of the earliest and most enduring project in this field of study is the my Personality project (Kosinski et al. 2013), which has attracted over 6 million Facebook users who have donated their digital footprints and responded to online questionnaires on a wide variety of psychometric measures, including Big Five personality traits, satisfaction with life, and intelligence.

Studies in this field share a common research design, broadly consisting of four steps: (1) User digital footprints are collected and mined for the automated extraction of multiple features; (2) Information about user individual characteristics is collected by means of different approaches (e.g., online survey, ecological momentary assessment via mobile apps); (3) Datasets combining the features extracted from digital traces and users' information are mined to explore associations between features and users' characteristics, and to train models aimed at predicting individuals' traits, typically using a machine-learning approach; and (4) Competing trained models are compared based on their accuracy in predicting users' characteristics on new independent datasets, leading to the identification of the best performing model. Most of existing studies employing this approach have explored the feasibility of mining digital footprints for the prediction of personality traits, as defined by the Big Five model (McCrae and Costa 1987; McCrae and John 1992) and the Dark Triad model (Paulhus and Williams 2002). The focus on personality is largely due to the importance of personality in predicting many life-course aspects for individuals, including academic success (e.g., Komarraju et al. 2009), job performance (e.g., Neal et al. 2012), financial decision making (Lauriola and Levin 2001; Bibby and Ferguson 2011), health and health-related behaviors (e.g., Soldz and Vaillant 1999; Bogg and Roberts 2004, 2013), subjective well-being (e.g., Hayes and Joseph 2003), and online behaviors (e.g., Matz et al. 2017).

Due to the current dominant position of Facebook over all existing social media platforms, the majority of existing studies exploring digital footprints for the prediction of personality traits have focused on the use of Facebook data. In this chapter, we present an overview of these studies discussing differences in the types of examined digital footprints, and the various analytical approaches used for mining them. Next, we refer to existing meta-analytical results to discuss the accuracy of prediction based on digital footprints. Finally, we discuss ethical issues related to this field of study, in particular with respect to privacy violations. Popularity of social media platforms can drastically change over time, resulting in a significant decline in user activity (e.g., Myspace, Ribeiro and Faloutsos 2015), and eventually, this may apply to Facebook. Still, regardless of the social media platform under focus, current findings indicate the potential of the analysis of social media data for research in *psychoinformatics* is expected to increase in the future as technology improves, and new methodologies are developed for the analysis of digital footprints (Hinds and Joinson 2019).

7.2 Facebook Digital Footprints and Their Use for the Study of Personality

Facebook provides developers with access to the digital footprints of consenting Facebook users, which can be accessed via a specifically-devised application programming interface (API), the Facebook Graph API (Facebook for developers 2019a). Following the Cambridge Analytica scandal, Facebook has introduced stricter requirements for accessing user data (Facebook for developers 2019b). Starting from August 2018, in order to be granted extended login permissions (e.g., access to user posts, or Likes), Facebook requires all apps to undergo a review process requiring developers to explain in details how they plan to use and manage user data; further, developers are required to pass a business verification procedure, ultimately limiting data access to business companies. Non-business developers (e.g., researchers) can still access user data (excluding user posts) if they pass an individual verification procedure. Downloadable user data includes personal information (e.g., name, age, gender, hometown), as well as posts, likes, pictures, and videos shared by the users on their Facebook wall. Access to posts, likes, pictures, and videos, include the possibility to download the actual user-generated content, as well as attached metadata (e.g., day and time of posting, received likes and comments). In the following sections, we present a brief literature overview of the studies that analyzed the connection between different types of Facebook digital footprints and personality, and describe some of the approaches used to mine collected data for prediction purposes.

Demographics and activity statistics. Facebook provides researchers with a vast array of user demographic information, such as age, gender, geographical location, and information about activity on Facebook, such as number of friends, and frequency and time of online posting. Furthermore, based on the examination of users' feed data, it is possible to compute summary statistics of users' specific online behaviors, such as the number of times the user has updated his/her Facebook status, uploaded photos or videos, sent or accepted a friend request, the number of events he/she attended, or the number of times he/she has been tagged in a photo.

Studies have shown significant links between activity statistics and personality, and specifically, the Big 5 traits have been shown to be significantly associated with users' behaviors on social media. For example, individuals with high Extraversion have been characterized by higher levels of activity on social media, and have a greater number of friends (Kosinski et al. 2014) than introverted individuals. Individuals with high Conscientiousness appear to be cautious in managing their social media profiles; they post fewer pictures, and engage in less group activity on social media (Kosinski et al. 2014). Furthermore, individuals with high openness tend to have larger networks than individuals low on the trait (Quercia et al. 2012). Studies varies in the type of analyses employed to investigate the links between extracted features, and personality. Gosling and colleagues (2011) studied bivariate associations between count statistics of a wide range of user activity information, including the number of photos, number of wall posts, the total number of friends, and personality ratings provided by both the user itself and an external observer. Findings showed that both self-report and observer-rated Extraversion scores had positive associations with the number of user wall posts, uploaded photos, and the size of the friendship network, while self-report openness was positively related with the number of user online. In turn, Wald and colleagues (2012) extracted 31 features from users' profile information and wall post activity -including age, gender, relationship status, and the number of friends, photos, interests, and comments—to test them as attributes for personality prediction using a set of machine learning algorithms. When examining the predictive power of single features, number of friends emerged as the stronger predictor of individual differences in agreeableness. Trained machine-learning models combining all features showed good accuracy (75% of correctly classified individuals) in detecting individuals with high openness scores (over the 90th percentile), while prediction on other traits was less satisfactory (<0.65% of accuracy).

Facebook likes. Facebook gives its users the possibility to "like" Facebook pages created by groups, companies, public figures, or external websites. Likes represent a mechanism used by Facebook users to express their positive association with specific web pages, comments, photos, and offline activities among others (Youyou et al. 2015). By accessing user Likes data through the Graph API (*user_likes* authorization), the following information can be obtained: the name of each page liked by the user, the category each page was registered in by their creators (e.g., *musician/band*; *Media/News Company*; *Italian Restaurant*), and a timestamp indicating when each page was liked by the user.

Several links exist the frequency of "Like" behaviors and users' personality. For example, individuals scoring high on conscientiousness tend to express less "Likes" on Facebook (Kosinski et al. 2014), while individuals high on openness tend to "Like" more content found on social media (Bachrach et al. 2012). Likes can also be mined to obtain information about users' interests and preferences as regards brands, politics, music, etc. However, because of the massive number of existing Facebook pages (>42 million pages), when examined at the page-level, user Likes data usually generates very large sparse logical matrices (i.e., matrix in which each row represents a user and each column represents a specific page), even at small sample sizes. For this reason, examining user Likes data for personality prediction generally requires the implementation of some form of dimensionality reduction method on the predictors set (i.e., digital footprints, in this case Likes). For example, Kosinski and colleagues (2013) processed a large matrix consisting of an average of 170 liked pages per 58,466 Facebook users using singular value decomposition (SVD), retaining a smaller subset 100 SVD components for performing personality predictions using logistic regression. When applied to Facebook Likes, the emerging SVD components may be interpreted as reflecting latent users' interests and preferences emerging from the co-occurrence of Likes to similar pages, e.g., persons who "likes" the official Facebook page of Harry Potter and Lord of the Rings movies tend to score similarly on a SVD component reflecting an interest in fantasy novels or movies. Using this approach, Kosinski and colleagues found remarkable accuracy prediction; in particular concerning the 'Openness' trait, for which score predictions based on user's Likes was found to be roughly as informative as using self-report personality scores (Kosinski et al. 2013). Using SVD can improve performance albeit at the expense of the interpretability of results, since information about user specific preferences is lost in the process of producing SVD components, and interpretation of emerging component is not always straightforward. Further, because of the sparsity of Likes data, this approach is only viable using large datasets (Kosinski et al. 2016). Another analytical approach which may help prevent overfitting problems when performing personality prediction using large datasets is the least absolute shrinkage and selection operator algorithm, or LASSO regression (Tibshirani 1996). Using the LASSO approach, prediction is initially performed using all available Likes for the user, but only Likes that contribute significantly to the overall prediction are included in final model. Because of the large number of features used to perform prediction, interpretation of results is problematic and it is typically limited to the predictive accuracy of trained models over personality scores. Using this analytic approach, Youyou and colleagues (2015) demonstrated that score predictions of Big 5 traits derived from the analysis of Facebook-Likes can be more accurate than personal judgments of a user's friends, relatives, and even spouse. Furthermore, as shown by Torfarson and colleagues (2017) using the same analytical approach, prediction accuracy over personality seems to increase proportionally with the number of user Likes analyzed, with only 20 Likes needed to obtain personality scores as accurate as those provided by users' spouse.

An alternative approach that can help face the problem of the sparsity of Facebook Likes consists in analyzing collected data at the category-level, as opposed to the page-level. In doing this, Facebook pages that are registered in Facebook under the same category (e.g., "Retail company" category: *Amazon.com, ebay,it, Macy's*; Musician/Band category: *Frank Sinatra, Adele, Kraftwerk*) are counted in a single category indicator. Using this approach, Baik and colleagues (2016) examined Likes data by recoding 8.355 distinct Likes pages into 183 categories. This procedure allowed them to perform linear regression analyses with no variable selection, while also preserving interpretability of results. Using this approached they were able to predict users' extraversion with average accuracy ($r = 42$), and provide some insight on the association between the personality trait and specific user interests (e.g., extroverts were more likely to show interest in hotels, sports, and shopping/retail, whereas introvert users showed interested in musicians, bands, and games/toy categories).

Texts. The *user_posts* Graph API authorization allows access to user posting activity, including personal status updates and comments, and the number of Likes received on users' posts.

Studies have shown significant links between the Big Five personality traits and features extracted from Facebook texts (e.g., Hall and Caton 2017; Schwartz et al. 2013). For example, Extraversion has been shown to be positively associated with the frequency of use of words about family and friends, and positive emotions (Schwartz et al. 2013) and to be negatively associated with use of words indicating cognitive processes (insight words, Hall and Caton 2017; words indicating tentativeness, causation, inhibition, Schwartz et al. 2013). Coherent with findings about depression and language use (Eichstaedt et al. 2018). Neuroticism has been linked with increased use of words indicating use of 1st person singular pronouns, negative emotions, and coarse language (Schwartz et al. 2013); in turn the Agreeableness trait has been linked to increased positive emotion words (Hall and Caton 2017; Schwartz et al. 2013). Furthermore, Garcia and Sikstrom (2014) explored associations between the Dark Triad personality traits, i.e., Machiavellianism, Narcissism, and Psychopathy, and textual features extracted from Facebook texts. Findings showed that Psychopathy was the personality trait most easily predictable from the semantic content of status updates. Results also showed that individuals with high levels on the Psychopathy and Narcissism traits posted more negative words in their Facebook posts, and published more "atypical" content when compared to individuals with low scores on these traits.

For the purpose of personality prediction, most studies have extracted features from texts using two text analyses approaches: the more traditional *closed-vocabulary* analysis and the recently emerging *open-vocabulary* analysis. Closed-vocabulary analysis has a long history in psychological science and can be viewed as theory-driven approach that consists of scoring language data according to predetermined semantic categories. One of the most popular instruments to apply this kind of approach is the Linguistic Inquiry and Word Count (LIWC) software, which has been developed over the past 20 years to measure multiple dimensions by computing the relative frequency of word categories (Pennebaker et al. 2015; Tausczik and Pennebaker 2010). LIWC allows the scoring of text documents based on a set of predetermined categories ranging from parts of speech (e.g., use of pronouns, numbers, punctuation), emotional expression (e.g., positive or negative emotions, anger, sadness), cognitive processes (e.g., insight, discrepancy), to social processes (e.g., friends, family), and personal concerns (e.g., body, death, money, occupation).

In contrast, the more recent open-vocabulary analysis employs data driven analytic approaches to explore the distribution of topics, words, and phrases naturally occurring in analyzed texts; thus producing results that are not limited by predetermined categories (Schwartz and Ungar 2015). For this reason, the open vocabulary approach is particularly suited for the analyses of alternative forms of communication, such as use of abbreviations (e.g., OMG, IMHO, NSFW), as well as pictorial symbols (e.g., emoticons, emoji), which represent a significant component of modern computer-mediated communication. Some of the most used approaches for performing open vocabulary analysis on Facebook texts are Latent Semantic Analysis (Dumais 2004) and Latent Dirichlet Allocation (Blei et al. 2003) algorithms. Both approaches have been used to infer the semantic and topical content of Facebook texts in studies exploring the link between personality and language use (e.g., Schwartz et al. 2013; Garcia and Sikström 2014). Despite the advantages linked to their use, these approaches generally require large datasets and emerging results are typically harder to interpret than those provided by closed-vocabulary analyses. Table 7.1 and Fig. 7.1 provide an illustrative example on how results from closed- and open-vocabulary analyses are usually presented when examining their association with personality scores. Presented results are based on the analyses collected on a sample of 296 adult Facebook users from Italy (Female $= 67\%$, Age: $M = 28.44$, $SD = 7.38$, previously unpublished results), and show the correlation between Neuroticism (Ten Item Personality Inventory, Gosling et al. 2003) and LIWC closed-vocabulary features (Table 7.1) and LDA open-vocabulary features (Fig. 7.1) computed on users' status updates. Topics emerging from LDA analyses are depicted using word clouds in which words that are more strongly related to the topic are depicted with larger font size and darker tones. Findings using LIWC and LDA topics are coherent in showing that Neuroticism correlates with increased negative emotionality in Facebook posts, and appear to be negatively related to the expression of positive emotions. However, each approach can provide different insights on the specific forma or semantic language features associated with the examined trait. Based on presented results, it is worthy to note that, similarly to what it is usually observed in the literature, effect-size of correlations between personality and both closed- and open-vocabulary features, is generally quite low.

Table 7.1 Personality and Facebook language: closed-vocabulary (LIWC) correlates of neuroticism (n = 296, Correlations significant at p < 0.05)	LIWC category	r
	I. Standard linguistic dimensions	
	Negations	0.15
	II. Psychological processes	
	Affective processes	
	Optimism	−0.15
	Negative emotions	0.20
	Anger	0.12
	Sadness	0.28
	Cognitive processes	
	Possibility	0.18
	Certainty	−0.13
	Social processes	
	Family	−0.15
	Personal concerns	
	Money and financial issues	−0.14
	Body states, symptoms	0.12

Fig. 7.1 Personality and Facebook language: open-vocabulary (LDA) correlates of neuroticism (n = 296, Correlations significant at p < 0.05)

However, as shown by Schwartz and colleagues (2013), when all extracted features are combined using predictive models, predictive accuracy is expected to improve significantly. Further, when compared to closed-vocabulary, open-vocabulary analysis has proven to yield additional insights and more in depth information about the behavioral characteristics of personality types beyond those that can be obtained by a traditional closed-vocabulary approach, resulting in increased prediction accuracy over personality (Schwartz et al. 2013).

Visual data. Sharing of pictures and videos is an increasingly frequent online activity and in the last few years new social media platforms have flourished focusing on the sharing of pictures and videos (e.g., Instagram, Snapchat) among users. This new phenomenon has also witnessed an increase on Facebook, as users are constantly provided with new features allowing for the proliferation of pictures and videos in their overall generated content, indicating a progressive shift from textual based interactions, to an increase in sharing of videos and photographs. Given the relative novelty of this phenomenon, only a few studies have explored the possible relations between visual data collected from Facebook, and personality, usually focusing on the analysis of user profile pictures (e.g., Celli et al. 2014; Torfason et al. 2016; Segalin et al. 2017). Access to the *default*, *user_photos* and *user_videos* Graph API authorizations allow researchers to inspect and download users' profile pictures, uploaded pictures, and videos. Celli and colleagues processed user profile pictures using a *bag-of-visual-words* approach, extracting 4096 non-interpretable visual features to perform predictions over personality scores. Prediction accuracy over personality traits was generally poor. However, they were able to show that extrovert and emotionally stable users post more frequently pictures in which they smile, and in which other people are included. On the other hand, introverts tend to appear alone, while Neurotics tend to post images without humans, and their pictures often feature close-up faces. Torfarson and colleagues, on the other hand, extracted information about specific facial features (e.g., eyeglasses, smiling, wearing lipstick) based on models trained on the CelebA dataset (Liu et al. 2015); based solely on features extracted from users' profile pictures, they observed a correlation of 0.18 between observed and predicted scores for the Big 5 personality traits. Finally, Segalin and colleagues (2017) extracted a large set of variables representing aesthetics-based features of images (i.e., color, composition, textual properties, and content), byte-level features, and both visual word and concept features extracted using other approaches (e.g., pyramid histogram of visual words; convolutional neural networks). In the study extroverts and agreeable individuals showed an increased inclination to post warm colored pictures and to exhibit many faces in their profile pictures; in turn, individuals high on neuroticism were more likely to post pictures of indoor places. As in the aforementioned studies, prediction accuracy was limited (~60% of correctly classified individuals), but still higher than that obtained in the study when using human raters. Overall, existing findings indicate that the accuracy of predicting personality achieved using visual data is still relatively limited compared to that achieved using other types of digital footprints (Azucar et al. 2018).

7.3   Establishing the Accuracy of Personality Predictions

Studies exploring the predictive power of Facebook digital footprints over personality vary significantly in the methods employed to assess such associations. Some studies implement simple bivariate analyses (e.g., zero-order correlation, independent samples t-test) examining the strength of associations between personality scores and numerical variables representing features extracted from digital footprints (e.g., Gosling et al. 2011; Panicheva et al. 2016), while other studies investigate the accuracy of personality predictions based of the mining of large set of features using predictive models (e.g., Kosinski et al. 2013; Celli et al. 2014). Other studies present both kinds of analyses (e.g., Schwartz et al. 2013; Farnadi et al. 2016).

Another important difference emerging from these studies relates to the use of validation techniques to avoid overfitting problems and support generalizability of results. Amongst the most used validation methods are the *holdout* method, the *k-fold* validation method, and the *leave-one-out* method. When using the holdout method, collected data is randomly split in two datasets of unequal size, a larger training set and a smaller test set, consisting of mutually independent observations. Analyses are first performed on the training set, resulting in a set of parameters or coefficients describing the association between features extracted from digital footprints, and personality scores. Then, the developed models are applied on the smaller test set using the estimated parameters or coefficients: Accuracy of personality predictions informs about the expected performance of the model on new, unseen observations. The *k*-fold validation method is similar, in that it also involves randomly splitting the dataset in a training set and a test set, but the process is repeated k times, resulting in k sets of results which are averaged to produce a single estimation of accuracy. Finally, in the leave-one-out method, the split is repeated for as many observations present in the dataset, i.e., the dataset is iteratively split so that only one observation is used to the test accuracy of predictions, while the remaining observations are used to estimate model parameters or coefficients. In general, *using k-fold validation and leave-one-out methods* is preferable over the *simple holdout* method (Kohavi 1995). As regards the number of folds, authors have suggested either using k = 5 and k = 10 folds when performing cross-validation over k = 2 folds, as using a larger number of folds is expected to decrease bias in estimating prediction errors (Rodriguez et al. 2010); however, it should be noted that as increasing the number of folds is only feasible using large datasets, as the large sample condition needs to be achieved in each of the fold (Wong 2015).

Most of the studies examining the predictive power of digital footprints collected from Facebook used a k-fold method to validate personality predictions (e.g., Golbeck et al. 2011; Bachrach et al. 2012; Quercia et al. 2012; Kosinski et al. 2013; Farnadi et al. 2016, 2018; Baik et al. 2016; Thilakaratne et al. 2016) followed by the holdout method (e.g., Celli et al. 2014; Schwartz et al. 2013). It is worthy to note that some authors did not employ a cross-validation technique in their studies, but they analyzed data from their whole samples (e.g., Gosling et al. 2011; Garcia and Sikström 2014). Caution should be used in interpreting their findings due to the limitations cited above, i.e. overfitting and lack of generalizability.

7.4   Accuracy of Personality Predictions Based on Facebook Data

Two recent meta-analytic studies have examined the literature aiming at estimating the overall prediction accuracy of digital footprints collected on a wide range of social media platforms (e.g., Facebook, Twitter, Sina Weibo, and Instagram) over users' individual characteristics. Settanni and colleagues (2018) identified 38 papers investigating associations between digital footprints and a set of individual characteristics, including personality (e.g., traits from the Big Five and the Dark Triad personality models), psychological well-being (e.g., satisfaction with life, depression), and intelligence. Overall, based on findings from a subset of 18 studies, the estimated overall accuracy in predicting personality, computed as a meta-analytical correlation was moderate ($r = 0.34$). Further, the results of the meta-analysis showed that the overall accuracy in predicting personality was lower than the accuracy in predicting psychological well-being ($r = 0.37$), but higher than what was computed for the prediction of intelligence ($r = 0.29$). In turn, the meta-analysis by Azucar and colleagues (2018) examined literature focusing on studies presenting associations between digital footprints and traits from the Big Five model—i.e., Openness, Conscientiousness, Extraversion, Agreeableness, and Neuroticism. Based on prediction results presented in 16 independent studies, Extraversion appears to be associated with the highest overall prediction accuracy (r = 0.40), followed by Openness (r = 0.39), Conscientiousness (r = 0.35), Neuroticism (r = 0.33), and Agreeableness (r = 0.29). In the aforementioned meta-analyses, a set of meta-regression was conducted to recognize factors influencing prediction accuracy, finding that the accuracy improves when models include information about users' demographics and more than one type of digital footprints. In both meta-analyses, examined studies referred to a conceptualization of personality as a multi-dimensional construct; hence, reported results should not be interpreted as indicating accuracy in predicting discrete personality types.

Given the relevance of Facebook in the social media world, we re-estimated overall prediction accuracy reported in the two cited meta-analyses, including only those studies presenting predictions based on Facebook data. Very similar results emerged, with r values ranging from 0.33 (Neuroticism) to 0.43 (Extraversion), and an overall correlation equal to 0.34, 95% CI [0.24–0.44] indicating that the use of data extracted from differing social media platforms is not expected to have a significant effect on the accuracy of personality predictions. As noted by Kosinski and colleagues (2013) this correlation size corresponds to the "personality coefficient" (a Pearson correlation ranging from 0.30 to 0.40; Meyer et al. 2001; Roberts et al. 2007), which is the upper limit of correlations between behaviors and personality traits reported in past psychological studies. Among the examined traits, Extraversion appears to be most easily predictable based on the examination of digital footprint from Facebook, a finding which is compatible with findings emerging from studies exploring other sources of digital footprints (e.g., smartphone usage, Stachl et al. 2017). Still, upon inspecting these results, it is important to note that the average prediction accuracy reported by existing studies is still quite limited. For this reason, reliability of individual personality predictions obtained by mining digital footprint is still quite low, in particular when compared with that obtainable with traditional self-report assessments, limiting their use of predicted scores for assessment purposes.

7.5   Conclusions

In this chapter, we presented an overview of studies examining the feasibility of inferring individual differences in personality of Facebook users based on the analyses of their digital footprints (e.g., user demographics, texts, Facebook Likes, and pictures). Published studies vary significantly in employed analytical approaches, both in terms of methods used for extracting features from raw social media data, performing predictions, and validating results. Overall, the accuracy of personality predictions based on the mining of digital footprints extracted from Facebook appears to be moderate, and similar to that achievable using data collected on other social media platforms. Given the relatively recency of this area of research, and the rapid evolution of data mining techniques, we expect accuracy of personality prediction to improve in the near future. The ability to use digital footprints for the unobtrusive assessment of personality traits can represent a rapid, cost-effective alternative to surveys to reach large online populations, an approach which can be beneficial for academic, health-related, and commercial purposes pursued in the online environment (e.g., improve the efficacy of online health-related messages and interventions, enhance online recommender systems, improve user experiences, enhance efficacy of advertising by tailoring online message to personality attributes, including political messages, Matz et al. 2017).

7.5.1   Future Directions and Ethical Concerns

Meta-analyses conducted to determine the predictive power of social media data on psychological characteristics in general, and in particular on personality traits, revealed that collecting information from multiple sources (e.g. from pictures and text) and different social media platforms, permit to achieve greater predictive power. This means that in the near future, the aggregation of features extracted from different types of data and the inclusion of data from different social media platforms, or from other sources, such as wearables (e.g., iwatch, runkeeper, etc.) or mobiles, will probably lead to relevant improvements in the predictive power of these kinds of models. Furthermore, the progressive expansion and complexification of individuals' online activities will support the creation of an increasing number of datasets, easily available for the development of predictive models. It is foreseeable that the development of new and more efficient approaches to data collection, integration, and analysis (e.g., using deep learning algorithms) will contribute to making predictions more accurate and reliable, extending their reach well beyond the field of personality traits, towards the prediction of more specific characteristics, behaviors, and even biological features (Montag et al. 2017; Sariyska et al. 2018). The acquisition of these new capabilities will raise important ethical issues that cannot be underestimated.

Social media data may be used in ways that surpass what users intend, or understand, when they give consent to their collection. Apparently innocuous data points may be and have been used to reveal information that users might expect to stay private. These predictions can have negative consequences for social media users: First, predicted traits can be used to make decisions relative to single Facebook users, without their explicit consent to disclose such characteristics (e.g., in hiring procedures); Second, as recently highlighted by Matz and colleagues (2017), psychological targeting procedures might be developed and aimed at manipulating the behavior of large groups of people, without the individuals being aware of it.

Given the possibility of using raw data to infer relevant individual characteristics, the need is emerging for a more careful consideration of ethical challenges related to the use of data extracted from Facebook or other social media. It is worthy to note that, while research in medicine and psychology is routinely subjected to IRB ethical approvals, the same does not apply for computer science. Same as in clinical disciplines, computers scientists should also develop and apply ethical restrictions when they do research in this field, and research projects should adhere to the principle of beneficence: the good of research participants should be taken in high consideration and influence the assessment of risks versus benefits when planning a research. As recently noted in a Nature editorial (2018) on the Cambridge Analytica scandal, the fact that data are there should not be a sufficient reason to exploit these in order to conduct research.

References

Azucar D, Marengo D, Settanni M (2018) Predicting the Big 5 personality traits from digital footprints on social media: a meta-analysis. Pers Individ Dif 124:150–159. https://doi.org/10.1016/j.paid.2017.12.018

Bachrach Y, Kosinski M, Graepel T et al (2012) Personality and patterns of facebook usage. In: Proceedings of the 4th annual ACM web science conference. ACM, New York, NY, USA, pp 24–32

Baik J, Lee K, Lee S et al (2016) Predicting personality traits related to consumer behavior using SNS analysis. New Rev Hypermedia Multimed 22(3):189–206. https://doi.org/10.1080/13614568.2016.1152313

Bibby PA, Ferguson E (2011) The ability to process emotional information predicts loss aversion. Pers Individ Dif 51(3):263–266. https://doi.org/10.1016/j.paid.2010.05.001

Blei DM, Ng AY, Jordan MI (2003) Latent dirichlet allocation. J Mach Learn Res 3(Jan):993–1022

Bogg T, Roberts BW (2004) Conscientiousness and health-related behaviors: a meta-analysis of the leading behavioral contributors to mortality. Psychol Bull 130(6):887–919. https://doi.org/10.1037/0033-2909.130.6.887

Bogg T, Roberts BW (2013) The case for conscientiousness: evidence and implications for a personality trait marker of health and longevity. Ann Behav Med 45(3):278–288. https://doi.org/10.1007/s12160-012-9454-6

Cadwalladr C, Graham-Harrison E (2018) Revealed: 50 million Facebook profiles harvested for Cambridge Analytica in major data breach. The Guardian

Celli F, Bruni E, Lepri B (2014) Automatic personality and interaction style recognition from facebook profile pictures. In: Proceedings of the 22nd ACM international conference on multimedia. ACM, New York, NY, USA, pp 1101–1104

Choudhury MD, Gamon M, Counts S, Horvitz EJ (2013) Predicting Depression via Social Media. In: ICWSM

Curtis B, Giorgi S, Buffone AEK et al (2018) Can Twitter be used to predict county excessive alcohol consumption rates? PLoS ONE 13(4):e0194290. https://doi.org/10.1371/journal.pone.0194290

Dumais ST (2004) Latent semantic analysis. ARIST 38(1):188–230. https://doi.org/10.1002/aris.1440380105

Eichstaedt JC, Smith RJ, Merchant RM et al (2018) Facebook language predicts depression in medical records. Proc Natl Acad Sci U S A 115(44):11203–11208. https://doi.org/10.1073/pnas.1802331115

Facebook for developers (2019a) Graph API. https://developers.facebook.com/docs/graph-api. Accessed 30 Jul 2019

Facebook for developers (2019b) App Review. https://developers.facebook.com/docs/apps/review/. Accessed 30 Jul 2019

Farnadi G, Sitaraman G, Sushmita S et al (2016) Computational personality recognition in social media. User Model User-Adap Inter 26(2):109–142. https://doi.org/10.1007/s11257-016-9171-0

Farnadi G, Tang J, De Cock M, Moens M-F (2018) User profiling through deep multimodal fusion. In: Proceedings of the eleventh ACM international conference on web search and data mining. ACM, New York, NY, USA, pp 171–179

Garcia D, Sikström S (2014) The dark side of Facebook: semantic representations of status updates predict the dark triad of personality. Pers Individ Dif 67:92–96. https://doi.org/10.1016/j.paid. 2013.10.001

Golbeck J, Robles C, Turner K (2011) Predicting personality with social media. In: CHI '11 extended abstracts on human factors in computing systems. ACM, New York, NY, USA, pp 253–262

Gosling SD, Augustine AA, Vazire S et al (2011) Manifestations of personality in online social networks: self-reported facebook-related behaviors and observable profile information. Cyberpsychol Behav Soc Netw 14(9):483–488. https://doi.org/10.1089/cyber.2010.0087

Gosling SD, Rentfrow PJ, Swann WB (2003) A very brief measure of the big-five personality domains. J Res Pers 37(6):504–528. https://doi.org/10.1016/S0092-6566(03)00046-1

Hall M, Caton S (2017) Am I who I say I am? Unobtrusive self-representation and personality recognition on Facebook. PLoS ONE 12(9):e0184417. https://doi.org/10.1371/journal.pone.0184417

Hayes N, Joseph S (2003) Big 5 correlates of three measures of subjective well-being. Pers Individ Dif 34(4):723–727. https://doi.org/10.1016/S0191-8869(02)00057-0

Hinds J, Joinson A (2019) Human and computer personality prediction from digital footprints. Curr Dir Psychol Sci 28(2):204–211. https://doi.org/10.1177/0963721419827849

Kohavi R (1995) A study of cross-validation and bootstrap for accuracy estimation and model selection. In: Proceedings of international joint conference on AI. pp 1137–1145

Komarraju M, Karau SJ, Schmeck RR (2009) Role of the Big Five personality traits in predicting college students' academic motivation and achievement. Learn Individ Differ 19(1):47–52. https://doi.org/10.1016/j.lindif.2008.07.001

Kosinski M, Bachrach Y, Kohli P et al (2014) Manifestations of user personality in website choice and behaviour on online social networks. Mach Learn 95(3):357–380. https://doi.org/10.1007/s10994-013-5415-y

Kosinski M, Stillwell D, Graepel T (2013) Private traits and attributes are predictable from digital records of human behavior. Proc Natl Acad Sci U S A 110(15):5802–5805. https://doi.org/10.1073/pnas.1218772110

Kosinski M, Wang Y, Lakkaraju H, Leskovec J (2016) Mining big data to extract patterns and predict real-life outcomes. Psychol Methods 21(4):493–506. https://doi.org/10.1037/met0000105

Lauriola M, Levin IP (2001) Personality traits and risky decision-making in a controlled experimental task: an exploratory study. Pers Individ Dif 31(2):215–226. https://doi.org/10.1016/S0191-8869(00)00130-6

Liu Z, Luo P, Wang X, Tang X (2015) Deep learning face attributes in the wild. pp 3730–3738

Liu L, Preotiuc-Pietro D, Samani ZR, Moghaddam ME, Ungar LH (2016) Analyzing personality through social media profile picture choice. In: ICWSM, pp 211–220

Markowetz A, Błaszkiewicz K, Montag C et al (2014) Psycho-informatics: big data shaping modern psychometrics. Med Hypotheses 82(4):405–411. https://doi.org/10.1016/j.mehy.2013.11.030

Matz SC, Kosinski M, Nave G, Stillwell DJ (2017) Psychological targeting as an effective approach to digital mass persuasion. Proc Natl Acad Sci U S A 114(48):12714–12719. https://doi.org/10.1073/pnas.1710966114

McCrae RR, Costa PT (1987) Validation of the five-factor model of personality across instruments and observers. J Pers Soc Psychol 52(1):81–90. https://doi.org/10.1037/0022-3514.52.1.81

McCrae RR, John OP (1992) An introduction to the five-factor model and its applications. J Pers 60(2):175–215. https://doi.org/10.1111/j.1467-6494.1992.tb00970.x

Meyer GJ, Finn SE, Eyde LD et al (2001) Psychological testing and psychological assessment: a review of evidence and issues. Am Psychol 56(2):128

Montag C, Markowetz A, Blaszkiewicz K et al (2017) Facebook usage on smartphones and gray matter volume of the nucleus accumbens. Behav Brain Res 329:221–228. https://doi.org/10.1016/j.bbr.2017.04.035

Nature (2018) Cambridge Analytica controversy must spur researchers to update data ethics. Nature 555:559–560. https://doi.org/10.1038/d41586018038564

Neal A, Yeo G, Koy A, Xiao T (2012) Predicting the form and direction of work role performance from the Big 5 model of personality traits. J Organ Behav 33(2):175–192. https://doi.org/10.1002/job.742

Panicheva P, Ledovaya Y, Bogolyubova O (2016) Lexical, morphological and semantic correlates of the dark triad personality traits in Russian Facebook texts. In: 2016 IEEE artificial intelligence and natural language conference (AINL). pp 1–8

Paulhus DL, Williams KM (2002) The dark triad of personality: narcissism, machiavellianism, and psychopathy. J Res Pers 36(6):556–563. https://doi.org/10.1016/S0092-6566(02)00505-6

Pennebaker JW, Boyd RL, Jordan K, Blackburn K (2015) The development and psychometric properties of LIWC2015

Quercia D, Lambiotte R, Stillwell D et al (2012) The personality of popular Facebook users. In: Proceedings of the ACM 2012 conference on computer supported cooperative work. ACM, New York, NY, USA, pp 955–964

Ribeiro B, Faloutsos C (2015) Modeling website popularity competition in the attention-activity marketplace. In: Proceedings of the eighth ACM international conference on web search and data mining. ACM, New York, NY, USA, pp 389–398

Roberts BW, Kuncel NR, Shiner R, Caspi A, Goldberg LR (2007) The power of personality: the comparative validity of personality traits, socioeconomic status, and cognitive ability for predicting important life outcomes. Perspect Psychol Sci 2(4):313–345

Rodriguez JD, Perez A, Lozano JA (2010) Sensitivity analysis of k-fold cross validation in prediction error estimation. IEEE Trans Pattern Anal Mach Intell 32(3):569–575. https://doi.org/10.1109/TPAMI.2009.187

Sariyska R, Rathner E-M, Baumeister H, Montag C (2018) Feasibility of linking molecular genetic markers to real-world social network size tracked on smartphones. Front Neurosci 12. https://doi.org/10.3389/fnins.2018.00945

Schwartz HA, Eichstaedt JC, Kern ML et al (2013) Personality, gender, and age in the language of social media: the open-vocabulary approach. PLoS ONE 8(9):e73791. https://doi.org/10.1371/journal.pone.0073791

Schwartz HA, Ungar LH (2015) Data-driven content analysis of social media: a systematic overview of automated methods. Ann Am Acad Pol Soc Sci 659(1):78–94. https://doi.org/10.1177/0002716215569197

Segalin C, Celli F, Polonio L et al (2017) What your facebook profile picture reveals about your personality. In: Proceedings of the 25th ACM international conference on multimedia. ACM, New York, NY, USA, pp 460–468

Settanni M, Azucar D, Marengo D (2018) Predicting individual characteristics from digital traces on social media: a meta-analysis. Cyberpsychol Behav Soc Netw 21(4):217–228. https://doi.org/10.1089/cyber.2017.0384

Settanni M, Marengo D (2015) Sharing feelings online: studying emotional well-being via automated text analysis of Facebook posts. Front Psychol 6. https://doi.org/10.3389/fpsyg.2015.01045

Soldz S, Vaillant GE (1999) The Big Five personality traits and the life course: a 45-year longitudinal study. J Res Pers 33(2):208–232. https://doi.org/10.1006/jrpe.1999.2243

Stachl C, Hilbert S, Au J-Q et al (2017) Personality traits predict smartphone usage. Eur J Pers 31(6):701–722. https://doi.org/10.1002/per.2113

Statista (2019) Number of monthly active Facebook users worldwide as of 2nd quarter 2019 (in millions). https://www.statista.com/statistics/264810/number-of-monthly-active-facebook-users-worldwide/. Accessed 30 Jul 2019

Tausczik YR, Pennebaker JW (2010) The psychological meaning of words: LIWC and computerized text analysis methods. J Lang Soc Psychol 29(1):24–54. https://doi.org/10.1177/0261927X09351676

Thilakaratne M, Weerasinghe R, Perera S (2016) Knowledge-driven approach to predict personality traits by leveraging social media data. In: 2016 IEEE/WIC/ACM international conference on web intelligence (WI). pp 288–295

Tibshirani R (1996) Regression shrinkage and selection via the lasso. J Royal Stat Soc Ser B (Methodol) 58(1):267–288. https://doi.org/10.1111/j.2517-6161.1996.tb02080.x

Torfason R, Agustsson E, Rothe R, Timofte R (2016) From face images and attributes to attributes. In: Lai S-H, Lepetit V, Nishino K, Sato Y (eds) Computer vision—ACCV 2016. Springer International Publishing, pp 313–329

Torfason R, Agustsson E, Rothe R, Timofte R (2017) From face images and attributes to attributes. In: Lai S-H, Lepetit V, Nishino K, Sato Y (eds) Computer vision—ACCV 2016. Springer International Publishing, pp 313–329

Wald R, Khoshgoftaar T, Sumner C (2012) Machine prediction of personality from Facebook profiles. In: 2012 IEEE 13th international conference on information reuse integration (IRI). pp 109–115

Wong T-T (2015) Performance evaluation of classification algorithms by k-fold and leave-one-out cross validation. Pattern Recognit 48(9):2839–2846. https://doi.org/10.1016/j.patcog.2015.03.009

Youyou W, Kosinski M, Stillwell D (2015) Computer-based personality judgments are more accurate than those made by humans. Proc Natl Acad Sci U S A 112(4):1036–1040. https://doi.org/10.1073/pnas.1418680112

Chapter 8
Orderliness of Campus Lifestyle Predicts Academic Performance: A Case Study in Chinese University

Yi Cao, Jian Gao and Tao Zhou

Abstract Different from the western education system, Chinese teachers and parents strongly encourage students to have a regular lifestyle. However, due to the lack of large-scale behavioral data, the relation between living patterns and academic performance remains poorly understood. In this chapter, we analyze large-scale behavioral records of 18,960 students within a Chinese university campus. In particular, we introduce orderliness, a novel entropy-based metric, to measure the regularity of campus lifestyle. Empirical analyses demonstrate that orderliness is significantly and positively correlated with academic performance, and it can improve the prediction accuracy on academic performance at the presence of diligence, another behavioral metric that estimates students' studying hardness. This work supports the eastern pedagogy that emphasizes the value of regular lifestyle.

Keywords Computational social science · Orderliness · Academic performance · Human behavior

8.1 Introduction

Asian traditional culture considers regularity as an important and valuable personal trait. Therefore, in addition to being diligent, parents and teachers in most Asian countries ask students to live disciplined and regular lives. Accordingly, a hardworking and self-disciplined student is usually recognized as a positive model. To maintain large-size classes, teachers in Far East Asia create the highly disciplined

Y. Cao · J. Gao · T. Zhou (✉)
CompleX Lab Web Sciences Center, University of Electronic Science
and Technology of China, Chengdu 611731, People's Republic of China
e-mail: zhutou@ustc.edu

Y. Cao
e-mail: caoyi318@qq.com

J. Gao
e-mail: gaojian08@hotmail.com

© Springer Nature Switzerland AG 2019   125
H. Baumeister and C. Montag (eds.), *Digital Phenotyping and Mobile Sensing*,
Studies in Neuroscience, Psychology and Behavioral Economics,
https://doi.org/10.1007/978-3-030-31620-4_8

classes (Ning et al. 2015; Baumann and Krskova 2016), while western teachers rarely emphasize discipline in class or regularity in life. In 2015, BBC broadcasted a documentary about an attempt of Chinese-style education in the UK (BBC 2015), where Chinese teachers and UK students were maladjusted to each other at the beginning but English pupils taught by Chinese teachers eventually got better scores than their peers in a series of exams.

Although eastern and western have different pedagogies, they face the same challenge in education management, that is, to uncover underlying ingredients affecting students' academic performance. Previous studies have demonstrated that educational achievement is related to health conditions (Santana et al. 2017; Hoffmann 2018), IQ (Duckworth and Seligman 2005) and even to molecular genetic markers (Okbay 2016; Selzam et al. 2017). For example, scientists identified 74 genome-wide significant loci associated with the years of schooling (Okbay 2016). Since students' mentality and behavior are more interventional, the majority of studies concentrate on psychological and behavioral issues (Conard 2006). Experiments have demonstrated correlations between academic performance and personality (Chamorro-Premuzic and Furnham 2003; Poropat 2014). In particular, conscientiousness is the best predictor of GPA, while agreeableness and openness are of weaker effects (Vedel 2014). Behaviors of students are also associated with their academic performance, for example, students with more class attendance (Credé et al. 2010; Kassarnig et al. 2018), longer time on study (Grave 2011; Cattaneo et al. 2017), good sleep habits (Taylor et al. 2013; Urrila et al. 2017) and more physical activity (Erwin et al. 2017) perform better on average.

This said, it is still debated in the scientific community if a regular lifestyle in general represents an important prerequisite for academic study or not. One of the reasons for this ongoing debate is that, statistical validation of these observations based on large-scale behavioral data remains lacking. Traditional framework relies on data coming from questionnaires and self-reports, which usually contains a small number of participants (Vedel 2014) and suffers from being biased by the tendency to answer in a social desirable fashion (Fisher 1993). Second, previous studies rarely isolate regularity in living patterns from diligence in studies. As a more regular studying pattern may be correlated with a longer studying time, it is hard to distinguish their independent effects on academic performance. So far, to our knowledge, a quantitative relationship between regularity in everyday life and academic achievement has not been demonstrated. Fortunately, rapid development of information technologies has made it possible to study students' activities in an unobtrusive way by collecting their digital records through smartphones (Wang et al. 2014), online course platforms (Brinton et al. 2016), campus WiFi (Zhou et al. 2016), and so on (Gao et al. 2019). These large-scale extracurricular behavioral data offer chances to quantify the regularity of campus lifestyle and explore its relation to academic performance.

In this chapter, we present a case study on the relation between students' campus lifestyles and their academic performance. Through campus smart cards, we have collected the digital records of 18,960 undergraduate students' daily activities including taking showers, having meals, entering/exiting library and fetching water. Accordingly, we proposed two high-level behavioral characters, orderliness

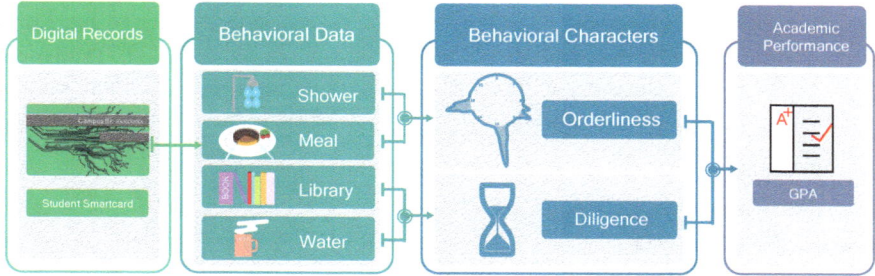

Fig. 8.1 Illustration of the methodology to reveal the relation between campus lifestyle and academic performance. Four types of behavioral records are collected by campus smart cards. Taking showers and having meals are used to measure orderliness, which represents the regularity level of campus life. And entering/exiting library and fetching water are used to measure diligence for which the reason is that cumulative occurrences of these behaviors in study places is naturally recognized as the total efforts taken on studying hardness. Correlations between behavioral features and academic performance are analyzed, and the predictive powers of orderliness and diligence are compared

and diligence (see Fig. 8.1 for the methodology). The orderliness factor is a novel entropy-based metric that measures the regularity of campus life, which is calculated based on temporal records of taking showers and having meals. The diligence factor is roughly estimated based on the cumulative occurrences of entering/exiting library and fetching water. Empirical analyses demonstrated that academic performance (GPA) is significantly correlated with orderliness and diligence. Moreover, orderliness can improve the prediction accuracy on academic performance at the presence of diligence. Some primary results have been published in a recent article (Cao et al. 2018), and the present chapter is an extension with more detailed analyses.

8.2 Data and Metrics

8.2.1 Data Description

In most Chinese universities, every student owns a campus smart card which is used for student identification and serves as the unique payment medium for many on-campus consumptions. For example, there are toll gates in shower rooms, where students have to keep the smart card inserted during the shower. Here, we introduce a specific case study in a Chinese university, the *University of Electronic Science and Technology of China* (UESTC), which provides on-campus dormitories to all undergraduate students and in principle does not allow students to live off-campus. Therefore, smart cards record large volume of behavioral data in terms of students' living and studying activities. Accordingly, we have collected digital records of $N = 18,960$ undergraduate students' daily activities from September 2009 to July 2015, covering the period from the beginning of their first year to the end of their third year. The data includes the purchase records for showers ($n = 3, 151, 783$) and meals ($n = 19, 015, 773$), the entry-exit records in library ($n = 3, 412, 587$) and fetching water records in teaching buildings ($n = 2, 279, 592$). GPAs of students in each semester are also collected.

8.2.2 Orderliness

We calculate orderliness based on two behaviors: taking showers in dormitories and having meals in cafeterias. Indeed, the meaning of orderliness is twofold, say timing and order. The happening times of the same kind of events should be close to each other, for example, having breakfast at about 8:00 is more regular than between 7:00 and 9:00. The temporal order of different events should also be regular, for instance, having meals following the order breakfast $\rightarrow$ lunch $\rightarrow$ supper $\rightarrow$ breakfast $\rightarrow$ lunch $\rightarrow$ supper is more regular than breakfast $\rightarrow$ supper $\rightarrow$ lunch $\rightarrow$ supper $\rightarrow$ breakfast $\rightarrow$ lunch. With these insights, we turn to the mathematical formula of orderliness. Considering a student's specific behavior within total n recorded actions happening at time stamps $\{t_1, t_2, \ldots, t_n\}$, where $t_i \in [00 : 01, 24 : 00]$ denotes the precise time with resolution in minutes. We first arrange all actions in the order of occurrence, namely, the i-th action happens before the j-th action if $i < j$, while we ignore the date information. Then, we divide one day into 48 time bins with a 30 minutes step (specifically, 0:01-0:30 is the 1st bin, 0:31-1:00 is the 2nd bin, ...), and map the time series $\{t_1, t_2, \ldots, t_n\}$ into a discrete sequence $\{t'_1, t'_2, \ldots, t'_n\}$, where $t'_i \in \{1, 2, \ldots, 48\}$. If a student's starting times of five consecutive showers are $\{21{:}05, 21{:}33, 21{:}13, 21{:}48, 21{:}40\}$, the corresponding binned sequence is $\mathcal{E} = \{43, 44, 43, 44, 44\}$. Next, we apply the actual entropy (Kontoyiannis et al. 1998; Xu et al. 2019) to measure the orderliness of any sequence $\mathcal{E}$. Formally, the actual entropy is defined as

$$S_{\mathcal{E}} = \left(\frac{1}{n} \sum_{i=1}^{n} \Lambda_i\right)^{-1} \ln n, \qquad (8.1)$$

where Λ_i represents the length of the shortest subsequence which starts from t'_i of $\mathcal{E}$ and has never appeared previously. Note that we set $\Lambda_i = n - i + 2$ if such subsequence cannot be found (Xu et al. 2019). Finally, we define orderliness as $O_{\mathcal{E}} = -S_{\mathcal{E}}$ and calculate regularized orderliness by normalizing $S_{\mathcal{E}}$ via Z-score (Kreyszig 2010):

$$O'_{\mathcal{E}} = \frac{O_{\mathcal{E}} - \mu_O}{\sigma_O} = \frac{\mu_S - S_{\mathcal{E}}}{\sigma_S}, \qquad (8.2)$$

where μ_O and σ_O are the mean and standard deviation of orderliness O, μ_S and σ_S are the mean and standard deviation of actual entropy S, and $O'_{\mathcal{E}}$ is the regularized orderliness for the student with binned sequence $\mathcal{E}$. The larger orderliness corresponds to higher regularity of a student's campus lifestyle.

8.2.3 Diligence

Diligence measures to what extent people take efforts to strive for achievements. As an important behavioral character, diligence is intuitively related to students' academic performance. Due to the lack of ground truth, however, it is difficult to quantify students' diligence. Here, we roughly estimate diligence based on two behaviors: entering/exiting the library, and fetching water in teaching buildings. Specifically, we use a student's cumulative occurrences of library entering/exiting and water fetching as a rough estimate of his/her diligence. Basically, self-studying and borrowing books are the most common purposes of going to the library, while attending courses usually take place in the teaching buildings. As teaching buildings have no check-in devices or entry terminals like the library, we use records of water fetching as the proxy.

8.3 Result

8.3.1 Validation of Behavioral Characters

Figure 8.2a and b present the distributions of actual entropies on taking showers and having meals, respectively. The broad distributions guarantee the discriminations of students with different orderliness. For student H with very high orderliness (at the 5th percentile) and student L with very low orderliness (at the 95th percentile), we notice that student H takes most showers around 21:00 while student L may take showers at any time in a day (Fig. 8.2c). We observe the similar discrepancy on having meals (Fig. 8.2d). Figure 8.2e and f present the distributions of cumulative occurrences for entering/exiting the library and fetching water. The two distributions are both broad, showing that the two diligence metrics can distinguish students with different levels of studying hardness.

We next explore the consistency and dependence of the two behavioral characters. As either orderliness or diligence is measured by two types of behavioral records, their intra correlations should be high if they are properly measured. That is, orderliness-meal should be correlated with orderliness-shower, and diligence-water should be correlated with diligence-library. Moreover, the effect of orderliness should be isolated from diligence, i.e., their inter correlations should be low. Figure 8.3 presents the Spearman's correlation matrix between each pair of behavioral features. The intra correlations are all positive and significant, with the correlation $r = 0.226$ between two orderliness metrics and $r = 0.262$ between two diligence metrics. Moreover, if orderliness provides additional information to diligence, the correlation between any pair of orderliness metric and diligence metric should be insignificant. As shown in Fig. 8.3, all inter correlations are close to 0, suggesting the absence of significant correlation between orderliness and diligence. These results validate the robustness of the two behavioral characters and demonstrate their independence.

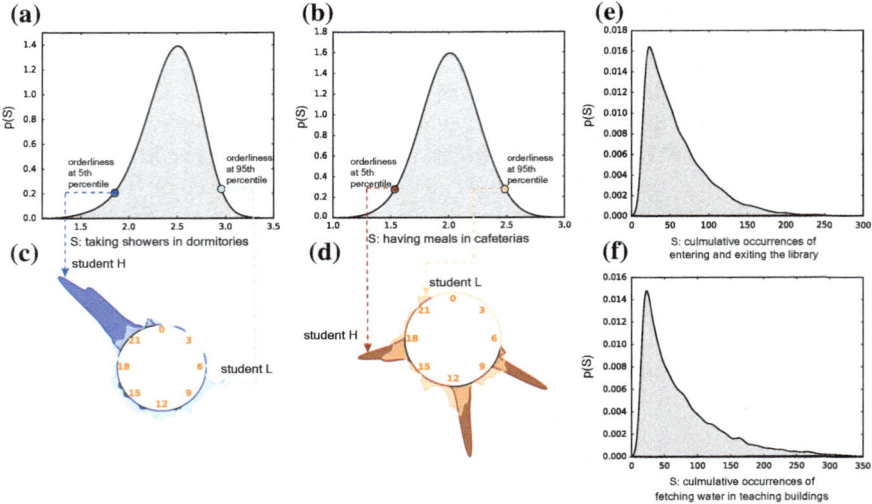

Fig. 8.2 Distributions of actual entropies and cumulative occurrences. Distributions $p(S)$ of students in taking showers (**a**) and having meals (**b**). The x-axis represents the actual entropies S, calculated in each semester. The broad distributions guarantee the discriminations of students with different orderliness. The behavioral clocks of two students at the 5th percentile and the 95th percentile are shown for taking showers (**c**) and having meals (**d**), where student H has high orderliness and student L has low orderliness. Distributions $p(C)$ of students in entering/exiting library (**e**) and fetching water (**f**), calculated in each semester. The broad distributions distinguish students with different diligence levels

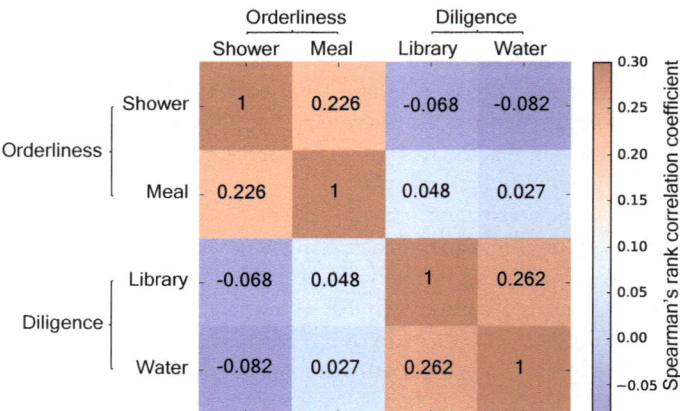

Fig. 8.3 Correlations among behavioral features. Shower and Meal are the two orderliness features, while Library and Water are the two diligence features. The color denotes the corresponding Spearman's rank correlation coefficient. Significance level: all p-values are less than 10^{-15}

8.3.2   Correlation Analysis

Students of higher orderliness and diligence are expected to have better grades. Here, we empirically assessed how students' orderliness and diligence are related to their academic performance (GPA). The regularized GPA for student i is defined as $G'_i = (G_i - \mu_G)/\sigma_G$, where G_i is his/her GPA, and μ_G and σ_G are the mean and standard deviation of G for all considered students.

Figure 8.4a and b present how regularized GPA is positively correlated to regularized orderliness-shower and regularized orderliness-meal, respectively. As the relationships between orderliness and GPA are not simply linear, we apply the well-known Spearman's rank correlation coefficient (Spearman 1904) to quantify the strength of correlation. Results show that the correlation between orderliness-meal and GPA is $r = 0.182$, and the correlation between orderliness-shower and GPA is $r = 0.157$, both with statistical significance $p < 0.0001$. Analogously, Fig. 8.4c and d present how regularized GPA is positively correlated to regularized diligence-library and regularized diligence-water, respectively. The correla-

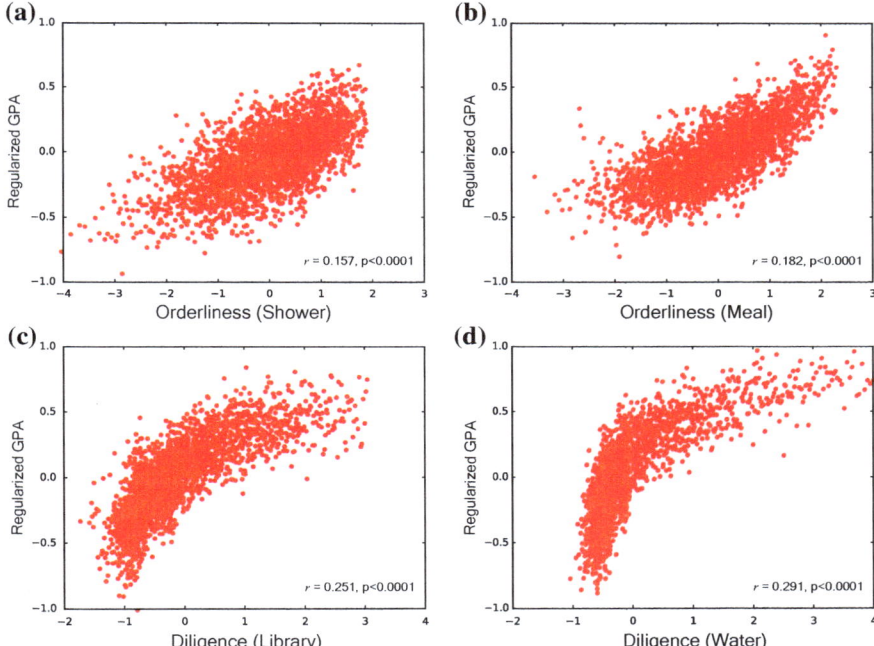

Fig. 8.4 Relations between behavioral features and academic performance. How regularized GPA is positively correlated with **a** regularized orderliness-shower, **b** regularized orderliness-meal, **c** regularized diligence-library, and **d** regularized diligence-water. The corresponding Spearman's rank correlation coefficients and the level of statistical significance are shown in each plot

tion between diligence-library and GPA is $r = 0.291$, and the correlation between diligence-water and GPA is $r = 0.251$.

As a summary, the two behavioral characters are significantly correlated to academic performance with correlations being about 0.2. The Spearman's rank correlations for diligence (library and water) are stronger than for orderliness (shower and meal), while eyeballing the data suggests the opposite (Fig. 8.4). As a robustness check, we additionally calculated the Pearson correlation coefficients. Results showed that correlations for diligence remain stronger than for orderliness. The visual discrepancy may be because the data points are dispersive. As orderliness is largely independent to diligence, the results suggest their independently potential effects on students' academic performance.

8.3.3 Predictive Analysis

The significant correlations between behavioral features and GPA imply that orderliness and diligence can be used as different feature classes to predict students' academic performance. Here, we predict the ranks of students' semester grades by applying a well-known supervised learning-to-rank algorithm named RankNet (Burges et al. 2005). Given a feature vector $\mathbf{x} \in \mathbb{R}^p$ of each student, RankNet tries to learn a scoring function $f : \mathbb{R}^p \to \mathbb{R}$, so that the predicted ranks according to f are as consistent as possible with the ground truth. The consistence is measured by the cross entropy between the actual probability and the predicted probability. Based on the scoring function, the predicted probability that a student i has a higher GPA than another student j (denoted as $i \rhd j$) is defined as $P(i \rhd j) = \sigma(f(\mathbf{x}_i) - f(\mathbf{x}_j))$, where $\sigma(z) = 1/(1 + e^{-z})$ is a sigmoid function.

We consider a simple regression function $f = \mathbf{w}^T \mathbf{x}$, where $\mathbf{w}$ is the vector of parameters. The cost function of RankNet is given by

$$\mathcal{L} = - \sum_{(i,j):i \rhd j} \log \sigma(f(\mathbf{x}_i) - f(\mathbf{x}_j)) + \lambda \Omega(f), \qquad (8.3)$$

where $\Omega(f) = \mathbf{w}^T \mathbf{w}$ is a regularized term. Given all students' feature vectors and their ranks, gradient decent is applied to minimize the cost function. The gradient of the lost function with respect to parameter $\mathbf{w}$ in f is

$$\frac{\partial \mathcal{L}}{\partial \mathbf{w}} = \sum_{(i,j):i \rhd j} (\sigma(f(\mathbf{x}_i) - f(\mathbf{x}_j)) - 1) \left(\frac{\partial f(\mathbf{x}_i)}{\partial \mathbf{w}} - \frac{\partial f(\mathbf{x}_j)}{\partial \mathbf{w}} \right) + \lambda \frac{\partial \Omega(f)}{\partial \mathbf{w}}. \qquad (8.4)$$

The prediction accuracy is evaluated by AUC value (Hanley and McNeil 1982), which is equal to the percentage of student pairs whose relative ranks are correctly predicted. The AUC value ranges from 0 to 1 with 0.5 being the random chance, therefore to which extent the AUC value exceeds 0.5 can be considered as the predictive power.

Table 8.1 AUC values for the GPA prediction. The abbreviations O, D and O + D stand for utilizing features on orderliness only, on diligence only and on the combination of orderliness and diligence, respectively

Features	Semester 2	Semester 3	Semester 4	Semester 5
O	0.618	0.617	0.611	0.597
D	0.630	0.655	0.663	0.668
O + D	0.668	0.681	0.685	0.683

We train RankNet based on the extracted orderliness and diligence features in one of the first four semesters and predict students' ranks of grades in the next semester. We use the abbreviations O, D and O+D to stand for utilizing features on orderliness only, on diligence only and on the combination of orderliness and diligence, respectively. Table 8.1 presents the results of AUC values under different feature combinations, where the column semester j represents the case in which we train the data of semester $j - 1$ and predict the ranks of grades in semester j. Obviously, both orderliness and diligence are predictive to academic performance, and orderliness can improve the prediction accuracy at the presence of diligence, showing its independent role in facilitating academic studying.

8.4 Discussion

Large parts of the Eastern world value regularity in campus lifestyle, while large parts of the Western world tend to provide a more unconstrained lifestyle to students. The disparity in educational philosophy between these different parts of the world may originate from their culture differences. Yet, the core question is whether regularized campus lifestyle is helpful to achieve higher academic performance. To answer this question, we presented the data-driven case study based on large-scale behavioral records of students' living and studying activities in a Chinese university campus (Cao et al. 2018). Specifically, we calculated orderliness based on temporal records of taking showers and having meals, which is not directly related to studying activities. Empirical analyses show that academic performance is significantly and positively correlated with orderliness. Moreover, orderliness can remarkably improve the accuracy of academic performance prediction even at the presence of diligence, suggesting the independent predictive power of orderliness.

Our work not only provides a quantitatively understanding of the relationships between students' behavioral patterns and academic performances, but probably also takes a significant step towards better educational management. On the one hand, education administrators could design personalized teaching and caring programs for individuals with different behaviors. For example, recent works have discussed the prediction of course failures and dropping out for K12 education (Kindergarten and the 1–12 grades) (Jayaprakash et al. 2014; Lakkaraju et al. 2015), and thus teachers could pay more attentions in advanced to those students who may develop difficulties in studying.

On the other hand, education managers can detect students' undesirable abnormal behaviors from traced data (e.g., Internet use disorder (IUD) Montag and Reuter 2017; Brand et al. 2016; Peterka-Bonetta et al. 2019) and implement interventions in time. IUD is negatively correlated with academic performance (Akhter 2013; Khan et al. 2016), and IUD is among the most important reasons resulting in the failure of college study in China. Two issues need to be discussed in the context of IUD, formerly also known as Internet addiction. Firstly, the sharp fall of exam performance or even failure of many courses appears about one or two semesters after developing IUD. Secondly, it requires a long time (usually a few months) for a student to rebuild learning ability after proper treatment of IUD.

Therefore, a student's academic performance would not drop immediately, while IUD immediately will impact on the student's behaviors (e.g., absence from classes). Students suffering from IUD demonstrated largely different behavioral patterns compared to those not suffering from IUD, for example, students with IUD have irregular bedtimes and dietary behavior (Kim et al. 2010), and their diligence and orderliness usually dramatically decline. Accordingly, it might be possible to establish models being able to predict whether students are more prone to develop IUD, and thus those problem students can be helped as soon as possible.

Even though traditional questionnaire surveys are limited by sample sizes and suffering from response biases such as tendencies to answer in a social desirable way (Paulhus and Vazire 2007), these two methodologies can complement and benefit each other. The use of unobtrusive digital records is helpful in improving the quality of questionnaires (Montag et al. 2016), meanwhile the assessment of psychological characteristics related to a target behavior can be also complemented by self-report questionnaires, e.g., assessing conscientiousness, which would have been of interest also in the present work. Indeed, it has been shown that is possible to infer self-reported personality and other private attributes from available online information such as Facebook-Likes (Kosinski et al. 2013; Youyou et al. 2015). Moreover, it is promising to establish a causal link between behavioral features and academic performance through designing controlled experiments. We are expecting psychologists and computer scientists to work together on such a promising research endeavor in the near future (Gao et al. 2019).

As we know from our culture, Chinese universities value disciplinary behaviors (Baumann and Krskova 2016). However, whether orderliness will be of same positive quality for academic study performance in other countries remains an open question. On the one hand, orderliness relies in our work on campus activities while students in other countries may live off-campus or spend a considerable portion of time doing part-time jobs, resulting perhaps in lower orderliness (but such differences between "West" and "East" need to be systematically evaluated on an empirical level and we explicitly state this to be a working hypothesis). On the other hand, it is difficult to isolate orderliness from the capacity to follow the teacher's advice (whatever that is) in and out of classes. Although previous studies have shown that better classroom discipline leads to better academic performance (Ning et al. 2015), whether a student's capacity to follow advice is related to her/his achievements is not clear or may be also attributable to other person characteristics including intelligence.

We hope that recent works leveraging large-scale behavioral data analysis and machine learning techniques also find its way into pedagogical sciences (Kassarnig et al. 2018; Wang et al. 2014; Gao and Zhou 2016). Indeed, uncovering factors that affect educational outcome play a significant role in future quantitative and personalized education management and could help to improve the schooling process.

References

Akhter N (2013) Relationship between Internet addiction and academic performance among university undergraduates. Educ Res Rev 8:1793–1796

Baumann C, Krskova H (2016) School discipline, school uniforms and academic performance. Int J Educ Manag 30:1003–1029

BBC (2015) Would Chinese-style education work on British kids? BBC News Magazine (4 August) https://www.bbc.com/news/magazine-33735517

Brand M, Young KS, Laier C, Wölfling K, Potenza MN (2016) Integrating psychological and neurobiological considerations regarding the development and maintenance of specific Internet-use disorders: an interaction of Person-Affect-Cognition-Execution (I-PACE) model. Neurosci Biobehav Rev 71:252–266

Brinton CG, Buccapatnam S, Chiang M, Poor HV (2016) Mining MOOC clickstreams: video-watching behavior versus in-video quiz performance. IEEE Trans Signal Process **64**, 3677–3692

Burges C, Shaked T, Renshaw E, Lazier A, Deeds M, Hamilton N, Hullender G (2005) Learning to rank using gradient descent. In: Proceedings of the 22nd international conference on machine learning. ACM Press, New York, pp 89–96

Cao Y, Gao J, Lian D, Rong Z, Shi J, Wang Q, Wu Y, Yao H, Zhou T (2018) Orderliness predicts academic performance: behavioural analysis on campus lifestyle. J R Soc Interface **15**, 20180210 (2018)

Cattaneo MA, Oggenfuss C, Wolter SC (2017) The more, the better? the impact of instructional time on student performance. Educ Econ 25:433–445

Chamorro-Premuzic T, Furnham A (2003) Personality predicts academic performance: evidence from two longitudinal university samples. J Res Pers 37:319–338

Conard MA (2006) Aptitude is not enough: how personality and behavior predict academic performance. J Res Pers 40:339–346

Credé M, Roch SG, Kieszczynka UM (2010) Class attendance in college: a meta-analytic review of the relationship of class attendance with grades and student characteristics. Rev Educ Res 80:272–295

Duckworth AL, Seligman ME (2005) Self-discipline outdoes IQ in predicting academic performance of adolescents. Psychol Sci 16:939–944

Erwin H, Fedewa A, Ahn S (2017) Student academic performance outcomes of a classroom physical activity intervention: a pilot study. Int Electron J ElemTary Educ 4:473–487

Fisher RJ (1993) Social desirability bias and the validity of indirect questioning. J Consum Res 20:303–315

Gao J, Zhou T (2016) Big data reveal the status of economic development. Journal of University of Electronic Science and Technology of China 45, 625–633

Gao J, Zhang YC, Zhou T (2019) Computational socioeconomics. Phys Rep 817:1–104

Grave B (2011) The effect of student time allocation on academic achievement. Educ Econ 19:291–310

Hanley JA, McNeil BJ (1982) The meaning and use of the area under a receiver operating characteristic (ROC) curve. Radiology 143:29–36

Hoffmann I, Diefenbach C, Gräf C, König J, Schmidt MF, Schnick-Vollmer K, Blettner M, Urschitz MS (2018) Chronic health conditions and school performance in first graders: a prospective cohort study. PLoS ONE 13, e0194846

Jayaprakash SM, Moody EW, Lauría EJ, Regan JR, Baron JD (2014) Early alert of academically at-risk students: an open source analytics initiative. J Learn Anal 1:6–47

Kassarnig V, Mones E, Bjerre-Nielsen A, Sapiezynski P, Dreyer Lassen D, Lehmann S (2018) Academic performance and behavioral patterns. EPJ Data Sci 7:1–16

Khan MA, Alvi AA, Shabbir F, Rajput TA (2016) Effect of Internet addiction on academic performance of medical students. J Islam Int Med Coll 11:48–51

Kim Y, Park JY, Kim SB, Jung IK, Lim YS, Kim JH (2010) The effects of Internet addiction on the lifestyle and dietary behavior of Korean adolescents. Nutr Res Pract 4:51–57

Kontoyiannis I, Algoet PH, Suhov YM, Wyner AJ (1998) Nonparametric entropy estimation for stationary processes and random fields, with applications to English text. IEEE Trans Inf Theory 44:1319–1327

Kosinski M, Stillwell D, Graepel T (2013) Private traits and attributes are predictable from digital records of human behavior. P Natl Acad Sci USA 110:5802–5805

Kreyszig E (2010) Advanced engineering mathematics. Wiley, Hoboken, New Jersey

Lakkaraju H, Aguiar E, Shan C, Miller D, Bhanpuri N, Ghani R, Addison KL (2015) A machine learning framework to identify students at risk of adverse academic outcomes. In: Proceedings of the 21th ACM SIGKDD international conference on knowledge discovery and data mining, ACM Press, New York, pp 1909–1918

Montag C, Duke É, Markowetz A (2016) Toward psychoinformatics: computer science meets psychology. Comput Math Methods Med 2016:2983685

Montag C, Reuter M (2017) Internet addiction. Springer International Publishing, Cham, Switzerland

Ning B, Van Damme J, Van Den Noortgate W, Yang X, Gielen S (2015) The influence of classroom disciplinary climate of schools on reading achievement: a cross-country comparative study. Sch Eff Sch Improv 26:586–611

Okbay A et al (2016) Genome-wide association study identifies 74 loci associated with educational attainment. Nature 533:539–542

Paulhus DL, Vazire S (2007) The self-report method. In: Robins RW, Fraley RC, Krueger RF (eds) Handbook of research methods in personality psychology. The Guilford Press, New York and London, pp 224–239

Peterka-Bonetta J, Sindermann C, Sha P, Zhou M, Montag C (2019) The relationship between Internet use disorder, depression and burnout among Chinese and German college students. Addict Behav 89:188–199

Poropat AE (2014) Other-rated personality and academic performance: evidence and implications. Learn Individ Differ 34:24–32

Santana CCA, Hill JO, Azevedo LB, Gunnarsdottir T, Prado WL (2017) The association between obesity and academic performance in youth: a systematic review. Obes Rev 18:1191–1199

Selzam S, Krapohl E, Stumm SV, O'Reilly PF, Rimfeld K, Kovas Y, Dale PS, Lee JJ, Plomin R (2017) Predicting educational achievement from DNA. Mol Psychiatry 22:267–272

Spearman C (1904) The proof and measurement of association between two things. Am J Psychol 15:72–101

Taylor DJ, Vatthauer KE, Bramoweth AD, Ruggero C, Roane B (2013) The role of sleep in predicting college academic performance: Is it a unique predictor? Behav Sleep Med 11:159–172

Urrila AS, Artiges E, Massicotte J, Miranda R, Vulser H, Bézivin-Frere P, Lapidaire W, Lemaître H, Penttilä J, Conrod PJ, Garavan H, Martinot MP, Martinot J (2017) Sleep habits, academic performance, and the adolescent brain structure. Sci Rep-UK 7:41678

Vedel A (2014) The big five and tertiary academic performance: a systematic review and meta-analysis. Pers Individ Differ 71:66–76

Wang R, Chen F, Chen Z, Li T, Harari G, Tignor S, Zhou X, Ben-Zeev D, Campbell AT (2014) StudentLife: assessing mental health, academic performance and behavioral trends of college students using smartphones. In: Proceedings of the 2014 ACM international joint conference on pervasive and ubiquitous computing. ACM Press, New York, pp 3–14

Xu P, Yin L, Yue Z, Zhou T (2019) On predictability of time series. Phys A 523:345–351

Youyou W, Kosinski M, Stillwell D (2015) Computer-based personality judgments are more accurate than those made by humans. P Natl Acad Sci USA 112:1036–1040

Zhou M, Ma M, Zhang Y, Sui K, Pei D, Moscibroda T (2016) EDUM: classroom education measurements via large-scale WiFi networks. In: Proceedings of the 2016 ACM international joint conference on pervasive and ubiquitous computing. ACM Press, New York, pp 316-327

Part III

Digital Phenotyping and Mobile Sensing in Health Sciences

Chapter 9
Latest Advances in Computational Speech Analysis for Mobile Sensing

Nicholas Cummins and Björn W. Schuller

Abstract The human vocal anatomy is an intricate anatomical structure which affords us the ability to vocalise a large variety of acoustically rich sounds. As a result, any given speech signal contains an abundant array of information about the speaker in terms of both the intended message, i.e., the linguistic content, and insights into particular states and traits relating to the speaker, i.e., the paralinguistic content. In the field of computational speech analysis, there are substantial and ongoing research efforts to disengage these different facets with the aim of robust and accurate recognition. Speaker states and traits of interest in such analysis include affect, depressive and mood disorders and autism spectrum conditions to name but a few. Within this chapter, a selection of state-of-the-art speech analysis toolkits, which enable this research, are introduced. Further, their advantages and limitations concerning mobile sensing are also discussed. Ongoing challenges and possible future research directions in relation to the identified limitations are also highlighted.

9.1 Introduction

Human speech is produced by an exceptionally complex and intricate interaction between our cognitive and neuromuscular systems (Fitch 2000; Levelt et al. 1999). As a result, speech is a highly sensitive output system; physiological, pathological and biochemical changes easily affect our speech production in ways that are audible and thus objectively measurable using intelligent signal analysis and machine learning methodologies. Indeed, there are a plethora of papers in the relevant literature highlighting the benefits of using speech as an objective marker for a wide range of

N. Cummins (✉) · B. W. Schuller
ZD.B Chair of Embedded Intelligence for Health Care and Wellbeing, University of Augsburg, Augsburg, Germany
e-mail: nicholas.cummins@ieee.org

B. W. Schuller
GLAM—Group on Language, Audio & Music, Imperial College London, London, UK
e-mail: bjoern.schuller@imperial.ac.uk

© Springer Nature Switzerland AG 2019   141
H. Baumeister and C. Montag (eds.), *Digital Phenotyping and Mobile Sensing*,
Studies in Neuroscience, Psychology and Behavioral Economics,
https://doi.org/10.1007/978-3-030-31620-4_9

emotional, pathological and mental health conditions, e.g., (Bone et al. 2017; Cummins et al. 2015; Schuller et al. 2018a, b; Schuller 2017). This work is centred in a field of research known as *computational paralinguistics* (Schuller and Batliner 2013), which is the extraction and analysis of the phenomena embedded into a speech signal. This information includes short-term speaker states such as one's current instantaneous level of arousal or valence, or longer-term speaker traits such as if a speaker is currently suffering a mood disorder or similar condition. Within computational paralinguistics, frequently studied pathological and mental health conditions include unipolar and bipolar depression (Cummins et al. 2015; Ringeval et al. 2018), autism spectrum conditions (Ringeval et al. 2016; Schuller et al. 2013), and Parkinson's disease (Orozco-Arroyave et al. 2016; Schuller et al. 2015).

The richness of information available in a speech signal is further complemented by the ease at which it can be collected remotely and unobtrusively. With the ongoing growth of the *Internet-of-Things* (IoT), microphone embedded smart devices and wearable technologies are at the point of ubiquity in modern society (Jankowski et al. 2014). This growth has enhanced researchers and clinicians, ability to collect data relating to, not only speech (Cunningham et al. 2017; Hagerer et al. 2017; Marchi et al. 2016; Tsiartas et al. 2017), but a wide array of bio- and behavioural markers. Such data can, in turn, be used to aid (early) detection and remotely monitor a wide range of conditions (Istepanian and Al-Anzi 2018).

The aim of this chapter is two-fold. The first aim is to introduce the reader to a range of intelligent audio signal processing toolkits. These toolkits, the OPENSMILE feature extraction tool (Sect. 9.2), the OPENXBOW crossmodal Bag-of-Words toolkit (Sect. 9.3), the DEEPSPECTRUM (Sect. 9.4) and AUDEEP (Sect. 9.5) Python-based toolkits for feature representation learning, as well as the END2YOU toolkit for multimodal end-to-end profiling (Sect. 9.6), allow users to easily and quickly extract rich and relevant information from speech and audio signals. Where relevant, the use of this toolkits in analysing other behavioural- and bio-signals is also highlighted. These toolkits are introduced as they are all open source and are widely prevalent in the relevant literature. Further, they have all been used as baseline systems within the popular *Computational Paralinguistics Challenge* (COMPARE) and *Audio/Visual Emotion Challenge* (AVEC) workshops, see (Ringeval et al. 2018; Schuller et al. 2018a, b) for the 2018 challenge papers, among other challenges, and can, therefore, be regarded as standards in the field of audio signal processing.[1] The second aim is to discuss current challenges associated with speech-based mobile sensing and, to this end, highlight possible future research direction (Sect. 9.7). The chapter finishes with a brief concluding statement (Sect. 9.8).

---

[1]Nicholas Cummins is a co-developer of the DEEPSPECTRUM and AUDEEP toolkits. Björn W. Schuller is a co-developer of all five toolkits.

9.2 OPENSMILE

OPENSMILE[2] is a well-established research tool in speech, music and audio processing (Eyben et al. 2010, 2013). OPENSMILE enables users to, in real-time, extract large—knowledge driven—audio feature spaces. Feature extraction, in terms of intelligent signal processing, is the extraction of information relevant to the task at hand. For speech analysis, the extracted information is generally prosodic-acoustic in nature. A wide set of speech features can be extracted using openSMILE including *prosodic* features, such as loudness and pitch; *voice quality* features, such as jitter and shimmer; and *spectral* features, such as Mel-/Bark-/Octave-spectra, mel-frequency cepstral coefficients and spectral shape descriptors. All speech-related features extractable through OPENSMILE are grounded in highly researched and well-documented theories which, due to room limitations will not be discussed herein. For further information relating to speech and audio feature extraction, the interested reader is referred to (O'Shaughnessy 1999; Quatieri 2002).

OPENSMILE itself, is a cross-platform toolkit capable of operating in *Windows*, *Linux*, *Mac* and *Android* environments. It can receive a wide set of inputs including *audio* (.wav), or previously extracted features in *comma separated value* (.csv) or text (.txt) formats. The toolkit, as well as extracting a wide set of audio and speech *low-level-descriptors* (LLDs)—features—from a given input, also provides support for the post-processing of these LLDs such as through smoothing filters and standardisation and normalisation. The extraction of most speech and audio LLDs is conducted in very short analysis windows (typically 25–40 ms in length), paralinguistic effects on the other hand, are often more evident in the evolution of these features over time (Schuller and Batliner 2013). In this regard, OPENSMILE also supports feature summarisation (over a chunk of time or the course of an utterance) using statistical functionals. Finally, OPENSMILE has a range of output options including playback (.wav), .csv, and .txt files. Support is also provided (both an input and output) for common machine learning platforms such as the *.arff* format for the WEKA platform (Hall et al. 2009); LIBSVM (Chang and Lin 2011); and HTK (Young et al. 2002).

As well as enabling the extraction of numerous audio features, a core strength of openSMILE is its ability to extract this information in real-time; as enabled by the platform's unique modular architecture. Data flow is handled by a central memory component which essentially manages a combination of data processing components each configured to perform a particular task. Each component has permission to write to one memory location but has permission to read from other component's memory locations. This permission set-up enables a highly efficient incremental computation procedure in which a specific task, needed in multiple steps in a feature extraction pipeline, is only performed once (Fig. 9.1).

As already mentioned, the OPENSMILE toolkit has been used for baselines within the COMPARE and AVEC workshops. Within these challenges, participants are supplied with a common dataset and have to perform a specific classification or regression task on this data; OPENSMILE is used to provide participants with a baseline

---

[2]https://www.audeering.com/technology/opensmile/.

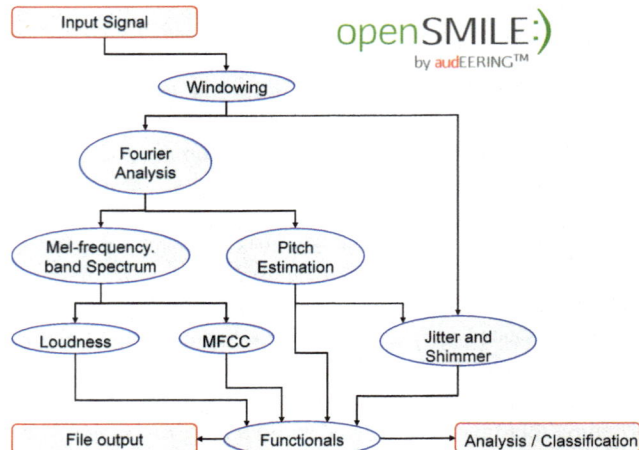

Fig. 9.1 An example of an incremental computation procedure enabled by OPENSMILE's unique modular architecture. In this example shows the information flow between different processing components needed to extract supra-segmental representation of loudness and pitch, both prosodic features; Mel Frequency Cepstral Coefficients (MFCC), a spectral feature representation; and both the Jitter and Shimmer voice quality features. Note, figure adapted from processing (Eyben et al. 2013)

feature set. The exact make-up of this feature set has evolved over the course of the challenges, and has now settled into two commonly used representations; the 6,373 dimensional 'paralinguistic omnibus' feature set known as the *Interspeech Computational Paralinguistics Challenge features set* (COMPARE) (Eyben et al. 2013), and the 88 dimensional 'tailor-made' for emotion recognition feature set known as the *extended Geneva Minimalistic Acoustic Parameter Set* (EGEMAPS) (Eyben et al. 2016). The scripts needed to implement the extraction of both the COMPARE and EGEMAPS, as well as scripts for a variety of other standard speech and audio features is provided with the OPENSMILE software.

9.3 OPENXBOW

OPENXBOW is a Java-based program for generating *bag-of-words* representations from either acoustic LLDs, transcriptions of natural speech, or visual features such as facial action units (Schmitt and Schuller 2017), physiological features or any other kind of time series feature data. The bag-of-words approach originated from natural language processing, where documents are classified based on a histogram representation of linguistic features such as the actual words present in the document. A *Bag-of-Audio-Words* (BoAW) on the other hand, involves quantisation of acoustic LLDs with respect to an audio word from a previously learnt codebook. BoAW

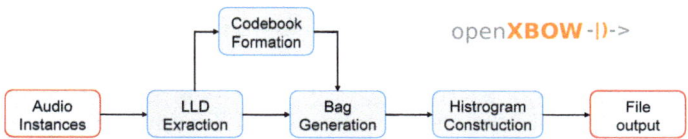

Fig. 9.2 A generalised overview of the key step involved when extracting a bag-of-audio-words feature representation using OPENXBOW. Note, figure adapted from (Schmitt and Schuller 2017)

representations have been shown to produce state-of-the-art results in a variety of audio classification tasks, in particular, continuous speech-based emotion prediction (Schmitt et al. 2016) and depression detection (Joshi et al. 2013). Visual features can be treated in a manner to form *Bag-of-Video-Words* (BoVW) representations, which have also been shown to be well suited to tasks such as emotion prediction (Ringeval et al. 2017) and sentiment analysis (Cummins et al. 2018a).

As already mentioned, a bagged feature representation is a sparse multi-dimensional feature space formed by the quantisation (bagging) of LLDs. This bagging procedure follows four keys steps (Fig. 9.2). First, is the extraction of the LLD features, using the openSMILE toolkit for instance. Second, a codebook is learnt from designated training data. In OPENXBOW the codebook can be performed by one of two methods, either by k-means clustering, or by a 'random' sampling of all LLDs in the training set. The default random sampling implemented in OPENXBOW is the initialisation step of the k-means clustering; the codebook entries are selected subsequently by a methodology which favours vectors which are farther away—as determined via Euclidean distance—from those already selected. The third step is the assignment of each frame-level LLD vector to words in the formed codebook. This assignment is achieved by identifying the vector in the codebook which returns the minimum Euclidean distance with the input vector. OPENXBOW also allows assignments to multiple codebook vectors; i.e., the quantisation is performed with respect to a predefined number of 'close' codebook entries again, as determined using the Euclidean distance. Finally, by counting the number of assignments for each word, a fixed length histogram (bag) representation of an audio clip is generated. This histogram represents the frequency of each identified word in a given input instance (Schmitt and Schuller 2017).

OPENXBOW supports two input and three output format types: *.csv* and *.arff* inputs and outputs are supported, while the LIBSVM file format is also supported but as an output only. OPENXBOW also offers a range of pre-processing options including normalisation and standardisation of the LLD's and post-processing options for normalising and reweighting the extracted histogram. The interested reader is referred to the OPENXBOW software repository for full details.[3]

In terms of mobile sensing, bagged feature representations offer several key advantages. Bagged representations are sparse by nature, with this property essentially

---

[3]https://github.com/openXBOW/openXBOW.

controlled by two parameters: the *codebook size* (*Cs*) which determines the dimensionality of the final feature vectors, and the *number of assignments* (*Na*) which determines the number of words assigned to an audio instance. Sparsity offers the advantages of being more computationally efficient, it costs less memory, and it is quicker to perform multiplication operation on sparse representations. Bagged feature spaces are also time-invariant; a single fixed length vector is generated regardless of the length of the input utterance; in its post-processing options, OPENXBOW allows users to normalise the extracted histogram with respect to the length of the input file. Further, recent research has shown that BoAW representations are more robust to noise in the LLD feature space (Cummins et al. 2017a); it is speculated that this is due to the quantisation step which allows this technique to have a degree of tolerance to small perturbations in the input data.

Finally, increasing consumer ethical awareness and legal frameworks, such the recently introduced *General Data Protection Regulation* (GDPR) in the European Union, has pushed issues relating to privacy to the forefront of current challenges in mobile and IoT based health applications (Kargl et al. 2019). In this regard, the OPENXBOW reduces the risks associated with recording, transmitting and analysing data. Primarily, the quantisation step undertaken when bagging features can be considered privacy conserving as the resulting time-invariant histogram of occurrences cannot be used to reconstruct the input space. Moreover, no critical information seems to be lost in this transformation, in general through tuning *Cs* and *Na* a BoAW representation can be found which outperforms more conventional speech and audio representations, e.g., Cummins et al. (2017b); Schuller et al. (2018a, b); Schuller et al. (2017).

9.4 DEEPSPECTRUM

Fuelled by high profile examples in popular media, interest in artificial intelligence applications, in particular, deep learning, has never been stronger. Deep learning itself is a particular subset of machine learning algorithms containing a vast set of interconnected nodes whose structure is inspired by the structure and function of neurons in the human brain. A deep learning model can be considered a multi-layered pipeline of non-linear transformations, capable of representing highly complex decision functions while maintaining a high degree of generalisability. Advances in deep learning have undoubtedly been responsible for dramatic increases in system robustness and accuracy in a number of audio and speech applications. However, due in part to securing adequate amounts of reliable training data, deep learning has not had the same dominating effect in speech-based health detection systems commonly associated with mobile sensing applications (Cummins et al. 2018b).

Deep learning can, however, be leveraged in these applications through techniques such as transfer learning and unsupervised feature extraction. The use of neural networks, or other machine learning paradigms, as a feature extractor is commonly known as *representation learning*. Handcrafted features, such as those extracted using

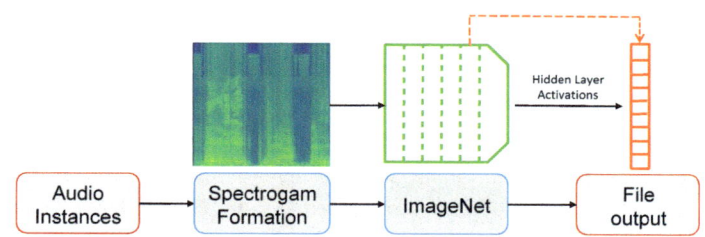

Fig. 9.3 An overview of extracting DEEPSPECTRUM features. Audio samples are converted to spectrogram images and then feed into a pre-train image classification Convolutional Neural Networks (CNNs); the activations of fully connected layers are then used as a feature representation for subsequent classification

OPENSMILE, have come about through engineers, or scientist somewhat subjectively select the relevant prosodic or acoustic property to extract; the extraction process is typically based on well-established theorems which can take decades to refine. Representation learning, on the other hand, is the development of techniques which enable machines to learn the discriminative characteristics of raw data or LLDs automatically. Learnt representations have many desirable properties, in particular, they can be easily adapted to suit a change in system requirements, such as different input data or classification targets (Bengio et al. 2013).

Convolutional Neural Networks (CNNs) have been continually proven adept at representation learning tasks. CNNs usually contain a combination of convolutional (filtering), pooling layers and non-linear activations which can learn a hierarchy of different feature representations, from broad in the initial layers to more task-specific in the later layers (LeCun et al. 2015). Indeed, it is standard practice in image processing to use the activations of different layers of so-called *imageNets* CNNs, trained on over a million images (Krizhevsky et al. 2012; Simonyan and Zisserman 2014), to perform feature extraction in other visual classification tasks. A somewhat recent fascinating result from computational speech analysis shows that imageNets can also produce meaningful speech, and audio feature representations for tasks such as emotion recognition and bipolar mood state, for example, (Cummins et al. 2017a; Ringeval et al. 2018) and audio features for tasks such as irregular heart sound detection (Ren et al. 2018). This approach, utilising pre-trained for image nets to extract audio feature representation from spectrogram images, is known as DEEPSPECTRUM feature extraction (Amiriparian et al. 2017a, b).

The DEEPSPECTRUM toolkit[4] is a python-based repository, for this extraction procedure. The core steps involved in DEEPSPECTRUM feature extraction are essentially choosing a suitable image representation of a speech segment and choosing which pre-trained image net to use as the feature extractor (Fig. 9.3). Currently, in terms of image formation, the DEEPSPECTRUM repository supports spectrogram, chromagram and mel-spectrum representations, while any image net in the Caffe-TensorFlow

---

[4]https://github.com/DeepSpectrum/DeepSpectrum.

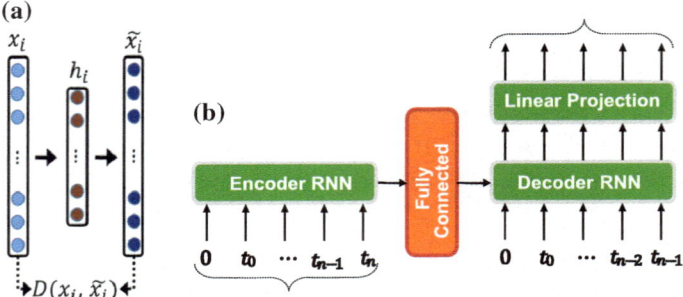

Fig. 9.4 AUDEEP's strengths lie in the recurrent sequence to sequence autoencoder technology implemented in the toolkit. The goal of autoencoders **a** is to create a robust representation of a data instance by passing it in through a neural network which attempts to recreate its input at its output. The inclusion of recurrent layers in the encoder and decoder parts if the network **b** enables the network to learn a fixed length representation of a variable length signal. Note, figure adapted from (Amiriparian et al. 2017a, b; Freitag et al. 2018)

toolkit,[5] including ResNet 50, VGG 16, GoogLeNet, or AlexNet, are supported. Deep spectrum features are given as the system output in either *.csv* or *.arff* formats. Fine-grained control and support, especially in relation to spectrogram formation, is available within the toolkit; the interested reader is referred to the repository for full details.

9.5 AUDEEP

The AUDEEP toolkit is a Python/tensorflow based repository for deep unsupervised representation learning from acoustic data.[6] One limitation of the CNN feature learning approach is difficulty in handling variable length data, as commonly found when working with audio data. The most common method for dealing with this is to divide audio files into overlapping chunks resulting in a lack of temporal continuity between the chunks. AUDEEP negates this limitation by implementing a recurrent sequence to sequence autoencoder for deep unsupervised representation learning (Amiriparian et al. 2017a, b; Freitag et al. 2018). Autoencoders are a specific type of neural networks which are designed to reconstruct its input data at its output while attempting to learn a robust 'compressed' representation of the data in its hidden layers (Fig. 9.4a). When implemented with recurrent neural networks, which can encode temporal data, it is possible to learn a fixed length representation of a variable length signal. Within AUDEEP, the sequence to sequence autoencoder is augmented with a fully connected layer between a multilayered *encoder* RNN, and another multilayered *decoder* RNN (Fig. 9.4b). Once trained, the activations of this fully connected

---

[5]https://github.com/ethereon/caffe-tensorflow.

[6]https://github.com/auDeep/auDeep.

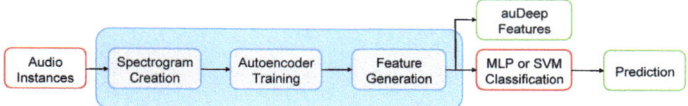

Fig. 9.5   To create AUDEEP feature representations, spectrograms are first extracted from raw audio files. Then, a recurrent sequence to sequence autoencoder is trained on the extracted spectrograms and the learnt representation of each instance is extracted as a feature vector. If the user has supplied instance labels, a classifier can then be trained and evaluated on the extracted features. Note, figure adapted from (Amiriparian et al. 2017a, b; Freitag et al. 2018)

layer are used as a feature representation. AUDEEP representations have been used in tasks such as acoustic scene classification (Amiriparian et al. 2017a, b), irregular heart sound detection (Amiriparian et al. 2018), atypical and self-assessed affect detection as well as infant cry recognition (Schuller et al. 2018a, b).

Taking a high-level overview of the feature extraction processes, the AUDEEP toolkit has four key steps (Fig. 9.5). First, the input data is converted into spectrogram or mel-spectrum based representations. Subsequently, the deep recurrent sequence to sequence autoencoder is trained on these spectra. Features can then be generated by passing the spectra back through the network and extracting the activations of the fully connected layer. Finally, AUDEEP utilises the Scikit-Learn toolkit (Pedregosa et al. 2011), to implement a *Multilayer Perceptron* (MLP) or *Support Vector Machine* (SVM). Alternatively, the user can choose to output the learnt representations as either a *.csv* or a *.arff* file. AUDEEP's command line interface offers a fine level of control over each step, with a large selection of related hyperparameters to tune.

The large amounts of hyperparameters, and the need to train deep neural networks themselves with numbers of parameters up into the millions, is a current limitation of applying AUDEEP in mobile sensing applications. However, the same can currently be stated of practically any deep learning system; put simply, the potential of DNNs has yet to be fully exploited in embedded systems as many state-of-the-art neural networks paradigms require too much memory to fit in on-chip storage (Han et al. 2015b; Zhu and Gupta 2017). Further, they have high computation demands to run efficiently in embedded devices (Chen et al. 2017; Han et al. 2016; Lane et al. 2016). This is discussed further in Sect. 9.7.

9.6   END2YOU

A recent exciting development in deep neural network technologies is the advancement of *end-to-end learning*. End-to-end learning takes the feature representation paradigm a step further by learning the entire classification pipeline directly from *raw* data instances (Trigeorgis et al. 2016). In doing so, hand-engineered features are removed entirely from an analysis pipeline; instead, the network learns its own robust feature representation specific for the task it is being trained on. End-to-end

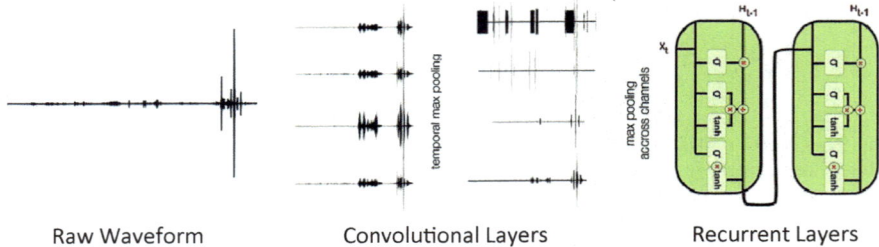

Raw Waveform   Convolutional Layers   Recurrent Layers

Fig. 9.6 An example of an end-to-end deep learning system pipeline. First, the raw data is divided into blocks of 40 ms, the raw instances are then fed into convolutional layers which learn a suitable feature representation, and recurrent layers are used to capture relevant temporal dynamics from the learnt features. Note, figure adapted from (Trigeorgis et al. 2016)

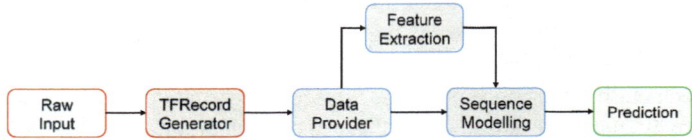

Fig. 9.7 The END2YOU pipeline consists of the *.tfrecord* generator, the data provider that feeds either the data through a feature extraction model (Convolutional Neural Network) into a sequencing model (Recurrent Neural Network) or, if desired, directly into the sequencing model, and finally the prediction model to generate the final output. Note, figure adapted from (Tzirakis et al. 2018)

systems typically consist of *convolutional* layers to learn robust feature representations followed by *recurrent* layers to leverage the temporal dynamics inherent in time-series data such as speech (Fig. 9.6). A fully connected layer can then be used to map between the output of the recurrent layers to the output score space. In terms of mobile sensing, notwithstanding current issues relating to deploying deep neural nets on embedded devices, is they offer a potential solution to alleviate privacy concerns by deploying an entire classification tool on a smart device.

The END2YOU toolkit,[7] implemented in python and based on Tensorflow, provides users with the capability of performing end-to-end learning on audio and visual systems (Tzirakis et al. 2018). Taking a generalised overview of the system, the workflow consists of two phases (Fig. 9.7). First, the raw signals are converted into Tensorflow's *.tfrecord* format; after that, it is possible to train and evaluate a model. During this second phase, a data provider unit reads the *.tfrecords* files, and feeds this information into the convolutional layers for feature extraction; this extracted information is then passed on to the recurrent layers for sequence modelling with the final step being the mapping to the prediction space using a fully connected network. Note that, it is also possible to pass the raw *.tfrecords* directly into a recurrent network.

The default audio system provided in END2YOU is composed of a 2-block of convolution max-pooling layers; the first layer has 40 filters of size 20 with a pooling

---

[7]https://github.com/end2you/end2you.

of size 2; the second layer has 40 filters of size 40 with a pooling layer of size 10. The default visual feature extractor is the ResNet-50 network (He et al. 2016). For both modalities, the default recurrent system is a 2-layer *Gated Recurrent Unit* network with 64 units. For further details on the toolkit, the interested reader is referred to (Tzirakis et al. 2018).

End-to-end learning has been used in a diverse array of tasks such as emotion detection (Trigeorgis et al. 2016), snore sound recognition, cold, and flu detection (Schuller et al. 2017), as well as irregular heart sound, or typical and self-assessed affect and crying detection (Schuller et al. 2018a, b). It has also been used to profile physiological signals, such as electrocardiogram and electrodermal activity for emotion detection (Keren et al. 2017). On large emotion detection databases, end-to-end learning has achieved state-of-the-art performances (Trigeorgis et al. 2016). However, the advantages of end-to-end learning are not as clear when tested on smaller datasets, and often unbalanced in terms of class distributions, typically found in health sensing applications. It has been speculated this effect is related to the training of an end-to-end model essentially on the statistics available in raw data representations. In this regard, smaller and unbalanced datasets may contain insufficient variation, especially concerning the underrepresented class, for robust end-to-end modelling (Schuller et al. 2017).

9.7 Challenges and Future Work Directions

Due to the richness of health information embedded into speech signals, paralinguistic analysis should be considered a core information stream in any health-based mobile sensing platform. However, to herald in the next generation of speech-enabled smart devices, future work directions need to focus around enabling deep learning approaches in smart and embedded devices. This challenge is highly non-trivial as current state-of-the-art deep learning solutions are large and computationally demanding models. Furthermore, for many somatic/physical and mental health conditions that may be of interest in a mobile sensing platform, there are challenges related to collecting and labelling sufficient amounts of data to adequately train deep learning solutions.

Low Resource Neural Networks: The contemporary speech processing approaches discussed in the preceding sections, DEEPSPECTRUM, AUDEEP and END2YOU, are based on neural networks. Such systems, which are capable of producing state-of-the-art results, have connection numbers measuring in the millions, potentially require hundreds of megabytes and create substantial data movement operation to support their computation. This makes them difficult to operate in the low resource and low power settings commonly associated with mobile sensing applications (Chen et al. 2017; Han et al. 2016; Lane et al. 2016). A growing research direction within neural networks is the development of approaches which can take in a large network and optimise it until it is executable on a low resource device (Table 9.1). Many of these approaches focus on reducing the *memory footprint*, how

Table 9.1 A comparison of
the advantages and
disadvantage of different
approaches to create a
(computationally) low
resource neural network

Approach	Advantages	Disadvantages
Network pruning	Applicable on pre-trained networks	Loss of precision
Mathematical optimisation	Applicable on pre-trained networks	Complexity of optimisation approaches
Knowledge distillation	A purpose built, smaller network is learnt	Increase in training time and effort
Spiking neural networks	Highly energy efficient	Require specialist hardware
Reconfigurable chips	Overcome memory bottlenecks	Require specialist hardware

much memory is required to store and run a network, and the *computational complexity*, the number of required calculations and their precision, of a network while at the same time preserving its level of accuracy.

One such approach is *network pruning* which not only reduces the size of the model, but also as it reduced the number of free learning parameters can counteract overfitting (Cheng et al. 2017). The general assumption of this approach is that neural nets have a lot of redundant weights which do not considerably contribute to the performance of the network. Pruning removes such weights entirely; weight sharing between connections with similar weights can also be used to reduce the footprint of the network (Chen et al. 2015). The pruning of large dense models has even been shown to improve the performance while reducing the model size by up to 80% (Zhu and Gupta 2017). Another promising approach is *low precision* neural networks (Fig. 9.8); these are networks in which the associated parameter values are not stored in a high precision—float 32 for example—but rather in a simplified quantised representation (Gupta et al. 2015). For example, activations can be stored in either a binary, -1 or $+1$, or ternary -1, 0 or $+1$ format as an integer value. These approaches have been shown to considerably reduce memory and computation costs whilst maintaining reasonable accuracy (Alemdar et al. 2017; Han et al. 2015a).

A range of other approaches can be found in the relevant literature. *Mathematical optimisation* and *knowledge distillation* are two such approaches. Optimisation techniques such as low-rank approximations of weight vectors (Nakkiran et al. 2015) have been shown to decrease network size by 75%. Thereby, knowledge distillation approaches aim to train a smaller 'student' network which is able to perform the same task as a larger 'teacher' network (Cheng et al. 2017). Furthermore, a range of specialised hardware solutions are being explored such as neuromorphic computing running so-called *spiking neural networks* (Schuman et al. 2017) and reconfigurable chips such as *Field-Programmable Gate Arrays* (FPGA) which aim to overcome memory read/write bottlenecks common in conventional computing architecture

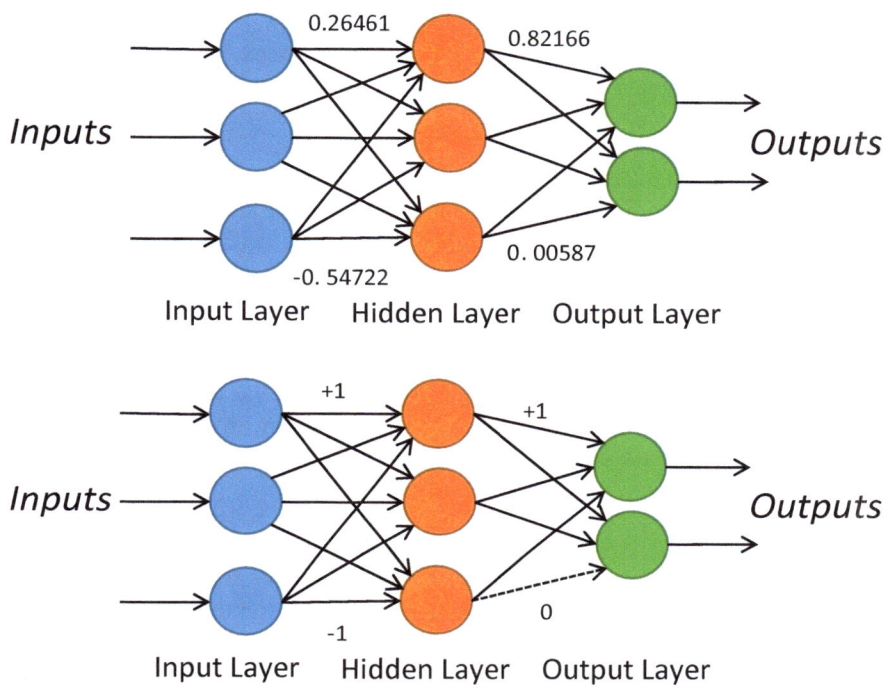

Fig. 9.8 An example of quantising weights in a neural network to form a low precision ternary neural network

(Ota et al. 2017). Google is also developing Tensor processing units specifically programmed for their TensorFlow framework (Jouppi et al. 2018), and are others developing *Neuro Processing Units* (NPUs).

As can be inferred from the above, low-resource network solutions are an active and growing area of research with many promising directions. Novel optimisation techniques and network topologies have shown promise in a range of learning tasks, but to date have yet to be fully explored for mobile sensing applications. In achieving such networks, researchers in mobile sensing applications cannot focus purely on system accuracy as their only evaluation metric—aspects such as runtime efficiency, noise robustness, generalisation, and energy consumption need to be considered as well during system design and development. Finally, developing networks increases the likelihood of systems being able to run offline, increasing user privacy and reducing energy consumption concerns associated with transmission bandwidth, both of which are core considerations for a robust mobile sensing unit.

Data Sparsity Challenges: A commonly occurring theme associated with speech databases for machine learning efforts into somatic/physical and mental health conditions is data sparsity. The corpora used to develop such systems are often small, both in terms of the total amount of speech data available and in the number of speakers present, and imbalance meaning they often contain more speech of individuals in a mild health state and noticeable less speech from individuals in a severe state. Such conditions make it difficult to train a model, especially deep learning models, which are well capable of generalising onto unseen data. However, ongoing research into *intelligent labelling* and *data augmentation* paradigms have the potential to alleviate these challenges.

Conventional data labelling paradigms require a large amount of time and resources to produce sufficiently reliable labels for machine learning. However, techniques such as semi-supervised learning, active learning, and cooperative learning have been shown to reduce these efforts (Zhang et al. 2017). These approaches leverage a smaller set of labelled data to annotate a larger dataset using machine learning techniques with minimal human involvement. Such methods have been shown to aid a wide range of speech-based classification techniques—in particular emotion and social signal recognition (Zhang et al. 2017). However, many of the advantages of many of these approaches have only been displayed in 'in laboratory' settings, and further experiments and advancements are needed to realise their suitability for mobile sensing platforms.

In this regard, future works should focus on integration, based around label confidence measures, in adaptive systems. Systems should be equipped with adequate confidence measures to perform co-operative adaptive learning; self-labelling data the system is highly confident about and interacting with the user to label data instances it has low confidence about but thinks are relevant to label and then utilising both sources of data to update its learning parameters to better match the user's need. Such systems should also focus on approaches, such as *few-shot networks*, that can efficiently solve new learning tasks requiring only a few instances of training data (Snell et al. 2017; Triantafillou et al. 2018).

Another potential method to address data sparsity concerns is data augmentation using techniques such as *Generative Adversarial Networks* (GANs) (Goodfellow et al. 2014; Han et al. 2018; Salimans et al. 2016) to generate new samples (Donahue et al. 2018; Saito et al. 2018). GANs consists of two neural networks: a *generative model* (generator) and a *discriminative model* (discriminator) which are set to compete against each other in a zero-sum game. During this game, the objective of the generator is to convert input noises from a simple distribution into realistic samples to fool the discriminator, while the objective of the discriminator is to distinguish between generated samples as being either '*real*' or '*fake*' (Fig. 9.9). The overall objective of the entire GAN network is to compel both models to continuously improve their methods until the generator is able to synthesise realistic data instances. Results published in (Deng et al. 2017), highlight the promise of GANs to mobile sensing applications. The authors demonstrated that GAN-based methods could be used to synthesise new training instances to aid a speech-based classification system to detect if a child was typically developing or had a developmental disorder such as an Autism spectrum condition.

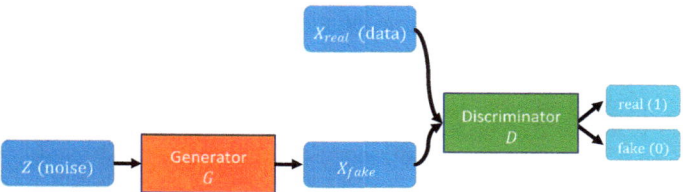

Fig. 9.9 The framework of *Generative Adversarial Network* (GAN); the objective of the generator is to fool the discriminator into misclassifying the generated samples while the discriminator is trained to accurately distinguish whether a given sample has been produced by the generator or drawn from a real data distribution. Training the networks concurrently in an adversarial setting allows the generator eventually produce realistic samples

9.8 Conclusion

As in most areas of intelligent signal analysis, deep learning has an unquestionable impact in computational speech analysis. While toolkits such as OPENSMILE and OPENXBOW are widely used due to their proven ability to generate robust audio and speech representations, there is a growing research interest in deep learning solutions, particularly for representation learning solutions that automatically and objectively learn meaningful feature representations. Results gained with the DEEPSPECTRUM, AUDEEP and END2YOU toolkits highlight the potential of representation learning across a range of computational paralinguistic tasks.

The toolkits discussed in this chapter, in combination with ubiquitous computing devices, place speech signals as a core modality for consideration in mobile health care solutions. However, to realise this, potential research efforts are needed to overcome challenges relating to the massive computational costs associated with current state-of-the-art neural networks and the data sparsity challenge associated in securing adequate data to train highly reliable and robust models. A combination of low resource deep learning paradigms, intelligent labelling solutions, and state-of-the-art data augmentation has the potential to foster in a new generation of mobile and intelligent patient-driven health care devices.

Acknowledgements This research has received funding from the Innovative Medicines Initiative 2 Joint Undertaking under grant agreement No. 115902. This joint undertaking receives support from the European Union's Horizon 2020 research and innovation programme and EFPIA.

References

Alemdar H, Leroy V, Prost-Boucle A, Pétrot F (2017) Ternary neural networks for resource-efficient AI applications. In: 2017 international joint conference on neural networks (IJCNN). Anchorage, AK, USA, pp 2547–2554

Amiriparian S, Freitag M, Cummins N, Schuller B (2017) Sequence to sequence autoencoders for unsupervised representation learning from audio. In: Proceedings of the detection and classification of acoustic scenes and events 2017 workshop. IEEE, Munich, Germany, pp 17–21

Amiriparian S, Gerczuk M, Ottl S, Cummins N, Freitag M, Pugachevskiy S, Schuller B (2017) Snore sound classification using image-based deep spectrum features. In: Proceedings of INTER-SPEECH 2017, 18th annual conference of the international speech communication association. ISCA, Stockholm, Sweden, pp 3512–3516. https://doi.org/10.21437/Interspeech.2017-434

Amiriparian S, Schmitt M, Cummins N, Qian K, Dong F, Schuller B (2018) Deep unsupervised representation learning for abnormal heart sound classification. In: Proceedings of the 40th annual international conference of the IEEE engineering in medicine and biology society, EMBC 2018. IEEE, Honolulu, HI, pp 4776–4779. https://doi.org/10.1109/EMBC.2018.8513102

Bengio Y, Courville A, Vincent P (2013) Representation learning: a review and new perspectives. IEEE Trans Pattern Anal Mach Intell 35:1798–1828

Bone D, Lee CC, Chaspari T, Gibson J, Narayanan S (2017) Signal processing and machine learning for mental health research and clinical applications. IEEE Signal Process Mag 34:189–196

Chang C-C, Lin C-J (2011) LIBSVM: a library for support vector machines. ACM Trans Intell Syst Technol 2:1–27

Chen W, Wilson J, Tyree S, et al (2015) Compressing neural networks with the hashing trick. In: 32nd international conference on machine learning. ICML, Lille, France, pp 2285–2294

Chen Y, Krishna T, Emer JS, Sze V (2017) Eyeriss: an energy-efficient reconfigurable accelerator for deep convolutional neural networks. IEEE J Solid-State Circuits 52:127–138. https://doi.org/10.1109/JSSC.2016.2616357

Cheng Y, Wang D, Zhou P, Zhang T (2017) A survey of model compression and acceleration for deep neural networks

Cummins N, Amiriparian S, Hagerer G, Batliner A, Steidl S, Schuller B (2017a) An image-based deep spectrum feature representation for the recognition of emotional speech. In: Proceedings of the 25th ACM international conference on multimedia, MM 2017. ACM, Mountain View, CA, pp 478–484

Cummins N, Amiriparian S, Ottl S, Gerczuk M, Schmitt M, Schuller B (2018a) Multimodal bag-of-words for cross domains sentiment analysis. In: Proceedings 43rd IEEE international conference on acoustics, speech, and signal processing, ICASSP 2018. IEEE, Calgary, Canada

Cummins N, Baird A, Schuller BW (2018b) The increasing impact of deep learning on speech analysis for health: challenges and opportunities. Methods Spec Issue Transl data Anal Heal informatics

Cummins N, Scherer S, Krajewski J, Schnieder S, Epps J, Quatieri TF (2015) A review of depression and suicide risk assessment using speech analysis. Speech Commun 71:10–49

Cummins N, Schmitt M, Amiriparian S, Krajewski J, Schuller B (2017b) You sound ill, take the day off: classification of speech affected by upper respiratory tract infection. In: Proceedings of the 39th annual international conference of the IEEE engineering in medicine and biology society, EMBC 2017. IEEE, Jeju Island, South Korea, pp 3806–3809

Cunningham S, Green P, Christensen H, Atria J, Coy A, Malavasi M, Desideri L, Rudzicz F (2017) Cloud-based speech technology for assistive technology applications (CloudCAST). Stud Health Technol Inform 242:322–329

Deng J, Cummins N, Schmitt M, Qian K, Ringeval F, Schuller B (2017) Speech-based diagnosis of autism spectrum condition by generative adversarial network representations. In: Proceedings of the 7th international digital health conference, DH '17. ACM, London, UK, pp 53–57

Donahue C, McAuley J, Puckette M (2018) Adversarial audio synthesis

Eyben F, Scherer K, Schuller B, Sundberg J, André E, Busso C, Devillers L, Epps J, Laukka P, Narayanan S, Truong K (2016) The Geneva minimalistic acoustic parameter set (GeMAPS) for voice research and affective computing. IEEE Trans Affect Comput 7:190–202

Eyben F, Weninger F, Schuller FGB (2013) Recent developments in openSMILE, the munich open-source multimedia feature extractor. In: Proceedings of the 21st ACM international conference on multimedia, MM '13. ACM, Barcelona, Spain, pp 835–838

Eyben F, Wöllmer M, Schuller B (2010) openSMILE: the munich versatile and fast open-source audio feature extractor. In: Proceedings of the 18th ACM international conference on multimedia, MM '10. ACM, Firenze, Italy, pp 1459–1462. https://doi.org/10.1145/1873951.1874246

Fitch WT (2000) The evolution of speech: a comparative review. Trends Cogn Sci 4:258–267

Freitag M, Amiriparian S, Pugachevskiy S, Cummins N, Schuller B (2018) auDeep: unsupervised learning of representations from audio with deep recurrent neural networks. J Mach Learn Res 18:1–5

Goodfellow I, Pouget-Abadie J, Mirza M, Xu B, Warde-Farley D, Ozair S, Courville A, Bengio Y (2014) Generative adversarial nets. In: Ghahramani Z, Welling M, Cortes C, Lawrence ND, Weinberger KQ (eds) Advances in neural information processing systems 27. Curran Associates, Inc., pp 2672–2680

Gupta S, Agrawal A, Gopalakrishnan K, Narayanan P (2015) Deep learning with limited numerical precision. In: 32nd international conference on machine learning. ICML, Lille, France, pp 1737–1746

Hagerer G, Cummins N, Eyben F, Schuller B (2017) "Did you laugh enough today?"—{D}eep neural networks for mobile and wearable laughter trackers. In: Proceedings of INTERSPEECH 2017, 18th annual conference of the international speech communication association. ISCA, Stockholm, Sweden, pp 2044–2045

Hall M, Frank E, Holmes G, Pfahringer B, Reutemann P, Witten IH (2009) The WEKA data mining software: an update. ACM SIGKDD Explor Newsl 11:10–18

Han J, Zhang Z, Cummins N, Schuller B (2018) Adversarial training in affective computing and sentiment analysis: recent advances and prospectives. IEEE Comput Intell Mag Spec Issue Comput Intell Affect Comput Sentim Anal

Han S, Liu X, Mao H, Pu J, Pedram A, Horowitz MA, Dally WJ (2016) EIE: efficient inference engine on compressed deep neural network. In: 2016 ACM/IEEE 43rd annual international symposium on computer architecture (ISCA). IEEE, Seoul, South Korea, pp 243–254. https://doi.org/10.1109/ISCA.2016.30

Han S, Mao H, Dally WJ (2015a) Deep compression: compressing deep neural networks with pruning, trained quantization and Huffman coding

Han S, Pool J, Tran J, Dally W (2015b) Learning both weights and connections for efficient neural network. In: Cortes C, Lawrence ND, Lee DD, Sugiyama M, Garnett R (eds) Advances in neural information processing systems 28. Curran Associates, Inc., pp 1135–1143

He K, Zhang X, Ren S, Sun J (2016) Deep residual learning for image recognition. In: The IEEE conference on computer vision and pattern recognition (CVPR). Las Vegas, NV, pp 770–778

Istepanian RSH, Al-Anzi T (2018). m-Health 2.0: new perspectives on mobile health, machine learning and big data analytics. Methods

Jankowski S, Covello J, Bellini H, Ritchie J, Costa D (2014) The internet of things: making sense of the next mega-trend

Joshi J, Goecke R, Alghowinem S, Dhall A, Wagner M, Epps J, Parker G, Breakspear M (2013) Multimodal assistive technologies for depression diagnosis and monitoring. J Multimodal User Interfaces 7:217–228

Jouppi NP, Young C, Patil N, Patterson D (2018) A domain-specific architecture for deep neural networks. Commun ACM 61:50–59. https://doi.org/10.1145/3154484

Kargl F, van der Heijden RW, Erb B, Bösch C (2019) Privacy in Mobile Sensing. In: Baumeister H, Montag C (eds) Mobile sensing and digital phenotyping in psychoinformatics. Springer, Berlin

Keren G, Kirschstein T, Marchi E, Ringeval F, Schuller B (2017) End-to-end learning for dimensional emotion recognition from physiological signals. In: Proceedings 18th IEEE international conference on multimedia and expo, ICME 2017. IEEE, Hong Kong, PR China, pp 985–990

Krizhevsky A, Sutskever I, Hinton GE (2012) ImageNet classification with deep convolutional neural networks. In: Pereira F, Burges CJC, Bottou L, Weinberger KQ (eds) Advances in neural information processing systems. Curran Associates, Inc., pp 1097–1105

Lane ND, Bhattacharya S, Georgiev P et al (2016) DeepX: a software accelerator for low-power deep learning inference on mobile devices. 2016 15th ACM/IEEE International Conference on Information Processing in Sensor Networks (IPSN). Austria, Vienna, pp 1–12

LeCun Y, Bengio Y, Hinton G (2015) Deep learning. Nature 521:436–444. https://doi.org/10.1038/nature14539

Levelt WJM, Roelofs A, Meyer AS (1999) A theory of lexical access in speech production. Behav Brain Sci 22:1–38

Marchi E, Eyben F, Hagerer G, Schuller BW (2016) Real-time tracking of speakers' emotions, states, and traits on mobile platforms. Proceedings INTERSPEECH 2016, 17th annual conference of the international speech communication association. ISCA, San Francisco, CA, pp 1182–1183

Nakkiran P, Alvarez R, Prabhavalkar R, Parada C (2015) Compressing deep neural networks using a rank-constrained topology. Proceedings INTERSPEECH 2015, 16th annual conference of the international speech communication association. ISCA, Dresden, Germany, pp 1473–1477

O'Shaughnessy D (1999) Speech communications: human and machine, 2nd edn. Wiley-IEEE Press, Piscataway, NJ

Orozco-Arroyave JR, Hönig F, Arias-Londoño JD, Vargas-Bonilla JF, Daqrouq K, Skodda S, Rusz J, Nöth E (2016) Automatic detection of Parkinson's disease in running speech spoken in three different languages. J Acoust Soc Am 139:481–500

Ota K, Dao MS, Mezaris V, De Natale FGB (2017) Deep learning for mobile multimedia: a survey. ACM Trans Multimed Comput Commun Appl 13, 34:1–34:22. https://doi.org/10.1145/3092831

Pedregosa F, Varoquaux G, Gramfort A, Michel V, Thirion B, Grisel O, Blondel M, Prettenhofer P, Weiss R, Dubourg V et al (2011) Scikit-learn: machine learning in Python. J Mach Learn Res 12:2825–2830

Quatieri TF (2002) Discrete-time speech signal processing: principles and practice. Prentice Hall, Upper Saddle River, NJ

Ren Z, Cummins N, Pandit V, Han J, Qian K, Schuller B (2018) Learning image-based representations for heart sound classification. In: Proceedings of the 2018 international conference on digital health, DH'18. ACM, Lyon, France, pp 143–147. https://doi.org/10.1145/3194658.3194671

Ringeval F, Marchi E, Grossard C, Xavier J, Chetouani M, Cohen D, Schuller B (2016) Automatic analysis of typical and atypical encoding of spontaneous emotion in the voice of children. Proceedings INTERSPEECH 2016, 17th annual conference of the international speech communication association. ISCA, San Francisco, CA, pp 1210–1214

Ringeval F, Schuller B, Valstar M, Cowie R, Kaya H, Schmitt M, Amiriparian S, Cummins N, Lalanne D, Michaud A, et al (2018). AVEC 2018 workshop and challenge: bipolar disorder and cross-cultural affect recognition. In: Proceedings of the 2018 on audio/visual emotion challenge and workshop. pp 3–13

Ringeval F, Schuller B, Valstar M, Gratch J, Cowie R, Scherer S, Mozgai S, Cummins N, Schmitt M, Pantic M (2017). AVEC 2017: real-life depression, and affect recognition workshop and challenge. In: Proceedings of the 7th annual workshop on audio/visual emotion challenge, AVEC '17. ACM, Mountain View, CA, pp 3–9

Saito Y, Takamichi S, Saruwatari H (2018) Statistical parametric speech synthesis incorporating generative adversarial networks. IEEE/ACM Trans Audio Speech Lang Process 26:84–96. https://doi.org/10.1109/TASLP.2017.2761547

Salimans T, Goodfellow I, Zaremba W, Cheung V, Radford A, Chen X, Chen X (2016) Improved techniques for training GANs. In: Lee DD, Sugiyama M, Luxburg UV, Guyon I, Garnett R (eds) Advances in neural information processing systems. Curran Associates Inc, Barcelona, Spain, pp 2234–2242

Schmitt M, Ringeval F, Schuller B (2016) At the border of acoustics and linguistics: bag-of-audio-words for the recognition of emotions in speech. Proceedings INTERSPEECH 2016, 17th annual conference of the international speech communication association. ISCA, San Francisco, CA, pp 495–499

Schmitt M, Schuller B (2017) openXBOW—introducing the Passau open-source crossmodal bag-of-words toolkit. J Mach Learn Res 18:3370–3374

Schuller B (2017) Can affective computing save lives? Meet Mobile Health. IEEE Comput Mag 50:40

Schuller B, Batliner A (2013) Computational paralinguistics: emotion, affect and personality in speech and language processing. Wiley

Schuller B, Steidl S, Batliner A, Bergelson E, Krajewski J, Janott C, Amatuni A, Casillas M, Seidl A, Soderstrom M, Warlaumont A, Hidalgo G, Schnieder S, Heiser C, Hohenhorst W, Herzog M, Schmitt M, Qian K, Zhang Y, Trigeorgis G, Tzirakis P, Zafeiriou S (2017) The INTERSPEECH 2017 computational paralinguistics challenge: addressee, cold and snoring. In: Proceedings INTERSPEECH 2017, 18th annual conference of the international speech communication association. ISCA, Stockholm, Sweden, pp 3442–3446. https://doi.org/10.21437/Interspeech.2017-43

Schuller B, Steidl S, Batliner A, Hantke S, Hönig F, Orozco-Arroyave JR, Nöth E, Zhang Y, Weninger F (2015) The INTERSPEECH 2015 computational paralinguistics challenge: degree of nativeness, Parkinson's and eating condition. Proceedings INTERSPEECH 2015, 16th annual conference of the international speech communication association. ISCA, Dresden, Germany, pp 478–482

Schuller B, Steidl S, Batliner A, Marschik P, Baumeister H, Dong F, Pokorny FB, Rathner E-M, Bartl-Pokorny KD, Einspieler C, Zhang D, Baird A, Amiriparian S, Qian K, Ren Z, Schmitt M, Tzirakis P, Zafeiriou S (2018) The INTERSPEECH 2018 computational paralinguistics challenge: atypical and self-assessed affect, crying and heart beats. In: Proceedings INTERSPEECH 2018, 19th annual conference of the international speech communication association. ISCA, Hyderabad, India, pp 122–126. https://doi.org/10.21437/Interspeech.2018-51

Schuller B, Steidl S, Batliner A, Vinciarelli A, Scherer K, Ringeval F, Chetouani M, Weninger F, Eyben F, Marchi E, Mortillaro M, Salamin H, Polychroniou A, Valente F, Kim S (2013) The INTERSPEECH 2013 computational paralinguistics challenge: social signals, conflict, emotion, autism. In: Proceedings INTERSPEECH 2013, 14th annual conference of the international speech communication association. ISCA, Lyon, France, pp 148–152

Schuller B, Weninger F, Zhang Y, Ringeval F, Batliner A, Steidl S, Eyben F, Marchi E, Vinciarelli A, Scherer K, Chetouani M, Mortillaro M (2018b) Affective and behavioural computing: lessons learnt from the first computational paralinguistics challenge. Comput, Speech Lang, p 32

Schuman CD, Potok TE, Patton RM, Birdwell JD, Dean ME, Rose GS, Plank JS (2017) A survey of neuromorphic computing and neural networks in hardware

Simonyan K, Zisserman A (2014) Very deep convolutional networks for large-scale image recognition

Snell J, Swersky K, Zemel R (2017) Prototypical networks for few-shot learning. In: Guyon I, Luxburg UV, Bengio S, Wallach H, Fergus R, Vishwanathan S, Garnett R (eds) Advances in neural information processing systems 30. Curran Associates, Inc., pp 4077–4087

Triantafillou E, Larochelle H, Snell J, Tenenbaum J, Swersky KJ, Ren M, Zemel R, Ravi S (2018) Meta-learning for semi-supervised few-shot classification

Trigeorgis G, Ringeval F, Brückner R, Marchi E, Nicolaou M, Schuller B, Zafeiriou S (2016) Adieu features? End-to-end speech emotion recognition using a deep convolutional recurrent network. In: Proceedings 41st IEEE international conference on acoustics, speech, and signal processing, ICASSP 2016. IEEE, Shanghai, PR China, pp 5200–5204

Tsiartas A, Albright C, Bassiou N, Frandsen M, Miller I, Shriberg E, Smith J, Voss L, Wagner V (2017) Sensay analyticstm: a real-time speaker-state platform. 2017 IEEE international conference on acoustics, speech and signal processing, ICASSP '17. IEEE, New Orleans, LA, pp 6483–6582

Tzirakis P, Zafeiriou S, Schuller B (2018) End2You—the imperial toolkit for multimodal profiling by end-to-end learning

Young S, Evermann G, Gales M, Hain T, Kershaw D, Liu X, Moore G, Odell J, Ollason D, Povey D et al (2002) The HTK book. Cambridge Univ Eng Dep 3:175

Zhang Z, Cummins N, Schuller B (2017) Advanced data exploitation in speech analysis—an overview. IEEE Signal Process Mag 34:107–129. https://doi.org/10.1109/MSP.2017.2699358

Zhu M, Gupta S (2017) To prune, or not to prune: exploring the efficacy of pruning for model compression

Chapter 10
Passive Sensing of Affective and Cognitive Functioning in Mood Disorders by Analyzing Keystroke Kinematics and Speech Dynamics

Faraz Hussain, Jonathan P. Stange, Scott A. Langenecker, Melvin G. McInnis, John Zulueta, Andrea Piscitello, Bokai Cao, He Huang, Philip S. Yu, Peter Nelson, Olusola A. Ajilore and Alex Leow

Life is too sweet and too short to express our affection with just our thumbs. Touch is meant for more than a keyboard.
—Kristin Armstrong, Olympic cyclist

F. Hussain · J. Zulueta · O. A. Ajilore · A. Leow
Collaborative Neuroimaging Environment for Connectomics, University of Illinois, Chicago, USA
e-mail: farazh@uic.edu

J. Zulueta
e-mail: oajilore@uic.edu

O. A. Ajilore
e-mail: oajilore@uic.edu

A. Leow
e-mail: weihliao@uic.edu

J. P. Stange (✉)
Cognition and Affect Regulation Lab, University of Illinois, Chicago, USA
e-mail: jstange@uic.edu

S. A. Langenecker
University Neuropsychiatric Institute, University of Utah, Salt Lake City, UT, USA
e-mail: s.langenecker@hsc.utah.edu

M. G. McInnis
Heinz C. Prechter Bipolar Research Program, University of Michigan, Ann Arbor, MI, USA
e-mail: mmcinnis@med.umich.edu

A. Piscitello
Department of Electronics, Information and Bioengineering, Politecnico di Milano, Milan, Italy
e-mail: andrea1.piscitello@mail.polimi.it

B. Cao
Video Understanding Team, Applied Machine Learning, Facebook, Menlo Park, CA, USA
e-mail: caobokai@fb.com

© Springer Nature Switzerland AG 2019   161
H. Baumeister and C. Montag (eds.), *Digital Phenotyping and Mobile Sensing*,
Studies in Neuroscience, Psychology and Behavioral Economics,
https://doi.org/10.1007/978-3-030-31620-4_10

Abstract Mood disorders can be difficult to diagnose, evaluate, and treat. They involve affective and cognitive components, both of which need to be closely monitored over the course of the illness. Current methods like interviews and rating scales can be cumbersome, sometimes ineffective, and oftentimes infrequently administered. Even ecological momentary assessments, when used alone, are susceptible to many of the same limitations and still require active participation from the subject. Passive, continuous, frictionless, and ubiquitous means of recording and analyzing mood and cognition obviate the need for more frequent and lengthier doctor's visits, can help identify misdiagnoses, and would potentially serve as an early warning system to better manage medication adherence and prevent hospitalizations. Activity trackers and smartwatches have long provided exactly such a tool for evaluating physical fitness. What if smartphones, voice assistants, and eventually Internet of Things devices and ambient computing systems could similarly serve as fitness trackers for the brain, without imposing any additional burden on the user? In this chapter, we explore two such early approaches—an in-depth analytical technique based on examining meta-features of virtual keyboard usage and corresponding typing kinematics, and another method which analyzes the acoustic features of recorded speech—to passively and unobtrusively understand mood and cognition in people with bipolar disorder. We review innovative studies that have used these methods to build mathematical models and machine learning frameworks that can provide deep insights into users' mood and cognitive states. We then outline future research considerations and close by discussing the opportunities and challenges afforded by these modes of researching mood disorders and passive sensing approaches in general.

10.1 Introduction

Mood disorders take a sizable toll on the world's population, affecting more than 1 in 20 people annually and nearly 1 out of every 10 people over the course of their lifetime (Steel et al. 2014). Bipolar disorder, which alone accounts for at least 1% of years lived with disability globally (GBD 2017), is a mood disorder that causes patients to alternate between manic episodes of abnormally elevated mood and energy levels, and depressive episodes marked by diminished mood, interest, and energy (APA 2013). Compared to major depressive disorder (MDD), bipolar disorder can be harder to diagnose, and even when an accurate diagnosis is made, it is often

H. Huang · P. S. Yu
Department of Computer Science, University of Illinois, Chicago, USA
e-mail: hehuang@uic.edu

P. S. Yu
e-mail: psyu@cs.uic.edu

P. Nelson
College of Engineering, University of Illinois, Chicago, USA
e-mail: nelson@uic.edu

delayed. The depressive episodes in both disorders share the same diagnostic criteria, and it is known that individuals suffering from bipolar disorder on average spend more time in the depressive phase than in mania. In particular, bipolar disorder type II, a subtype which is differentiated by attenuated levels of mania-like symptoms (termed hypomania) is difficult to diagnose by non-specialists as it can be challenging to distinguish from recurring unipolar depression. The presence of mood episodes with mixed features, i.e., those that exhibit characteristics of both mania and depression, can further complicate the process of diagnosis (Phillips and Kupfer 2013).

10.1.1   Current State of Diagnosis and Monitoring of Bipolar Disorder

Clinical approaches to diagnosing and monitoring bipolar disorder usually start with careful history-taking by the clinician (detailed interviews with patients and their family members as well as probing for a family history of the disorder), followed by the frequent use of self- and clinician-administered rating scales that assess for a history of possible mania or hypomania in patients with depression. Even with these tools at their disposal, it is often difficult for clinicians to ascertain whether any noted changes in mood, sleep, or energy are within normal ranges—or whether they are evidence of, say, a manic/hypomanic episode (Wolkenstein et al. 2011). Achieving inter-rater reliability between administered assessments and scales poses its own challenges.

After a correct diagnosis has been made, monitoring of symptoms commonly relies upon self-reports that may include mood charting and self-ratings or clinician-rated scales. These scales can only assess the severity of symptoms experienced by the patients and cannot actually screen for mania or hypomania; patients in manic states also may not be cognizant of their manic symptoms, casting doubt on the validity of some of these assessments (NCCMH 2018).

Ecological momentary assessments (EMA) have been used for supplementary monitoring in mood disorders with varying degrees of success (Ebner-Priemer and Trull 2009; Asselbergs et al. 2016; Kubiak and Smyth 2019). Asselbergs and colleagues reported that the clinical utility of self-report EMA is too often limited by the heavy response burden that is imposed upon respondents—which can result in large dropout rates after an initial period of activity—and furthermore, that the predictive models constructed using unobtrusive EMA data were inferior to existing benchmark models.

In recent years, other techniques including neuroimaging (Phillips et al. 2008; Leow et al. 2013; Ajilore et al. 2015; Andreassen et al. 2018) and genomics (Hou et al. 2016; Ikeda et al. 2017) have also been used in attempts to discover biomarkers for bipolar disorder. Although they may not currently be feasible either for diagnosis or for monitoring on an individual level, in the near future we may begin finding immense value in these and related methods beyond their immediate research applications.

In addition to its affective components, bipolar disorder also influences cognitive ability (APA 2013). Among the most severely impaired domains of cognition are attention, working memory, and response inhibition (Bourne et al. 2013). These provide another avenue to further aid in distinguishing a possible diagnosis of bipolar disorder from other mood disorders and assessing its course and treatment.

10.1.2 Passive Sensing in Physical Health

Smartwatches, fitness trackers, and associated physical health and fitness apps in general have to a large extent enabled and encouraged users to self-manage chronic medical conditions and attempt to take better care of their physical health (Anderson et al. 2016; Canhoto and Arp 2017; Messner et al. 2019). The Apple Watch, for instance—which uses photoplethysmography to passively sense atrial fibrillation—and the associated Apple Heart Study (Turakhia 2018) have already been credited with saving several lives by alerting enrolled users to the onset of life-threatening conditions and directing them to seek immediate medical attention (Feng 2018; Perlow 2018).

10.1.3 What About Passive Sensing for Mental Health?

Portable sensors to track the health of the rest of the body have so far proven easier to develop than those that can track brain health. As yet, there are no portable functional magnetic resonance imaging (fMRI) scanners or brain-computer interfaces (BCI) that can be used to unobtrusively analyze brain functioning—although science fiction has proposed examples of each in the form of, respectively, cowboy hats that conduct brain scans to map wearers' cognition in television shows such as Westworld (Avunjian 2018) and biomechanical computer implants called neural lace in author Iain M. Banks' series *The Culture* (Banks 2002, 2010)—which science may in fact someday deliver instead in the shape of the startup Openwater's fMRI-replacing ski hats that are purportedly being designed to use infrared holography to scan oxygen utilization by the wearer's brain (Jepsen 2017; Clifford 2017) and implantable electronic circuits capable of neural communication such as those being developed by Neuralink and others (Fu et al. 2016; Chung et al. 2018; Sanford 2018).

Until these nascent technologies reach maturity, there is a need for passive sensing tools that can bridge the divide and perhaps eliminate the need for more onerous means of sensing altogether. Smartphones are already ubiquitous enough and offer a wide array of sensors, which when used in concert with mHealth and digital phenotyping tools, offer a greater degree of precision medicine tools to users, researchers, and healthcare providers than ever before. Indeed, the very use of smartphones, and mobile social networking apps in particular, has been found to be associated with structural and functional changes in the brain (Montag et al. 2017); the corollary that smartphone usage patterns can be used to quantify the presence of established biomarkers has also been explored by Sariyska and colleagues (2018) in their preliminary study examining the feasibility of probing molecular genetic variables corresponding to individual differences in personality and linked social traits, in this case a variant of the promoter gene coding for the oxytocin receptor, and simultaneously surveying their real world behavior as reflected by the myriad different ways and purposes for which they used their phones over the course of the day.

The proliferation of touchscreen smartphones with software keyboards has, at least for the time being, tilted the balance of telecommunications in favor of typed rather than spoken messages (Shropshire 2015). Combined with the data provided by a phone's accelerometer, gyroscope, and screen pressure sensors, keystroke dynamics can be used to build mathematical models of a person's mood and cognition based only on how, and not what, they type.

Voice itself, of course, remains a valuable instrument for gaining insight into the speaker's mood state, and will only continue to become more so as the tide eventually turns toward speech-based interactions with both intelligent voice assistants and other human users of connected devices. Using similar statistical modeling and machine learning techniques, the acoustic features of speech are just as well-suited for analysis as typing kinematics (Cummings and Schuller 2019).

As more and more computing comes to be offloaded from personal devices to Internet of Things (IoT) devices and the cloud, and ambient computing becomes the norm, we expect that techniques like keystroke analysis will be supplanted by speech meta-feature analysis, facial emotional recognition (for more information on FER software, see Chapter 3 by Geiger and Wilhelm in this book), and altogether novel passive mood sensing tools. For the present time, being aware of the increasing ubiquity of algorithms and their influence on data analytics, digital architectures and digital societies (Dixon-Román 2016), as well as mindful of the absence of a codified analog for the Hippocratic Oath in the current practice of artificial intelligence in medicine as well as other applications (Balthazar et al. 2018), we nevertheless stand to learn a great deal from leveraging currently used input methods to derive models for sensing users' inner states.

10.2 Mobile Typing Kinematics

In the first known study of its kind, researchers from the University of Illinois at Chicago (UIC), the University of Michigan, the Politecnico di Milano, Tsinghua University and Sun Yat-sen University used passively obtained mobile keyboard usage metadata to predict changes in mood state with significant degrees of accuracy. The team recruited subjects from the Prechter Longitudinal Study of Bipolar Disorder at the University of Michigan as part of the BiAffect-PRIORI consortium for its pilot study based on an Android mobile keyboard and associated app. After winning the grand prize in the Mood Challenge supported by Apple and sponsored by the New Venture Fund of Robert Wood Johnson Foundation, UIC is currently conducting a full-scale study on the iOS platform using an app based on the open source ResearchKit mobile framework, enrolling both people with bipolar disorder as well as healthy controls from the general population.

The BiAffect study (https://www.biaffect.com/) involves the installation of a companion app containing a custom keyboard that is cosmetically similar to the stock system keyboard. The app includes mood surveys; self-rating scales; and active tasks such as a the go/no-go task and the trail-making test (part B) to measure reaction time, response inhibition, and set-shifting as part of executive functioning—all overlapping domains of cognition identified by Bourne and colleagues (2013) to be the most affected in bipolar disorder.

All data collected by the app and keyboard are first encrypted and then transmitted and stored on secure study servers; these were hosted at UIC for the Android pilot app, whereas study management services are being supported by Sage Bionetworks for the ongoing iOS study with the data being hosted on their Synapse platform. The Android pilot phase, which has concluded data collection, involved the keyboard, trail making test, Hamilton Depression Rating Scale (HDRS), Young Mania Rating Scale (YMRS), and slider-based daily self-rating scales for mood, energy, impulsiveness, and speed of thoughts; the main iOS study included each of these [with the notable substitution of the clinician-rated HDRS and YMRS with the self-reported Patient Health Questionnaire (PHQ) and the Altman Mania Rating Scale, respectively] as well as a daily self-rating scale querying ability to focus, and the aforementioned reaction time task. Metadata collected for keyboard usage include timestamps associated with each keystroke, residence time on each key, intervals between successive keystrokes, and accelerometer readings over the course of all active typing sessions. The actual character corresponding to any given keypress is not recorded, apart from noting whether it was a backspace, alphanumeric, or symbol key. In addition to backspace usage, instances of autocorrection and autosuggestion invocations are also logged.

Table 10.1 summarizes the literature that has been published thus far based on analyses of data collected during the pilot phase of the study, which included 40 participants—between 9 and 20 of whose data were used for any given one depending on the number of days of metadata logged, diagnosis of the participant, and other requirements; up to 1,374,547 keystrokes and 14,237,503 accelerometer readings across 37,647 sessions were incorporated into some of the resulting models. Data collection for the main arm of the study is ongoing and has already resulted in over 8,000 cumulative hours of active typing sessions culled from across hundreds of users.

Zulueta and colleagues (2018) built mixed-effects linear models to correlate keyboard activity metadata during the week preceding when each pair of mood rating scales was administered to the corresponding HDRS and YMRS scores. A representative sampling of these metadata over several weeks from one study participant is illustrated in Fig. 10.1, while Fig. 10.2 compares the scores predicted by these models against actual scores for both mood scales. Autocorrect rates were positively correlated with depression scores, probably because error-awareness becomes impaired

Table 10.1 A summary of analyses published by researchers using data from the BiAffect study

Author	Analytical technique	Predictors used	Main findings
Zulueta et al. (2018)	Linear mixed-effects models (*Preferred over ANOVA in settings where measurements are made on clusters of related statistical units due to advantages in dealing with missing values*)	Average inter-key delay, backspace ratio, autocorrect rate, circadian baseline similarity, average accelerometer displacement, average session length, and session count	Keystroke activity was predictive of depressive, and to a lesser extent, manic symptoms. Specifically, accelerometer displacement, average inter-key delay, session count, and autocorrect rate were positively correlated with the HDRS scores, whereas accelerometer displacement was positively correlated and backspace rate negatively correlated with YMRS scores
Stange et al. (2018)	Multilevel models to evaluate predictiveness of instability metrics computed using the root mean square successive difference (*Specific models for each level of multilevel data, thereby modeling the non-independence of observations due to cluster sampling*)	Instability of EMA affective ratings and daily typing speed	Greater instability of mood during baseline EMA was predictive of future depressive symptoms, while instability of energy predicted future manic but not depressive symptoms. Instability of typing speed predicted prospective depressive but not manic symptoms. Models built using data gathered during only 5–7 days were as reliable and predictive as those assessing instability over longer time periods
Cao et al. (2017)	Comparison of late fusion based DeepMood LSTM-type GRU ML architecture with a multi-view RNN machine layer, factorization machine layer, or conventional fully connected layer against early fusion approaches (*Recurrent connections between machine learning layers allow modeling of nonlinear time series that, after training on sufficient data, can solve problems with prolonged temporal dependencies, such as linguistic, semantic, and topic inference tasks*)	Multiple representative views of the features of each typing session such as alphanumeric characters, special characters, and accelerometer values	Healthy people showed a wider range of variability in the time intervals between successive alphanumeric keypresses than people who were experiencing a mood disturbance. People in a manic state tend to hold down a keypress longer than people in a stable mood state, while depressed people pressed down on keys for shorter than average durations. The DMVM and DFM based architectures were the most predictive of depression scores, with prediction performances of 90.31% and 90.21%, respectively

(continued)

Table 10.1 (continued)

Author	Analytical technique	Predictors used	Main findings
Huang et al. (2018)	dpMood ML architecture based on early fusion, stacked CNNs to capture local typing dynamics and RNNs to capture temporal dynamics, and final predictions based on individual circadian calibrations (*More typically employed in computer vision models, CNNs outperform shallow architectures when predicting mental health related aspects from multiple data streams, while reaching at least comparable performance levels as predesignated architectures*)	Metadata for alphanumeric characters, including duration of keypress, time since last keypress, distance from the center of the last pressed key along both axes, and corresponding accelerometer values during active sessions	The proposed dpMood architecture incorporating CNNs, RNNs, early fusion and time-based calibration taken together outperformed any individual approach alone or in combination with just a few others. The integrated analysis of local patterns and temporal dependencies allowed for the isolation of variations in keyboard usage at different times of the day and from day to day over the course of the week, and the personalized calibration was sensitive enough to be able to distinguish between healthy controls and subjects with type I and type II bipolar disorder

Abbreviations EMA: ecological momentary assessment, *HDRS*: Hamilton Depression Rating Scale, *YMRS*: Young Mania Mania Rating Scale, *LSTM*: long short-term memory, *GRU*: gated recurrent unit, *ML*: machine learning, *DMVM*: DeepMood multi-view machine, *DFM*: DeepMood factorization machine, *DNN*: DeepMood neural network, *CNN* convolutional neural network, *RNN*: recurrent neural network

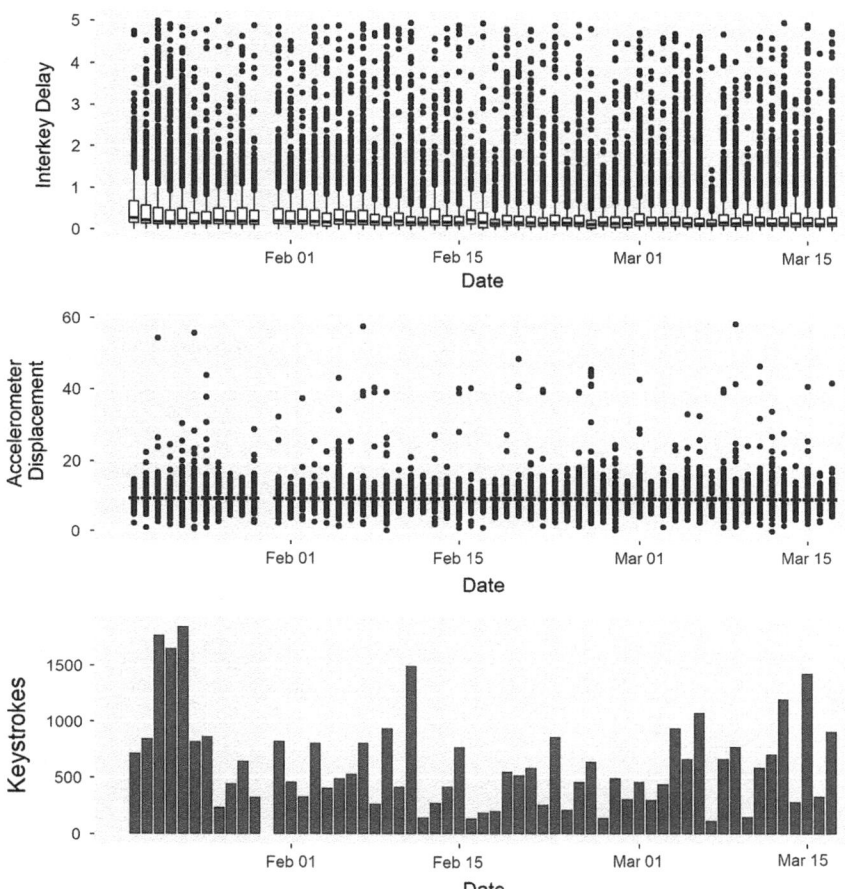

Fig. 10.1 An example of the deep personalized sensing possible with BiAffect showing the number of keystrokes, corresponding accelerometer readings, and the time between successive keypresses logged for an individual participant over the duration of the pilot study phase. Adapted from Zulueta et al. (2018)

when depressed (Fig. 10.3a). Backspace usage rate was found to be negatively correlated with higher mania scores, possibly because it is reflective of decreased self-monitoring and impaired response inhibition (Fig. 10.3b). Accelerometer activity was positively correlated with both depression and mania scores, possibly because study subjects were experiencing depression with mixed features or agitated/irritable depression. The trail making test is a standard neuropsychological assessment that measures processing speed and task-switching, which are both good indicators of cognitive functioning; Fig. 10.4 shows how typing kinematics data were just as predictive as trail making test results at establishing cognitive ability.

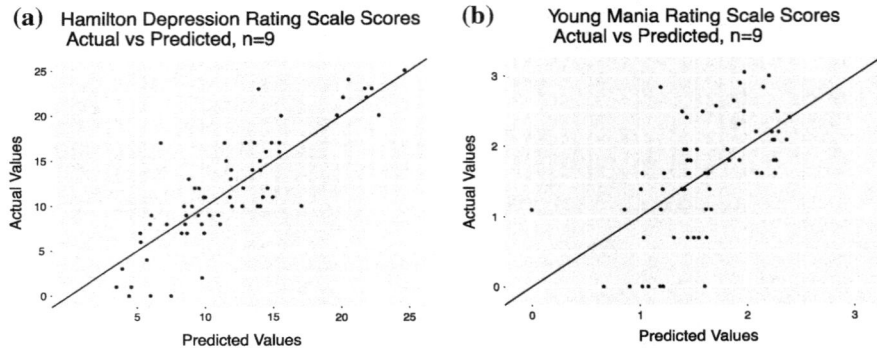

Fig. 10.2 Mixed effects modeling accounted for 63% of the variability of Hamilton Depression Rating Scale scores (Conditional $R^2 = 0.63$, Marginal $R^2 = 0.41$, $\chi^2_7 = 17.6$, $P = 0.014$). Ordinary least squares modeling accounted for 34% of the natural log of Young Mania Rating Scale scores (Multiple $R^2 = 0.34$, Adjusted $R^2 = 0.26$, $F_{7,56} = 4.1$, $P = 0.0011$). Adapted from Zulueta et al. (2018)

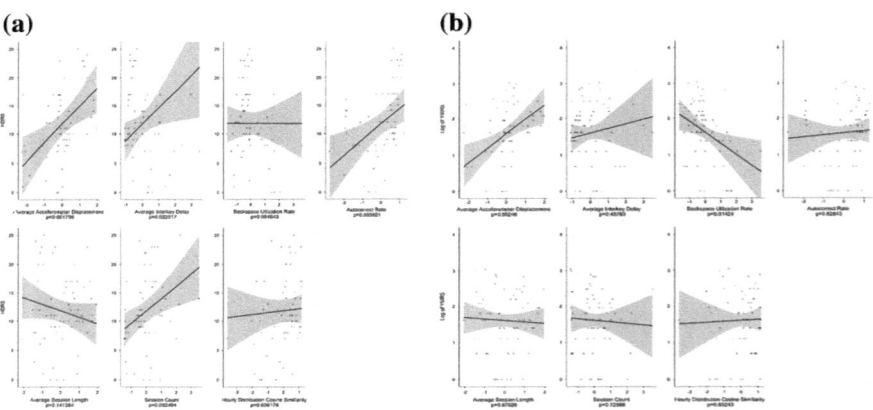

Fig. 10.3 Significant predictors for Hamilton Depression Rating Scale scores included accelerometer displacement ($P = 0.0017$), interkey delay ($P = 0.022$), autocorrect rate ($P = 0.0036$), and session count ($P = 0.0025$). Significant predictors for the natural log of the Young Mania Rating Scale scores include accelerometer displacement ($P = 0.003$) and backspace rate ($P = 0.014$). Adapted from Zulueta et al. (2018)

Stange et al. (2018) took a different approach by constructing multilevel models based on instability metrics calculated for EMA ratings and daily typing speeds (Fig. 10.5) using the root mean square of the successive differences (rMSSD)—a time-domain measure that takes into account the magnitude, frequency, and temporal order of intra-user fluctuations (Ebner-Priemer et al. 2009). Greater instability in baseline mood EMA ratings was significantly predictive of elevated future symptoms of both depression (Fig. 10.6a) and mania, whereas instability in energy ratings was

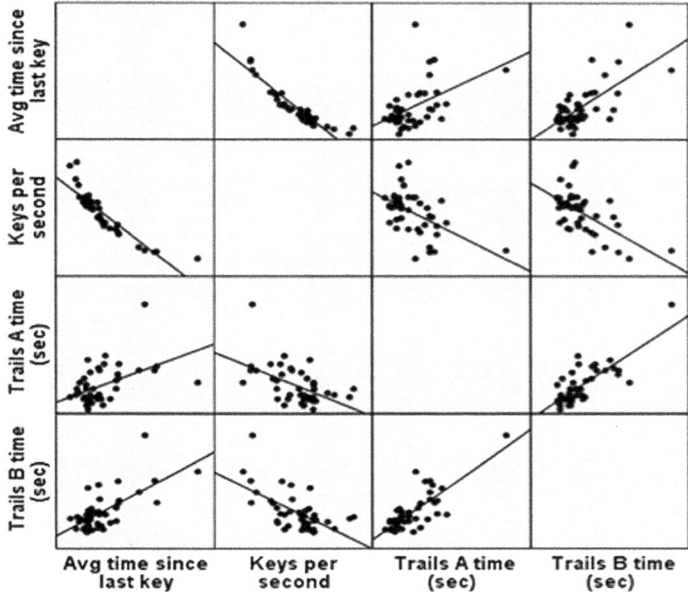

Fig. 10.4 Comparison of the predictiveness of keystroke data with that of trail making test results for assessing cognitive ability. Processing speed, as measured by trail taking test (part A) scores, was significantly correlated with average interkey delay (i.e., time since last key, $r = 0.5$, $p < 0.001$) and keys/second ($r = -0.54$, $p < 0.001$). Set shifting, as measured by trail taking test (part B) scores, was highly associated with average time since last key ($r = 0.68$, $p < 0.00001$) and keys/second ($r = -0.62$, $p < 0.00001$). Adapted from Zulueta et al. (2018)

predictive of future mania but not depression; other affective EMA ratings were not found to be significantly predictive of either. Typing speed instability was predictive of elevated prospective symptoms of depression (Fig. 10.6b) but not of mania. Interestingly, as little as one week of data provided levels of predictiveness comparable to data collected over durations of time longer than 5–7 days, perhaps because this time period is a representative enough snapshot to capture day-to-day typing variability (Fig. 10.7). Turakhia and colleagues (2019) have subsequently gone on to demonstrate the feasibility of exploiting variability in similar irregular noncontinuous datastreams to identify, predict, and prevent potential serious episodes—atrial flutters and fibrillations in the case of their app- and wearable-based study on cardiac arrhythmia.

Cao and colleagues (2017) were among the first to model keystroke dynamics data using deep learning. Their method, DeepMood, consisted of comparing the predictive performance of a multi-view machine layer architecture (Fig. 10.8) to that of other late fusion approaches such as factorization and conventional fully connected layers as well as early fusion strategies like tree boosting systems, linear support vector machines, and logistic ridge regression models. For the uninitiated, a review on current applications of deep neural networks in the field of psychiatry by Durstewitz

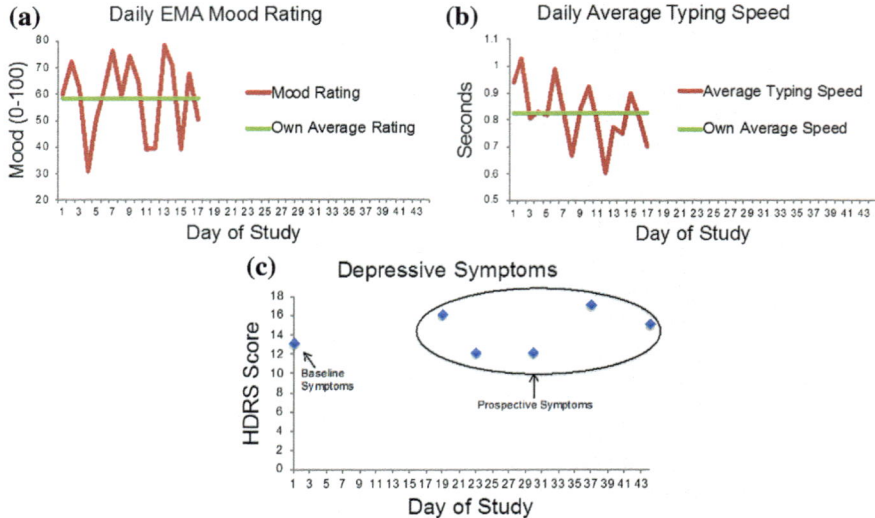

Fig. 10.5 An individual participant's **a** self-rated ecological momentary assessment scores, **b** passively collected daily typing speeds and **c** baseline and future course of depression symptom severity. Adapted from Stange et al. (2018) and reproduced with permission from the publisher

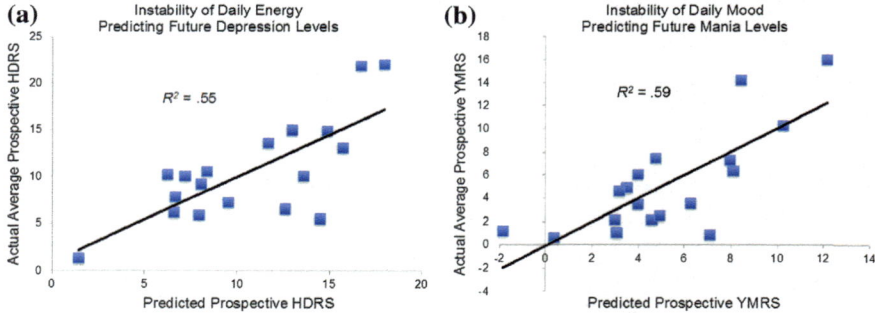

Fig. 10.6 Comparison of actual scores with those predicted by multilevel instability models for an individual participant's. **a** Hamilton Depression Rating Scale and **b** Young Mania Rating Scale. Adapted from Stange et al. (2018) and reproduced with permission from the publisher

et al. (2019) may serve as a primer. DeepMood's early fusion approaches align each of the data views—alphanumeric characters, special characters, and accelerometer values—with their associated timestamps (Fig. 10.9), and then immediately concatenate the multi-view time series per session. However, this does not take into proper account unaligned features in certain views, such as special characters, that do not have corresponding data points from other views like acceleration or inter-key distance. This shortcoming is addressed by the late fusion approach, in which each of the multi-view series is first modeled separately by a recurrent neural network

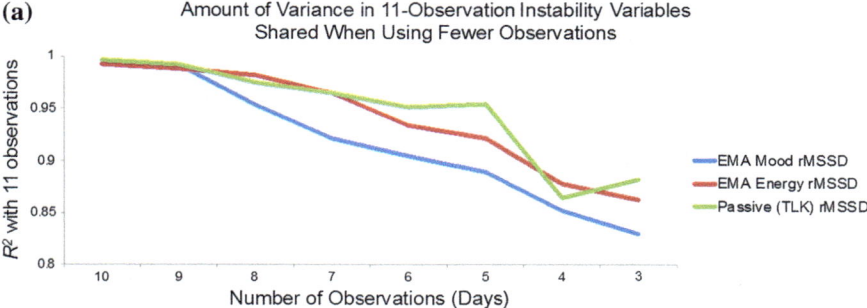

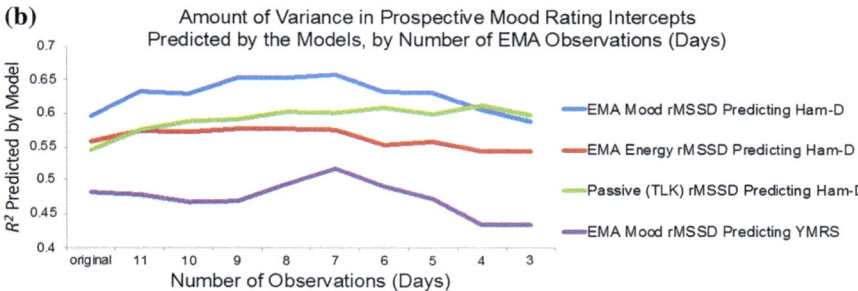

Fig. 10.7 **a** Reliability of active and passive assessments of instability depending on number of days of assessment. **b** Predictive utility of active and passive assessments of instability depending on number of days of assessment. Adapted from Stange et al. (2018) and reproduced with permission from the publisher

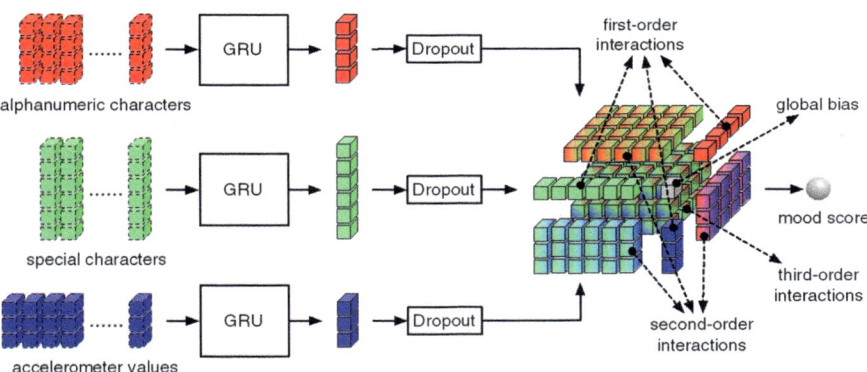

Fig. 10.8 DeepMood machine learning architecture with a multi-view machine layer for late data fusion. Adapted from Cao et al. (2017)

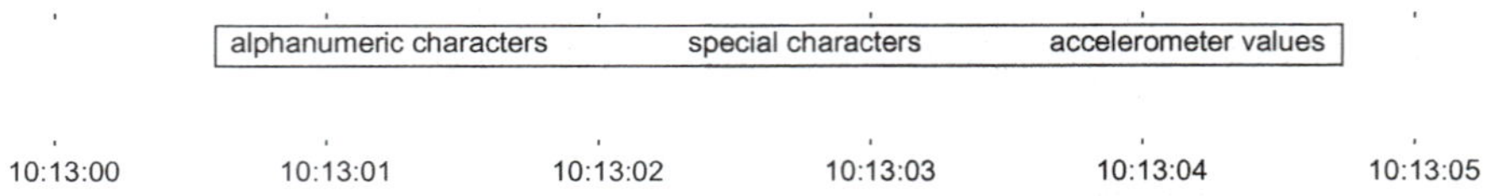

alphanumeric characters
special characters
accelerometer values
10:13:00
10:13:01
10:13:02
10:13:03
10:13:04
10:13:05

Fig. 10.9 A representative sample of the multi-view metadata collected in a time series. Adapted from Cao et al. (2017) and reproduced with permission from the publisher

Fig. 10.10 Comparison of the improvements in accuracy of different DeepMood architectural approaches over the course of successive training epochs. Adapted from Cao et al. (2017) and reproduced with permission from the publisher

Accuracy
1.0
0.9
0.8
0.7
0.6
0.5
0.4
0
100
200
300
400
500
Epoch
DMVM
DFM
DNN
XGB

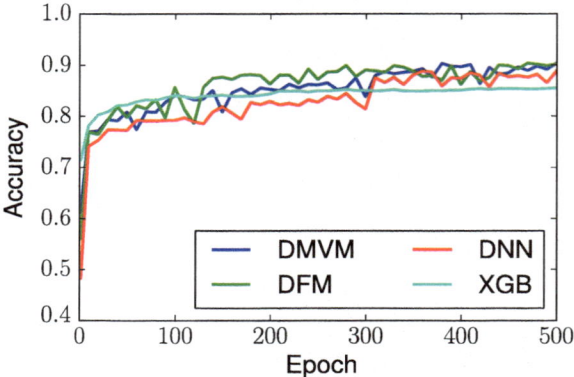

(RNN), and then fused in the next stage by analyzing first-, second-, and third-order interactions between each view's output vectors. Cao and colleagues established that their late fusion approach significantly outperformed early fusion in the ability to predict mood disturbances and their severity (Fig. 10.10), with the multi-view machines demonstrating the highest rate of accuracy at 90.31% followed by the factorization machines at 90.21%.

In a subsequent analysis, Huang et al. (2018) found that an early fusion approach integrating both convolutional and recurrent deep architectures and incorporating users' circadian rhythms allowed their model, dpMood, to attain even greater predictive performance as well as make more precise personalized mood predictions that took into fuller account an individual's biological clock and unique typing patterns. Their approach consisted of using convolutional neural networks (CNNs) that focused on temporal dynamics to analyze local features in typing kinematics over small periods of time, in conjunction with a special type of RNN called a gated recurrent unit (GRU) to model longer-term time-related dynamics (Fig. 10.11). GRUs address the vanishing gradient problem—the inherent inability of simpler RNNs to effectively learn those parameters that only cause very small changes in the neural network's output—and moreover have fewer parameters than comparable ameliorative approaches, allowing them to perform better on smaller datasets (Cho et al. 2014) such as the keystroke kinematics collected by BiAffect. This early fusion approach allowed for the alignment of features from multiple views to include additional information about temporal relationships between these data points that would otherwise

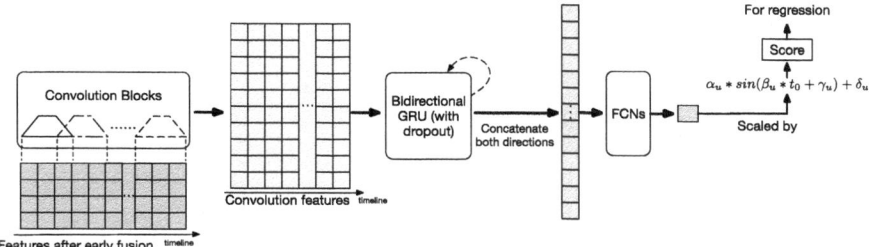

Fig. 10.11 dpMood machine learning architecture based on early data fusion stacked CNNs and GRUs, and time-based calibrations. Adapted from Huang et al. (2018) and reproduced with permission from the publisher

be lost in late fusion models. In the final analysis, the proposed dpMood architecture with the best predictive performance and the lowest regression error rate was the one that made combined use of both CNNs and RNNs to learn local patterns as well as temporal dependencies, learned each user's individual circadian rhythm, and retained accelerometer values that had no contemporaneous alphanumeric keypresses by filling the unaligned alphanumeric features with zero values instead of dropping unaligned accelerometer values altogether. Accelerometric and time-based analyses elucidated both daily (Figs. 10.12 and 10.13) and hourly (Fig. 10.14) variations in keyboard use, with the notably smaller Z-axis accelerations that help pinpoint when a phone is being typed on from a supine position having been observed more predominantly in the evenings (Fig. 10.14c) and on weekends (Fig. 10.13d). Modeling individuals' circadian rhythms as a sine function with parameters automatically learned by gradient descent algorithms and backpropagation resulted in one of these parameters conspicuously clustering based on the subjects' diagnoses, permitting dpMood to successfully classify users as participants with bipolar I disorder, those with bipolar II disorder, or healthy controls (Fig. 10.15). These sophisticated

Fig. 10.12 Distribution of daily typing hours visualized as a 7 day × 24 h matrix. Adapted from Huang et al. (2018) and reproduced with permission from the publisher

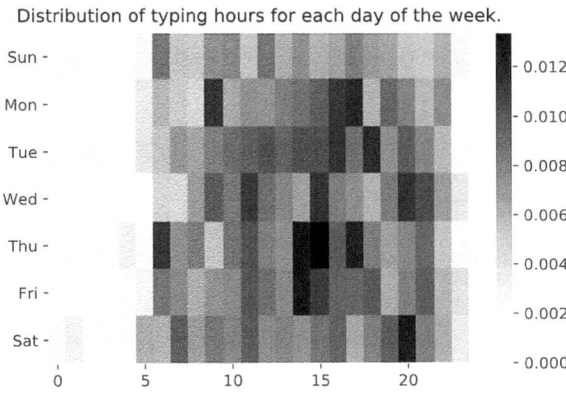

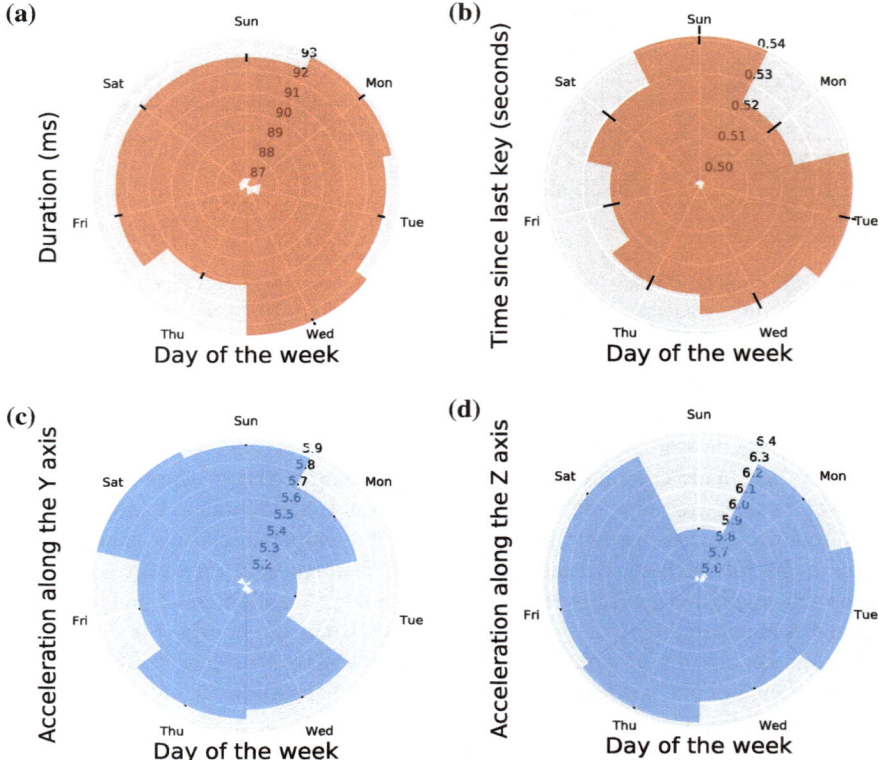

Fig. 10.13 Day-to-day fluctuations over the course of a week in **a** duration of a keypress, **b** time between successive keypresses, **c** acceleration along Y-axis, and **d** acceleration along Z-axis. Adapted from Huang et al. (2018) and reproduced with permission from the publisher

techniques can combine to provide extraordinarily insightful mood-sensing tools to users and precision medicine practitioners alike.

Preliminary analysis of study participants' performance on the go/no-go task has indicated that reaction times vary both within and between individuals (Fig. 10.16a) as well as continue to change over time (Fig. 10.16b); variations in daily typing patterns in BiAffect users have been found to correlate with their performance on the go/no-go task, and concurrent analyses of both data streams are now under way to examine their interrelationships and interactions with mood and cognition as well.

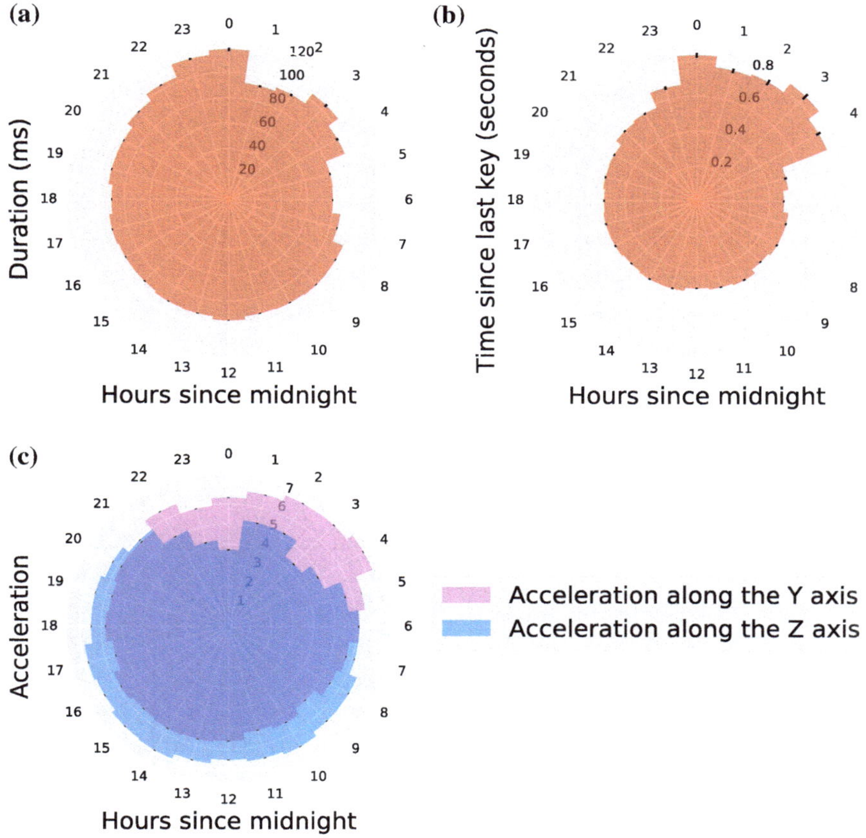

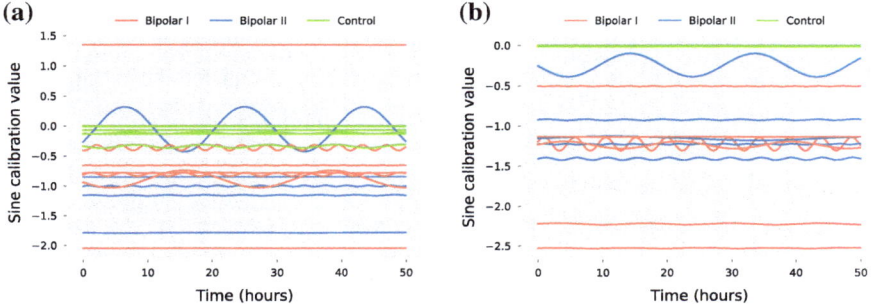

Fig. 10.14 Circadian rhythm mediated fluctuations in **a** duration of a keypress, **b** time between successive keypresses, and **c** acceleration along Y- and Z-axes. Adapted from Huang et al. (2018) and reproduced with permission from the publisher

Fig. 10.15 Visualizations of each individuals' calibration sine functions for **a** Hamilton Depression Rating Scale scores and **b** Young Mania Rating Scale scores. Adapted from Huang et al. (2018) and reproduced with permission from the publisher

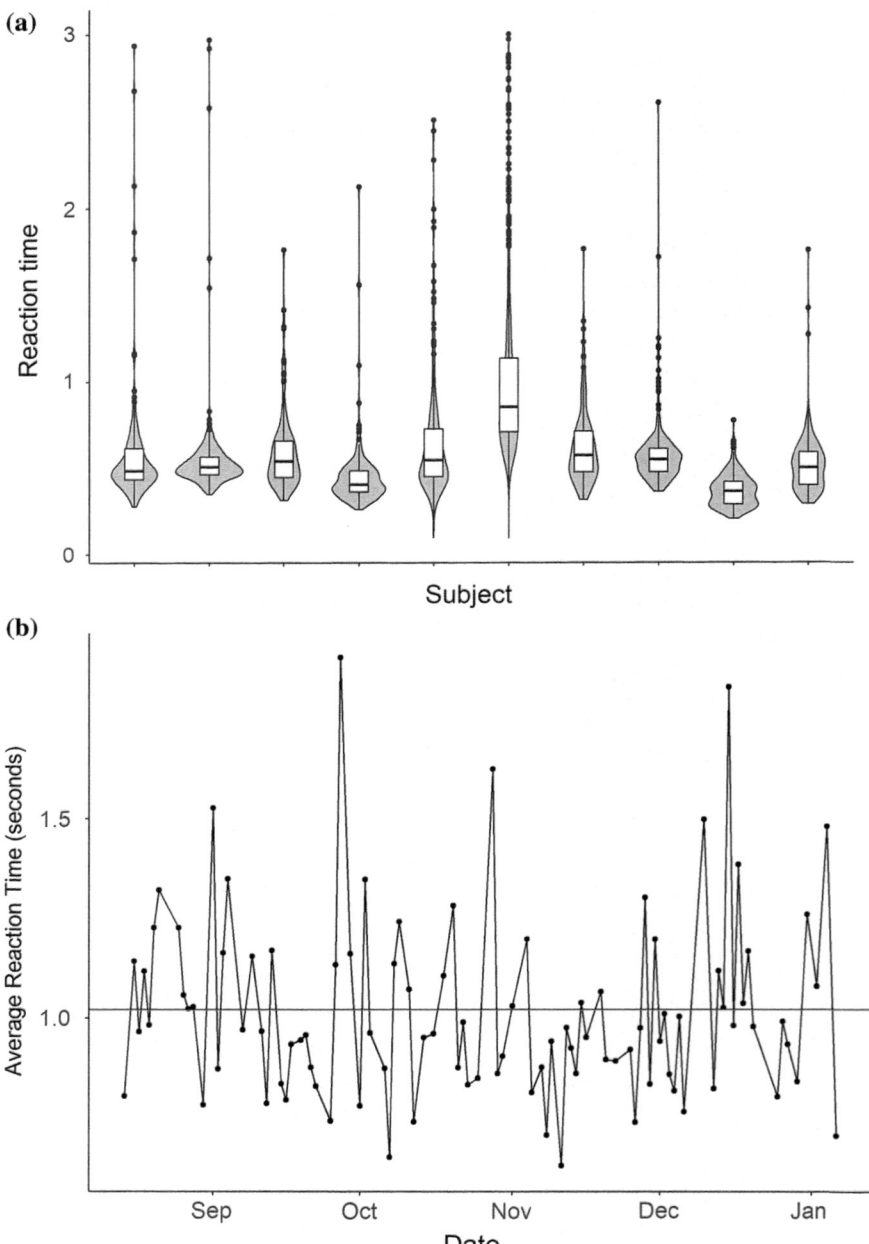

Fig. 10.16 **a** Go/no-go reaction time varies between and within individuals. **b** Average reaction time changes over the course of time

10.3 Speech Dynamics

Research on keystroke kinematics was inspired by the work of colleagues at the University of Michigan's Heinz C. Prechter Bipolar Research Program on the Predicting Individual Outcomes for Rapid Intervention (PRIORI) project, which is based on analyzing voice patterns in participants enrolled in the longest longitudinal research study of bipolar disorder; BiAffect aims to infer mood from typing metadata just as PRIORI does from the acoustic meta-features of speech. Participants were enrolled in the PRIORI study for an average of 16 to 48 weeks and were provided a rooted Android smartphone with a preinstalled secure recording application that captured audio of the participant's end of every phone call. Study staff called participants weekly to administer HDRS and YMRS mood assessments; these calls were labeled separately from personal calls. The dataset has accumulated over 52,000 recorded calls totaling above 4,000 h of speech from 51 participants with bipolar disorder and 9 healthy controls.

Karam et al. (2014) used a support vector machine (SVM) classifier to perform participant-independent modeling of segment- and low-level features extracted by the openSMILE audio signal processing toolkit, and were able to separate euthymic speech from hypomanic and depressed speech using an average of 5–8 judiciously selected features. In a later study, Gideon et al. (2016) used a declipping algorithm to approximate the original audio signal, and performed noise-robust segmentation to improve inter-device audio recording comparability. Rhythm features were classified using multi-task SVM analysis, then transformed into call-level features, and finally Z-normalized either globally or individually by subject. Declipping and SVM classification was found to increase the performance of manic but not depressive predictiveness, whereas segmentation and normalization significantly increased both. Khorram et al. (2016) captured subject-specific mood variations using i-vectors, and utilized a speaker-dependent SVM to classify both these i-vectors as well as rhythm features. Fusion of the subject-specific model—using unlabeled personal calls—with a population-general system enabled significantly improved predictive performance for depressive symptoms compared to the earlier approach used by Gideon and colleagues (2016). Khorram et al. (2018) went on to develop an 'in the wild' emotion dataset collating valence and activation annotations made by human raters drawing only upon the acoustic characteristics, and not the spoken content, of recordings from both personal and assessment calls.

Ongoing analyses, confounding challenges, and proposed solutions related to voice analysis have been outlined in a concise review by the PRIORI team (McInnis et al. 2017); their current focus is to isolate elements in the speech signal that are most strongly correlated with incipient disturbances in mood, enabling the development of on-device analytical systems without compromising limited mobile phone battery life.

10.4 Future Directions

The eventual goal of these projects is to be able to generate an early warning signal when changes in users' patterns of typing, speech, and behavior identify them to be at risk for an imminent manic or depressive episode. This would allow for just-in-time adaptive interventions that can circumvent or at least minimize the acuteness of the episode and any resulting cases of hospitalization, medication adjustment, or self-harm (Rabbi et al. 2019).

It has not escaped our attention that these passive sensing techniques can have applications in conditions other than bipolar disorder and indeed beyond just mood disorders; we have been investigating the use of a voice-enabled intelligent agent that are responsive to users' mood in order to provide emotionally aware education and guidance to patients with comorbid diabetes and depression (Ajilore 2018), as well as exploring the effectiveness of keystroke dynamics modeling in disparate conditions ranging from neurodegenerative processes such as Alzheimer's disease to cirrhotic sequelae such as hepatic encephalopathy.

The BiAffect keyboard has not only proven extremely adept at enabling digital phenotyping of its users' affective and cognitive states, but is also sensitive enough to their unique typing patterns that it can serve as an effective behavior-based biometric user identification and authentication tool. Sun et al. (2017) created DeepService, a multi-view multi-class deep learning method which is able to use data collected by the BiAffect keyboard to identify users with an accuracy rate of over 93% without using any cookies or account information. Until recently, the use of keystroke kinematics in hardware personal computer keyboards had been limited to similar continuous authentication applications, but physical keyboard sensing techniques are now expanding in scope to include identifying and measuring digital biomarkers as well (Samzelius 2016).

Mindful of the myriad potential concerns related to user privacy, data security and ethical implications inherent in the mass development and deployment of such applications, as well as in drawing conclusions based on findings generated using a relatively small number of smartphone users from a handful of geographic regions (Lovatt and Holmes 2017; Martinez-Martin and Kreitmair 2018), and remaining particularly cognizant of the clinical imperative to only use those methods informed by established transtheoretical frameworks—the overarching lack of which may have led to the current replication crisis in psychology and the medical sciences (Muthukrishna and Henrich 2019)—the research teams investigating BiAffect data streams have endeavored to adopt a deliberately paced approach that harmonizes the latest developments in cognitive science, psychological theory, nosology, and treatment with state-of-the-art deep learning techniques and statistical methods. By paying close attention to safeguarding the individual privacy and protected health information of its users, and by adopting the most transparent possible model of sharing research techniques and findings in order to prioritize the use of digital phenotyping data for ethical medical applications, the BiAffect platform has been built on the twin paradigms of open source and open science as an invitation to collaborators from around the world to replicate, validate, amend or correct our hypotheses.

Perhaps one day we will all sport brain scanning ski caps that tell us how we feel, and install BCI implants to communicate wordlessly with our gadgets and with one another, while our IoT devices infer our emotions by analyzing our behavior at a distance; in the meantime, there is already no dearth of data streams readily available for passively mining users' mood, cognition, and much more with greater preservation of privacy and potential for predictiveness.

Acknowledgements We are grateful to the Robert Wood Johnson Foundation, the Prechter Bipolar Research Fund, Apple, Luminary Labs, and Sage Bionetworks, all of whom have helped enable much of the research discussed in this chapter. Jonathan P. Stange was supported by grant K23MH112769 from NIMH.

References

Ajilore O (2018) A voice-enabled diabetes self-management program that addresses mood—the DiaBetty experience. In: American Diabetes Association's 78th Scientific Sessions, Orlando, FL, USA

Ajilore O, Vizueta N, Walshaw P et al (2015) Connectome signatures of neurocognitive abnormalities in euthymic bipolar I disorder. J Psychiatr Res 68:37–44

American Psychiatric Association (2013) Diagnostic and statistical manual of mental disorders (DSM-5®). American Psychiatric Publishing, Arlington, VA, USA

Anderson K, Burford O, Emmerton L (2016) Mobile health apps to facilitate self-care: a qualitative study of user experiences. PLoS ONE 11(5):e0156164

Andreassen O, Houenou J, Duchesnay E et al (2018) 121. Biological insight from large-scale studies of bipolar disorder with multi-modal imaging and genomics. Biol Psychiat 83(9):S49–S50

Asselbergs J, Ruwaard J, Ejdys M et al (2016) Mobile phone-based unobtrusive ecological momentary assessment of day-to-day mood: an explorative study. J Med Internet Res 18(3):e72

Avunjian N (2018) 'Westworld' cognition cowboy hats are a step up from a real science tool (inverse). USC Leonard Davis School of Gerontology. http://gero.usc.edu/2018/06/20/westworld-cognition-cowboy-hats-are-a-step-up-from-a-real-science-tool-inverse/

Balthazar P, Harri P, Prater A et al (2018) Protecting your patients' interests in the era of big data, artificial intelligence, and predictive analytics. J Am Coll Radiol 15(3, Part B):580–586

Banks IM (2002) Look to windward. Simon and Schuster

Banks IM (2010) Surface detail. Orbit

Bourne C, Aydemir Ö, Balanzá-Martínez V et al (2013) Neuropsychological testing of cognitive impairment in euthymic bipolar disorder: an individual patient data meta-analysis. Acta Psychiat Scand 128(3):149–162

Canhoto AI, Arp S (2017) Exploring the factors that support adoption and sustained use of health and fitness wearables. J Mark Manag 33(1–2):32–60

Cao B, Zheng L, Zhang C et al (2017) Deepmood: modeling mobile phone typing dynamics for mood detection. In: Proceedings of the 23rd ACM SIGKDD international conference on knowledge discovery and data mining. ACM, pp 747–755

Cho K, van Merrienboer B, Gulcehre C et al (2014) Learning phrase representations using RNN encoder-decoder for statistical machine translation. eprint arXiv:1406.1078

Chung JE, Joo HR, Fan JL et al (2018) High-density, long-lasting, and multi-region electrophysiological recordings using polymer electrode arrays. bioRxiv:242693

Clifford C (2017) This former Google[X] exec is building a high-tech hat that she says will make telepathy possible in 8 years. This former Google[X] exec is building a high-tech hat that she says will make telepathy possible in 8 years. https://www.cnbc.com/2017/07/07/this-inventor-is-developing-technology-that-could-enable-telepathy.html

Cummings N, Schuller BW (2019) Advances in computational speech analysis for mobile sensing. In: Baumeister H, Montag C (eds) Mobile sensing and psychoinformatics. Berlin, Springer, pp 141–159

Dixon-Román E (2016) Algo-Ritmo: more-than-human performative acts and the racializing assemblages of algorithmic architectures. Cult Studies? Crit Methodol 16(5):482–490

Durstewitz D, Koppe G, Meyer-Lindenberg A (2019) Deep neural networks in psychiatry. Mol Psychiatry

Ebner-Priemer UW, Eid M, Kleindienst N et al (2009) Analytic strategies for understanding affective (in)stability and other dynamic processes in psychopathology. J Abnorm Psychol 118(1):195–202

Ebner-Priemer UW, Trull TJ (2009) Ecological momentary assessment of mood disorders and mood dysregulation. Psychol Assess 21(4):463–475

Feng CH (2018) How a smartwatch literally saved this man's life and why he wants more people to wear one. South China Morning Post. https://www.scmp.com/lifestyle/health-wellness/article/2145681/how-apple-watch-literally-saved-mans-life-and-why-he-wants

Fu T-M, Hong G, Zhou T et al (2016) Stable long-term chronic brain mapping at the single-neuron level. Nat Methods 13:875

Gideon J, Provost EM, McInnis M (2016) Mood state prediction from speech of varying acoustic quality for individuals with bipolar disorder. In: 2016 IEEE international conference on acoustics, speech and signal processing (ICASSP), 20–25 March 2016, pp 2359–2363

Global Burden of Disease Collaborative Network (2017) Global Burden of Disease study 2016 (GBD 2016) results. Institute for Health Metrics and Evaluation (IHME) Seattle, United States

Hou L, Bergen SE, Akula N et al (2016) Genome-wide association study of 40,000 individuals identifies two novel loci associated with bipolar disorder. Hum Mol Genet 25(15):3383–3394

Huang H, Cao B, Yu PS et al (2018) dpMood: exploiting local and periodic typing dynamics for personalized mood prediction. Paper presented at the IEEE International Conference on Data Mining

Ikeda M, Takahashi A, Kamatani Y et al (2017) A genome-wide association study identifies two novel susceptibility loci and trans population polygenicity associated with bipolar disorder. Mol Psychiatr 23:639

Jepsen ML (2017) Open Water Internet Inc. Optical imaging of diffuse medium. U.S. Patent No. 9,730,649,

Karam ZN, Provost EM, Singh S et al (2014) Ecologically valid long-term mood monitoring of individuals with bipolar disorder using speech. In: 2014 IEEE international conference on acoustics, speech and signal processing (ICASSP), 4–9 May 2014, pp 4858–4862

Khorram S, Gideon J, McInnis MG et al (2016) Recognition of depression in bipolar disorder: leveraging cohort and person-specific knowledge. In: INTERSPEECH

Khorram S, Jaiswal M, Gideon J et al (2018) The PRIORI emotion dataset: linking mood to emotion detected in-the-wild. ArXiv e-prints

Kubiak T, Smyth JM (2019) Connecting domains—ecological momentary assessment in a mobile sensing framework. In: Baumeister H, Montag C (eds) Mobile sensing and psychoinformatics. Berlin, Springer, pp 201–207

Leow A, Ajilore O, Zhan L et al (2013) Impaired inter-hemispheric integration in bipolar disorder revealed with brain network analyses. Biol Psychiat 73(2):183–193

Lovatt M, Holmes J (2017) Digital phenotyping and sociological perspectives in a Brave New World. Addiction (Abingdon, England) 112(7):1286–1289

Martinez-Martin N, Kreitmair K (2018) Ethical issues for direct-to-consumer digital psychotherapy apps: addressing accountability, data protection, and consent. JMIR Ment Health 5(2):e32–e32

McInnis M, Gideon J, Mower Provost E (2017) Digital Phenotyping in bipolar disorder. Eur Neuropsychopharm 27:S440

Messner E-M, Probst T, O'Rourke T et al (2019) mHealth applications: potentials, limitations, current quality and future directions. In Baumeister H, Montag C (eds) Mobile sensing and psychoinformatics. Berlin, Springer

Montag C, Markowetz A, Blaszkiewicz K et al (2017) Facebook usage on smartphones and gray matter volume of the nucleus accumbens. Behav Brain Res 329:221–228

Muthukrishna M, Henrich J (2019) A problem in theory. Nat Hum Behav

National Collaborating Centre for Mental Health (2018) Bipolar disorder: the NICE guideline on the assessment and management of bipolar disorder in adults, children and young people in primary and secondary care. In: British Psychological Society, pp 39–40

Perlow J (2018) How Apple watch saved my life. ZDNet. https://www.zdnet.com/article/how-apple-watch-saved-my-life/

Phillips ML, Kupfer DJ (2013) Bipolar disorder diagnosis: challenges and future directions. Lancet 381(9878):1663–1671

Phillips ML, Ladouceur CD, Drevets WC (2008) A neural model of voluntary and automatic emotion regulation: implications for understanding the pathophysiology and neurodevelopment of bipolar disorder. Mol Psychiatr 13:833

Rabbi M, Klasnja P, Choudhury T et al (2019) Optimizing mHealth interventions with a bandit. In: Baumeister H, Montag C (eds) Mobile sensing and psychoinformatics. Berlin, Springer, pp 277–291

Samzelius J (2016) Neurametrix Inc. System and method for continuous monitoring of central nervous system diseases. U.S. Patent No. 15,166,064,

Sanford K (2018) Will this "neural lace" brain implant help us compete with AI? http://nautil.us/blog/-will-this-neural-lace-brain-implant-help-us-compete-with-ai

Sariyska R, Rathner E-M, Baumeister H et al (2018) Feasibility of linking molecular genetic markers to real-world social network size tracked on smartphones. Front Neurosci 12(945)

Shropshire C (2015) Americans prefer texting to talking, report says. Chicago Tribune. http://www.chicagotribune.com/business/ct-americans-texting-00327-biz-20150326-story.html

Stange JP, Zulueta J, Langenecker SA et al (2018) Let your fingers do the talking: passive typing instability predicts future mood outcomes. Bipolar Disord 20(3):285–288

Steel Z, Marnane C, Iranpour C et al (2014) The global prevalence of common mental disorders: a systematic review and meta-analysis 1980–2013. 43(2):476–493

Sun L, Wang Y, Cao B et al (2017) Sequential keystroke behavioral biometrics for mobile user identification via multi-view deep learning. Paper presented at the Joint European Conference on Machine Learning and Knowledge Discovery in Databases, 01 November 2017

Turakhia MP (2018) Moving from big data to deep learning—the case of atrial fibrillation. JAMA Cardiol 3(5):371–372

Turakhia MP, Desai M, Hedlin H et al (2019) Rationale and design of a large-scale, app-based study to identify cardiac arrhythmias using a smartwatch: the Apple heart study. Am Heart J 207:66–75

Wolkenstein L, Bruchmuller K, Schmid P et al (2011) Misdiagnosing bipolar disorder—do clinicians show heuristic biases? J Affect Disorders 130(3):405–412

Zulueta J, Piscitello A, Rasic M et al (2018) Predicting mood disturbance severity with mobile phone keystroke metadata: a biaffect digital phenotyping study. J Med Internet Res 20(7):e241

Chapter 11
Studying Psychopathology in Relation to Smartphone Use

Dmitri Rozgonjuk, Jon D. Elhai and Brian J. Hall

Abstract Smartphones allow for several daily life enhancements and productivity improvements. Yet, over the last decade the concern regarding daily life adversities in relation to excessive smartphone use have been raised. This type of behavior has been regarded as "problematic smartphone use" (PSU) to describe the effects resembling a behavioral addiction. In addition to other daily life adversities, research has consistently shown that PSU is related to various psychopathology constructs. The aim of this chapter is to provide an overview of some findings in PSU research regarding associations with psychopathology. We also discuss some of the theoretical explanations that may be helpful in conceptualizing PSU. We then take a look at self-reported PSU in relation to objectively measured smartphone use, and, finally, provide some insight into current findings and future opportunities in objectively measuring smartphone use in relation to psychopathology measures. This chapter may be useful as an introductory overview into the field of PSU research.

D. Rozgonjuk (✉)
Institute of Psychology, University of Tartu, Näituse 2, Tartu, Estonia
e-mail: dmroz@ut.ee

Center of IT Impact Studies, Johann Skytte Institute of Political Studies, University of Tartu, Tartu, Estonia

D. Rozgonjuk · J. D. Elhai
Department of Psychology, University of Toledo, Toledo, OH, USA

J. D. Elhai
Department of Psychiatry, University of Toledo, Toledo, OH, USA

Academy of Psychology and Behavior, Tianjin Normal University, No. 393, Binshui West Avenue, Xiqing District, China

B. J. Hall
Global and Community Mental Health Research Group, Department of Psychology, University of Macau, Taipa, Macao (SAR), People's Republic of China

Department of Health, Behavior and Society, Johns Hopkins Bloomberg School of Public Health, Baltimore, MD, USA

© Springer Nature Switzerland AG 2019   185
H. Baumeister and C. Montag (eds.), *Digital Phenotyping and Mobile Sensing*,
Studies in Neuroscience, Psychology and Behavioral Economics,
https://doi.org/10.1007/978-3-030-31620-4_11

Keywords Problematic smartphone use · Smartphone addiction · Internet addiction · Technological addictions · Psychopathology · I-PACE

11.1 Introduction

Smartphones provide many advantages to society that were unthinkable before the mid-to late 2000s. In addition to the traditional voice calls and text messages, smartphones are essentially small computers that provide access to the Internet and various applications with utility ranging from productivity enhancement to entertainment. These features, in fact, have the capability to facilitate important daily activities. Furthermore, with a smartphone, one may purchase their meals, clothing, or other necessary items without actually stepping into a physical store. These activities can be conducted almost anywhere, anytime.

In fact, because smartphones allow for efficient connectivity and even the replacement of many in-person, face-to-face daily-life activities, people may increasingly opt for using smartphones to execute their tasks. This preference may be associated with more frequent smartphone use as a result. Excessive engagement with smartphones, however, can be associated with several detrimental daily life conditions. In this chapter, we introduce problematic smartphone use, describe how smartphone use may be related to psychopathology, review how studies have implemented objectively measured smartphone use coupled with psychopathology constructs, and explore how smartphone features could be used in psychopathology research.

11.2 Problematic Smartphone Use and Smartphone Addiction

Soon after the initial emergence of smartphones, concern about people spending excessive time on their devices was raised. In fact, such concern was already raised during the pre-smartphone era with internet use (Young 1996; Young and Rogers 1998), and with earlier mobile phones (Bianchi and Phillips 2005). However, smartphone ownership quickly became prevalent in modern society. The extreme portability and easy accessibility may have lead some people to engage in excessive smartphone use.

Smartphone "addiction" is not recognized as a bona fide disorder. Nonetheless, it is an important construct that is studied in the scientific literature. Some research uses the term "smartphone addiction" (Kwon et al. 2013a), or similarly "proneness to smartphone addiction" (Rozgonjuk et al. 2016). More recently, researchers tend to avoid labeling this construct as an "addiction," instead using alternate labels such as smartphone overuse (Lee et al. 2017), and excessive smartphone use (Chen et al. 2016). Most recently, this construct was conceptualized and labeled as "problematic smartphone use" (PSU) to emphasize the adversity related to excessive smartphone use, and avoid addiction terminology (Panova and Carbonell 2018). PSU could be described by some emotional, cognitive, and behavioral difficulties associated with smartphone use. Frequently, the framework inspired by addiction research has been implemented, suggesting that PSU could be characterized by symptoms such as withdrawal (negative affect when separated from one's smartphone), tolerance (failed attempts to reduce smartphone use), and daily-life disturbances (impairment at work/school and/or health problems), among others (Kwon et al. 2013a). However, it should be noted that by using "problematic smartphone use" as the term describing similar conditions (e.g., smartphone addiction), there is still no consensus regarding using this term. One possible reason is that "problematic" could either describe a person being on the way from "healthy" to experiencing full blown psychopathological symptoms, or it could be the end condition in itself. Therefore, another used term is "smartphone use disorder" (Lachmann et al. 2018; Sha et al. 2019). With that being said, we will continue using the "problematic smartphone use" terminology in this text, while acknowledging that debate regarding vocabulary is ongoing.

How exactly is this *problematic* condition measured? Typically, self-report measures (questionnaires) were used. For example, measures such as the Smartphone Addiction Scale, developed by Kwon et al. (2013a, b), include symptom-based items adapted from substance use disorder scales and criteria, which assessed functional impairments associated with PSU. This questionnaire, and its shorter version (Kwon et al. 2013a, b), are probably the most commonly used contemporary instruments for measuring PSU, and have evidenced adequate psychometric properties (Demirci et al. 2014; Lopez-Fernandez 2017).

Mental health problems related to (self-reported) PSU are reported across different cultures, ranging from the Americas (Khoury et al. 2017; Elhai et al. 2018a), Europe (Lopez-Fernandez et al. 2017), and Asia (Kwon et al. 2013b). Apparently, concerns regarding PSU have been raised almost all over the world, further emphasizing the need for understanding this phenomenon. We now focus more specifically on mental health-related correlates of PSU.

11.3 How May Psychopathology Relate to Increased Smartphone Use?

To answer how and why psychopathology may be related to increased smartphone use and PSU, it would be reasonable to discuss several theories that conceptualize digital technology and media engagement. Specifically, we first discuss Uses and Gratifications Theory (UGT; Blumler 1979) and Compensatory Internet Use Theory (CIUT; Kardefelt-Winther 2014).

UGT posits that people actively seek out specific media to satisfy their psychological needs (Blumler 1979). These particular needs are the main drivers of certain types of media selection (Sundar and Limperos 2013; Rubin 2009). There are several gratifications, or need satisfactions, that people could obtain from their mobile phones, such as social need fulfillment, seeking information online, relaxation, and mobility (Sundar and Limperos 2013). Broadly, gratifications have been divided into three typologies: content (obtained from media content, e.g., reading news), process (from using the media, e.g., surfing the web), and social (media as a social environment, e.g., communicating with others) (Stafford et al. 2004). Derived from these gratifications, different media may have respective uses. In a study by van Deursen and colleagues (2015), smartphone use has broadly been categorized into process and social use, and shown that PSU was related to process, but not social smartphone use. On the other hand, there are also studies that show the prevalent role of social media use in developing PSU (Lopez-Fernandez et al. 2017; Rozgonjuk et al. 2018a; Sha et al. 2019, p. 201).

Another relevant tenet of UGT is that there are individual differences in engagement with media (Blumler 1979). In other words, levels of motivation to satisfy one's needs, but also differences in personality traits may drive some people to engage in higher levels of smartphone use. Regarding personality traits, it has recently been demonstrated that lower willpower may reflect the core vulnerability towards developing addictive tendencies towards digital technologies (Lachmann et al. 2017). Extraversion was shown to be positively and conscientiousness negatively correlated with social media engagement (Montag et al. 2015a, b). In addition, biological factors, such as age and sex of the smartphone user, may drive differences in engagement (Andone et al. 2016a). Another relevant factor of these individual differences may manifest in how people cope with experiencing negative affect. Here, CIUT is useful to conceptualize why some people engage in higher levels of digital technology use such as smartphones. According to CIUT, people may engage in excessive technology use in order to alleviate negative affect. In other words, some people who experience stressful life events cope with their negative emotions by using their smartphones (Kardefelt-Winther 2014). In fact, the SAS by Kwon et al. (2013a, b) (Wang et al. 2015; Zhitomirsky-Geffet and Blau 2016; Elhai et al. 2018b), includes items such as "Being able to get rid of stress with a smartphone" and "Feeling calm and cozy while using a smartphone", suggesting that emotion regulation may be a central motive for engaging in smartphone use. CIUT has been used to conceptualize psychopathology in relation to PSU (Wang et al. 2015; Zhitomirsky-Geffet and Blau 2016; Elhai et al. 2018c).

A more comprehensive approach explaining the associations between digital technology engagement and other psychological and environmental variables is the Interaction of Person-Affect-Cognition-Execution (I-PACE) model of specific Internet-use disorders by Brand and colleagues (2016). Based on the cognitive-behavioral model of pathological Internet use by Davis (2001), and further developed from Brand and colleagues (2014), the I-PACE takes the process approach regarding the development of digital technology addictions. According to the I-PACE model, individual differences in predisposing variables, such as personality traits, genetics, and psychopathology, drive the affective and cognitive responses to stimuli. These responses interact with the individual's coping and decision-making that results in the use of certain Internet-based media and platforms. In some cases, this may result in problematic digital technology use, such as PSU. In fact, some recent studies have used the I-PACE model as a theoretical framework in explaining the associations between daily-life adversities and PSU (Montag et al. 2016; Duke and Montag 2017; Carvalho et al. 2018).

Recent findings have consistently demonstrated that PSU is associated with increased severity of mental disorders, mainly mood-related. PSU severity is related to increased severity of depression and anxiety (Elhai et al. 2017), including social anxiety (Elhai et al. 2018b), PTSD symptoms (Contractor et al. 2017a), and stress (Wang et al. 2015). In addition, PSU has been associated with the so-called "transdiagnostic constructs". These are psychopathology-related characteristics that tend to overlap between different mental disorders (e.g., mood and anxiety disorders) and are considered to be core vulnerability factors in mental disorders (Krueger and Eaton 2015). Studies have found that PSU is associated with impulsivity (Contractor et al. 2017b), boredom proneness (Elhai et al. 2018d), procrastination (Rozgonjuk et al. 2018a), anger and worry (Elhai et al. 2019), and lower distress tolerance and mindfulness (Elhai et al. 2018d). Althogether, these studies suggest that PSU is related to psychopathology severity, and that smarphone use has been shown to have detrimental effects on daily-life social situations (Dwyer et al. 2018; Kushlev et al. 2019). Finally, works on the associations between (excessive) digital technology use and different cognitive functioning domains (attention, memory, delay of gratification) show that more engagement is typically associated with poorer cognitive functioning (Wilmer et al. 2017), possibly explaining the link between smartphone use and ADHD symptoms (Kushlev et al. 2016).

To summarize, people differ in their motives for using various digital technologies, including smartphones. Yet, prevailing explanations indicate that seeking gratifications, and alleviating negative affect, are a central component of technology engagement. Findings in PSU studies have consistently demonstrated that one's affect regulation and ability to cope with stressful situations could be one of the most important factors driving problematic technology use and in turn increase mental burden as theorized by vicious cycle models.

11.4 Objectively Measured Smartphone Use (OMSU)

Limitations of the research discussed above include that the majority relied on cross-sectional designs. Additionally, most of these studies used self-report measures rather than measuring smartphone use objectively.

Measuring smartphone use objectively, for example by using specific apps measuring smartphone use frequency and data, could allow us to overcome the two main limitations of smartphone use studies. It was previously demonstrated that self-reported smartphone use is typically inaccurately estimated when compared to objectively measured smartphone use (Boase and Ling 2013; Montag et al. 2015a, b; Andrews et al. 2015). Therefore, for the purpose of studying psychopathology's relations with digital technology engagement, using recorded behavioral measures may provide a more valid insight into smartphone engagement (Miller 2012; Yarkoni 2012; Andone et al. 2016b).

Probably the most straightforward approach to measure smartphone use is to measure actual time spent using smartphones and the number of phone checks/screen unlocks. These measures should provide information on (a) how much time one spends on their smartphone (usage duration), and (b) how frequently one initiates interactions with their smartphone (usage initiation frequency). Phone-checking and time spent using smartphones (reflected as active screen time) are shown to be two separate behavioral measures (Wilcockson et al. 2018; Rozgonjuk et al. 2018b). Whereas the former may indicate more active engagement with a smartphone (e.g., checking for messages), the latter could reflect more passive content consumption (e.g., watching videos, browsing social media sites). Based on the content of this chapter, one may have the following questions: (1) how is PSU related to OMSU, and (2) how are other psychopathological conditions associated with objectively measured behavior?

As mentioned earlier, actual smartphone use is not strongly correlated with self-reported smartphone use (Boase and Ling 2013; Andrews et al. 2015). Results from studying relations between PSU and objective behavioral smartphone use data are mixed. For instance, while Lin et al. (2017) found that both smartphone frequency (smartphone checking) and duration were associated with PSU severity, Rozgonjuk and colleagues (2018b) found that only duration but not frequency was associated with PSU. The latter findings are further supported by Lin et al. (2015) who found phone checking behavior related to PSU severity. These studies support the notion that phone checking behavior and time spent on smartphones are distinct behavioral measures, and that the behavioral manifestation of PSU could be screen time. However, we should note that Wilcockson et al. (2018) did not find a relationship between the measures of PSU and OMSU. In addition, evidence from a recent study suggested the possibility that genetic differences in predisposition for social behavior may be associated with more active phone usage (Sariyska et al. 2018).

Beyond this relatively straightforward approach, contextual data could provide additional insights into PSU's relations with actual smartphone use behavior. Over the last years, the feasibility of objectively observing smartphone use behavior improved significantly. One of the reasons is that it is now possible to retrieve and analyze various smartphone use logs stored on one's device. This has also provided the opportunity to analyze people's behavior from actually recorded, or objectively measured, data. For instance, these objectively measured phone use data have been applied for personality research, predicting either phone use behavior from (self-reported) personality measures, or vice versa (Chittaranjan et al. 2013; Montag et al. 2014; Stachl et al. 2017).

One of the aspects to consider in this context is also the distinction between the frequency and duration of specific app usage on one's smartphone. A study by Ahn and colleagues (2014) showed that different application categories and time of day may be indicative of problematic technology use. Specifically, differences in usage time and frequency of social networking sites (SNS) were associated with PSU. This seems logical, as one of the main uses of smartphones is to access and consume social media and SNS, and excessive social media use has been regarded as a vulnerability factor for developing PSU (Lopez-Fernandez et al. 2017). However, Montag et al. (2017) showed that objectively measured Facebook use duration in one's smartphone may be a more reliable measure in relation to other outcomes, such as the gray matter volume of nucleus accumbens (often regarded as the brain's reward center).

While the literature on OMSU and other constructs, including PSU, is increasing, there is still a lack of studies implementing recorded, or objectively measured, data to measure behavioral outcomes. Yet, we believe that the improving feasibility of retrieving these data will introduce a richer and larger body of research in the upcoming years.

11.5 Objectively Measured Smartphone Use (OMSU) in Relation to Psychopathology

We discussed some of the findings between both OMSU frequency and duration and PSU, but how are these OMSU data related to psychopathology? Probably the most studied relationship is between increased smartphone use and depression severity. Here, too, findings are mixed. For example Saeb et al. (2015), found that both smartphone use duration and frequency are positively associated with specific psychopathology severity (depression). A study conducted among bipolar disorder patients found that while a depressive state was related to more screen time, a manic state was related to more frequent smartphone use (Faurholt-Jepsen et al. 2016). In addition, Elhai and collegues (2018a, b, c, d) found that maladaptive emotion regulation, a transdiagnostic construct, was associated with higher baseline smartphone use duration. However, studies also found the opposite. Specifically, the aforementioned study by Elhai and collegues (2018a, b, c, d) also demonstrated that lower depression severity predicted increased smartphone use duration over one week; and Rozgonjuk et al. (2018c) found that smartphone use duration was not predicted by depression, but lower depression severity was predictive of higher phone checking behavior.

Literature suggests that OMSU is related to severity of mood-related disorders. Even within those relationships, the number of studies is relatively small and there are mixed results, mainly due to methodological differences (Dogan et al. 2017). We mentioned some studies that investigated depression and bipolar disorder. However, the role of anxiety has also been investigated, with it not being a significant predictor of OMSU (Rozgonjuk et al. 2018b). Research on other disorders in relation to OMSU is scarce.

In conclusion, the number of studies where PSU, psychopathology and OMSU have been of interest, is small. A majority of those studies have investigated how mood-related disorders, such as depression, are related to smartphone use behavior. These studies suggest that the duration of smartphone use is associated with higher intensity of depressive symptoms. It would be interesting to see how OMSU is related to psychopathology other than mood-related disorders.

11.6   Using Smartphone Features to Measure Psychopathology

Another less-explored avenue in smartphone use and psychopathology research is sensor data use. Contemporary smartphones include many sensors that could track the user's mobility and environmental conditions. These sensors could also be helpful in both measuring PSU and the relationship between smartphone use and psychopathology. Below we will outline some ideas that might be helpful in studying psychopathology by using one's smartphone sensors. In order to execute these ideas, one probably needs to use third-party applications that retrieve relevant sensor data and that makes data export feasible.

Contemporary smartphones typically include an ambient light sensor that measures light in the room (or outside); typically the purpose of this sensor is to adjust the smartphone's screen brightness according to current lighting conditions. It could also be used to capture light levels of the environment in which the person is spending time. There is evidence that depression severity is related to the perception of ambient light, with higher depression levels predicting dimmer light perception (Friberg and Borrero 2000). However, those findings relied on self-report, rather than objective measures. Ambient light sensor data from smartphones could provide further validation to these results. Additional research questions would be to test how frequently and for how long do depressed people spend time in dim environments while *using* their smartphone. This idea stems from findings where PSU and social media use are associated with poorer sleep quality (Woods and Scott 2016). Finally, light density could also inform us about people's sleep-wake cycles which are typically disturbed in depressive disorders (Dogan et al. 2017). This research might help in developing prevention and intervention approaches to the individuals in need.

Microphones in smartphones could help in studying the effects of noise on a person's well-being. Noise is related to poorer sleep quality and more annoyance (Basner et al. 2014); noise annoyance has also been linked to depression and anxiety (Beutel et al. 2016). Again, smartphones could assist research in that domain by recording the noise levels of a person's surroundings. This research could be helpful in studying the relationship between physical noise and psychopathology. Microphones could also be useful in stress recognition. For example, an application called StressSense (Lu et al. 2012) was developed for that function. This application is relatively intrusive, though, requiring recording audio and video conversations of the user. Another potential utilization of microphones could be in speech, language, and voice analysis (Rathner et al. 2018a, b; Cummings and Schuller 2019). For example, it is possible to infer a person's sentiment through word use (Rathner et al. 2018b), and similarly, psychopathological tendencies, such as depression, anxiety, and narcissism could be detected by the use of social words (Rathner et al. 2018a).

Weather could influence how people feel, with negative affect and/or fatigue being generally more experienced in colder and darker environments (Kööts et al. 2011). Contemporary smartphones may include a thermometer that tracks temperature of the surrounding environment, while a barometer provides data about atmospheric pressure, another factor shown to influence fatigue (Denissen et al. 2008) and migraine headaches (Kimoto et al. 2011). Again, the mentioned studies have either mainly relied on subjective self-report data and/or more aggregate meteorological data from a local or national meteorological centre. Thermometers and barometers embedded into one's smartphone may provide more accurate, temporal data on these environmental factors and their relations to mood changes and potential psychopathology.

Levels of physical activity and mobility could be measured with a smartphone's accelerometer. This sensor can be found in fitness trackers as well as in smartphones. Accelerometers provide data that could inform about one's number of steps walked/run, flights of stairs climbed, and miles traveled by foot during a given day. Smartphone apps based on this sensor tend to be quite accurate in providing information about the user's steps (Case et al. 2015). Little or no physical activity is generally associated with poorer mental health (Hiles et al. 2017). Using the data from accelerometers could help in that line of research. In addition, investigating smartphone user's gait patterns (gait acceleration and walking speed) over a period of time might also be indicative of a person's affective state and potential psychopathology. Some support for these relationships are provided in the scientific literature where more sadness and depression were associated with reduced walking speed (Michalak et al. 2009). Additionally, different wearables (e.g., a smartwatch) could complement smartphone data by providing additional measures such as heart rate. This, more direct index of physical activity could help in discriminating between a regular walk and a physical exercise session, further specifying the smartphone user's physical activity patterns.

Another feature that provides insight into one's mobility is the global positioning system (GPS). GPS utilizes satellite technology in order to calculate and pinpoint one's location. This technology also allows for investigating the smartphone user's mobility patterns by foot or vehicle. While GPS might provide more accurate data of one's mobility, another (but less accurate) method to track one's mobility is to investigate movement between cell towers. Moving between cell towers basically means that a person (and their smartphone) will be changing their geographical location between the reception areas of different cell phone signalling towers. For instance, in a study by Faurholt-Jepsen et al. (2016) with bipolar disorder patients, depressive states were related to less movement between cell towers, while more severe manic symptoms predicted more movement between cell towers. In other words, while depressive symptoms may be related to less mobility, psychological states including higher arousal could be manifested in more mobility. The idea that people who are in a depressive phase are expected to travel and stay outside less often is also demonstrated in a study by Gruenerbl et al. (2014). Location data could also provide more insight about a person's whereabouts and the time spent at a specific location, which could be associated with other addictive behavior comorbidities, like alcohol use in relation to proximity or duration in a bar.

Of course, the combination of the aforementioned utilities could provide a more accurate and valid representation of one's mental health condition. Elaborate machine learning algorithms that include smartphone usage frequency and duration, different sensors' data, and external factors (e.g., socio-demographic data, date and time of day of smartphone usage, additional self-reported measures, etc.) have been developed to predict smartphone users' negative affective states (Hung et al. 2016) and stress (Reimer et al. 2017) [and suicide—Nock].

In summary, in this section we described some of the opportunities of utilizing smartphone features and sensors to investigate affective states and potential psychopathology. While the list of sensors, and certainly the list of research questions, is not exhaustive, we find that there is an abundance of options for conducting research that include objective behavioral data retrieved from smartphones. Including objective behavioral data could provide a ground for replicating and validating previous research findings in psychopathology studies that mostly rely on self-report and cross-sectional study designs.

11.7 Concluding Remarks

Smartphones have enhanced people's everyday lives by providing means for ubiquitous communication, information consumption, and productivity enhancement. However, excessive engagement in smartphone use has been associated with psychopathology. People who experience stress may try to cope with negative emotions by seeking gratifications provided by smartphone use. This hypothesis is, to some extent, supported by findings from several studies where PSU was associated with psychopathology symptoms and other core vulnerabilities driving those forms of psychopathology. In addition to measuring smartphone use duration and frequency, other features of smartphones could be helpful in studying psychopathology.

This line of research could also provide some academic and clinical implications. As mentioned in this text, there is a relatively small number of studies that objectively measured smartphone use and correlated these results to measures of psychopathology and PSU. Objectively measuring smartphone use may specify the relations between engagement in digital technology use and mental health. This type of measurement could also be useful in clinical settings, as knowing who, how, and why people engage in more digital technology use may help in intervention and prevention of mental illness.

References

Ahn H, Wijaya ME, Esmero BC (2014) A systemic smartphone usage pattern analysis: focusing on smartphone addiction issue. Int J Multimed Ubiquitous Eng 9(6):9–14

Andone I, Błaszkiewicz K, Eibes M et al (2016a) How age and gender affect smartphone usage. In: Proceedings of the 2016 ACM international joint conference on pervasive and ubiquitous computing adjunct—UbiComp '16. ACM Press, Heidelberg, Germany, pp 9–12

Andone I, Błaszkiewicz K, Eibes M et al (2016b) Menthal: a framework for mobile data collection and analysis. In: Proceedings of the 2016 ACM international joint conference on pervasive and ubiquitous computing: adjunct. ACM, New York, NY, USA, pp 624–629

Andrews S, Ellis DA, Shaw H, Piwek L (2015) Beyond self-report: tools to compare estimated and real-world smartphone use. PLoS ONE 10(10):e0139004. https://doi.org/10.1371/journal.pone. 0139004

Basner M, Babisch W, Davis A et al (2014) Auditory and non-auditory effects of noise on health. Lancet 383(9925):1325–1332. https://doi.org/10.1016/S0140-6736(13)61613-X

Beutel ME, Jünger C, Klein EM et al (2016) Noise annoyance is associated with depression and anxiety in the general population-the contribution of aircraft noise. PLoS ONE 11(5):e0155357. https://doi.org/10.1371/journal.pone.0155357

Bianchi A, Phillips JG (2005) Psychological predictors of problem mobile phone use. Cyberpsychol Behav 8(1):39–51. https://doi.org/10.1089/cpb.2005.8.39

Blumler JG (1979) The role of theory in uses and gratifications studies. Commun Res 6(1):9–36. https://doi.org/10.1177/009365027900600102

Boase J, Ling R (2013) Measuring mobile phone use: self-report versus log data. J Comput Mediat Commun 18(4):508–519. https://doi.org/10.1111/jcc4.12021

Brand M, Young KS, Laier C et al (2016) Integrating psychological and neurobiological considerations regarding the development and maintenance of specific internet-use disorders: an interaction of person-affect-cognition-execution (I-PACE) model. Neurosci Biobehav Rev 71:252–266. https://doi.org/10.1016/j.neubiorev.2016.08.033

Brand M, Young KS, Laier C (2014) Prefrontal control and internet addiction: a theoretical model and review of neuropsychological and neuroimaging findings. Front Hum Neurosci 8. https://doi. org/10.3389/fnhum.2014.00375

Carvalho LF, Sette CP, Ferrari BL (2018) Problematic smartphone use relationship with pathological personality traits: Systematic review and meta-analysis. Cyberpsychology (Brno) 12(3). https:// doi.org/10.5817/cp2018-3-5

Case MA, Burwick HA, Volpp KG, Patel MS (2015) Accuracy of smartphone applications and wearable devices for tracking physical activity data. Jama 313(6):625–626. https://doi.org/10. 1001/jama.2014.17841

Chen J, Liang Y, Mai C et al (2016) General deficit in inhibitory control of excessive smartphone users: evidence from an event-related potential study. Front Psychol 7. https://doi.org/10.3389/ fpsyg.2016.00511

Chittaranjan G, Blom J, Gatica-Perez D (2013) Mining large-scale smartphone data for personality studies. Pers Ubiquitous Comput 17(3):433–450. https://doi.org/10.1007/s00779-011-0490-1

Contractor AA, Frankfurt SB, Weiss NH, Elhai JD (2017a) Latent-level relations between DSM-5 PTSD symptom clusters and problematic smartphone use. Comput Hum Behav 72:170–177. https://doi.org/10.1016/j.chb.2017.02.051

Contractor AA, Weiss NH, Tull MT, Elhai JD (2017b) PTSD's relation with problematic smartphone use: mediating role of impulsivity. Comput Hum Behav 75:177–183. https://doi.org/10.1016/j. chb.2017.05.018

Cummings N, Schuller BW (2019) Advances in computational speech analysis for mobile sensing. In: Baumeister H, Montag C (eds) Mobile sensing and psychoinformatics. Springer, Berlin, pp x–x

Davis RA (2001) A cognitive-behavioral model of pathological internet use. Comput Hum Behav 17(2):187–195. https://doi.org/10.1016/S0747-5632(00)00041-8

Demirci APK, Orhan APH, Demirdas APA et al (2014) Validity and reliability of the turkish version of the smartphone addiction scale in a younger population. Klinik Psikofarmakol Bülteni 24(3):226–234. https://doi.org/10.5455/bcp.20140710040824

Denissen JJA, Butalid L, Penke L, van Aken MAG (2008) The effects of weather on daily mood: a multilevel approach. Emotion 8(5):662–667. https://doi.org/10.1037/a0013497

Dogan E, Sander C, Wagner X et al (2017) Smartphone-based monitoring of objective and subjective data in affective disorders: where are we and where are we going? systematic review. J Med Internet Res 19(7):e262. https://doi.org/10.2196/jmir.7006

Duke É, Montag C (2017) Smartphone addiction, daily interruptions and self-reported productivity. Addict Behav Rep 6:90–95. https://doi.org/10.1016/j.abrep.2017.07.002

Dwyer RJ, Kushlev K, Dunn EW (2018) Smartphone use undermines enjoyment of face-to-face social interactions. J Exp Soc Psychol 78:233–239. https://doi.org/10.1016/j.jesp.2017.10.007

Elhai JD, Dvorak RD, Levine JC, Hall BJ (2017) Problematic smartphone use: a conceptual overview and systematic review of relations with anxiety and depression psychopathology. J Affect Disord 207:251–259. https://doi.org/10.1016/j.jad.2016.08.030

Elhai JD, Levine JC, O'Brien KD, Armour C (2018a) Distress tolerance and mindfulness mediate relations between depression and anxiety sensitivity with problematic smartphone use. Comput Hum Behav 84:477–484. https://doi.org/10.1016/j.chb.2018.03.026

Elhai JD, Tiamiyu M, Weeks J (2018b) Depression and social anxiety in relation to problematic smartphone use: the prominent role of rumination. Internet Res 28(2):315–332. https://doi.org/10.1108/IntR-01-2017-0019

Elhai JD, Tiamiyu MF, Weeks JW et al (2018c) Depression and emotion regulation predict objective smartphone use measured over one week. Pers Individ Dif 133:21–28. https://doi.org/10.1016/j.paid.2017.04.051

Elhai JD, Vasquez JK, Lustgarten SD et al (2018d) Proneness to boredom mediates relationships between problematic smartphone use with depression and anxiety severity. Soc Sci Comput Rev 36(6):707–720. https://doi.org/10.1177/0894439317741087

Elhai JD, Rozgonjuk D, Yildirim C et al (2019) Worry and anger are associated with latent classes of problematic smartphone use severity among college students. J Affect Disord 246:209–216. https://doi.org/10.1016/j.jad.2018.12.047

Faurholt-Jepsen M, Vinberg M, Frost M et al (2016) Behavioral activities collected through smartphones and the association with illness activity in bipolar disorder. Int J Methods Psychiatr Res 25(4):309–323. https://doi.org/10.1002/mpr.1502

Friberg TR, Borrero G (2000) Diminished perception of ambient light: a symptom of clinical depression? J Affect Disord 61(1):113–118. https://doi.org/10.1016/S0165-0327(99)00194-9

Gruenerbl A, Osmani V, Bahle G et al (2014) Using smart phone mobility traces for the diagnosis of depressive and manic episodes in bipolar patients. In: Proceedings of the 5th augmented human international conference. ACM, New York, NY, USA, pp 38

Hiles SA, Lamers F, Milaneschi Y, Penninx BWJH (2017) Sit, step, sweat: longitudinal associations between physical activity patterns, anxiety and depression. Psychol Med 47(8):1466–1477. https://doi.org/10.1017/S0033291716003548

Hung GC-L, Yang P-C, Chang C-C et al (2016) Predicting negative emotions based on mobile phone usage patterns: an exploratory study. JMIR Res Protoc 5(3):e160. https://doi.org/10.2196/resprot.5551

Kardefelt-Winther D (2014) A conceptual and methodological critique of internet addiction research: towards a model of compensatory internet use. Comput Hum Behav 31:351–354. https://doi.org/10.1016/j.chb.2013.10.059

Khoury JM, de Freitas AAC, Roque MAV et al (2017) Assessment of the accuracy of a new tool for the screening of smartphone addiction. PLoS ONE 12(5):e0176924. https://doi.org/10.1371/journal.pone.0176924

Kimoto K, Aiba S, Takashima R et al (2011) Influence of barometric pressure in patients with migraine headache. Intern Med 50(18):1923–1928. https://doi.org/10.2169/internalmedicine.50.5640

Kööts L, Realo A, Allik J (2011) The influence of the weather on affective experience. J Individ Differ 32(2):74–84. https://doi.org/10.1027/1614-0001/a000037

Krueger RF, Eaton NR (2015) Transdiagnostic factors of mental disorders. World Psychiatry 14(1):27–29. https://doi.org/10.1002/wps.20175

Kushlev K, Proulx J, Dunn EW (2016) "Silence your phones": smartphone notifications increase inattention and hyperactivity symptoms. In: Proceedings of the 2016 CHI conference on human factors in computing systems. ACM, New York, NY, USA, pp 1011–1020

Kushlev K, Hunter JF, Proulx J et al (2019) Smartphones reduce smiles between strangers. Comput Hum Behav 91:12–16. https://doi.org/10.1016/j.chb.2018.09.023

Kwon M, Kim D-J, Cho H, Yang S (2013a) The smartphone addiction scale: development and validation of a short version for adolescents. PLoS ONE 8(12):e83558. https://doi.org/10.1371/journal.pone.0083558

Kwon M, Lee J-Y, Won W-Y et al (2013b) Development and validation of a smartphone addiction scale (SAS). PLoS ONE 8(2):e56936. https://doi.org/10.1371/journal.pone.0056936

Lachmann B, Duke É, Sariyska R, Montag C (2017) Who's addicted to the smartphone and/or the internet? Psychol Pop Media Cult No Pagination Specified-No Pagination Specified. https://doi.org/10.1037/ppm0000172

Lachmann B, Sindermann C, Sariyska RY et al (2018) The role of empathy and life satisfaction in internet and smartphone use disorder. Front Psychol 9. https://doi.org/10.3389/fpsyg.2018.00398

Lee H-K, Kim J-H, Fava M et al (2017) Development and validation study of the smartphone overuse screening questionnaire. Psychiatry Res 257:352–357. https://doi.org/10.1016/j.psychres.2017.07.074

Lin Y-H, Lin Y-C, Lee Y-H et al (2015) Time distortion associated with smartphone addiction: identifying smartphone addiction via a mobile application (App). J Psychiatr Res 65:139–145. https://doi.org/10.1016/j.jpsychires.2015.04.003

Lin YH, Lin PH, Chiang CL et al (2017) Incorporation of mobile application (App) measures into the diagnosis of smartphone addiction. J Clin Psychiatry 78(7):866–872. https://doi.org/10.4088/JCP.15m10310

Lopez-Fernandez O (2017) Short version of the smartphone addiction scale adapted to Spanish and French: towards a cross-cultural research in problematic mobile phone use. Addict Behav 64:275–280. https://doi.org/10.1016/j.addbeh.2015.11.013

Lopez-Fernandez O, Kuss DJ, Romo L et al (2017) Self-reported dependence on mobile phones in young adults: a European cross-cultural empirical survey. J Behav Addict 6(2):168–177. https://doi.org/10.1556/2006.6.2017.020

Lu H, Frauendorfer D, Rabbi M et al (2012) StressSense: detecting stress in unconstrained acoustic environments using smartphones. In: Proceedings of the 2012 ACM conference on ubiquitous computing. ACM, New York, NY, USA, pp 351–360

Michalak J, Troje NF, Fischer J et al (2009) Embodiment of sadness and depression–gait patterns associated with dysphoric mood. Psychosom Med 71(5):580–587. https://doi.org/10.1097/PSY.0b013e3181a2515c

Miller G (2012) The smartphone psychology manifesto. Perspect Psychol Sci 7(3):221–237. https://doi.org/10.1177/1745691612441215

Montag C, Błaszkiewicz K, Lachmann B et al (2014) Correlating personality and actual phone usage. J Individ Differ 35(3):158–165. https://doi.org/10.1027/1614-0001/a000139

Montag C, Błaszkiewicz K, Lachmann B et al (2015a) Recorded behavior as a valuable resource for diagnostics in mobile phone addiction: evidence from psychoinformatics. Syst Res Behav Sci 5(4):434–442. https://doi.org/10.3390/bs5040434

Montag C, Błaszkiewicz K, Sariyska R et al (2015b) Smartphone usage in the 21st century: who is active on WhatsApp? BMC Res Notes 8(1):331. https://doi.org/10.1186/s13104-015-1280-z

Montag C, Markowetz A, Blaszkiewicz K et al (2017) Facebook usage on smartphones and gray matter volume of the nucleus accumbens. Behav Brain Res 329:221–228. https://doi.org/10.1016/j.bbr.2017.04.035

Montag C, Sindermann C, Becker B, Panksepp J (2016) An affective neuroscience framework for the molecular study of internet addiction. Front Psychol 7. https://doi.org/10.3389/fpsyg.2016.01906

Panova T, Carbonell X (2018) Is smartphone addiction really an addiction? J Behav Addict 7(2):252–259. https://doi.org/10.1556/2006.7.2018.49

Rathner EM, Djamali J, Terhorst Y et al (2018a) How did you like 2017? detection of language markers of depression and narcissism in personal narratives. In: Interspeech 2018. ISCA, pp 3388–3392

Rathner EM, Terhorst Y, Cummins N et al (2018b) State of mind: classification through self-reported affect and word use in speech. In: Interspeech 2018. ISCA, pp 267–271

Reimer U, Laurenzi E, Maier E, Ulmer T (2017) Mobile stress recognition and relaxation support with SmartCoping: user-adaptive interpretation of physiological stress parameters

Rozgonjuk D, Kattago M, Täht K (2018a) Social media use in lectures mediates the relationship between procrastination and problematic smartphone use. Comput Hum Behav 89:191–198. https://doi.org/10.1016/j.chb.2018.08.003

Rozgonjuk D, Levine JC, Hall BJ, Elhai JD (2018b) The association between problematic smartphone use, depression and anxiety symptom severity, and objectively measured smartphone use over one week. Comput Hum Behav 87:10–17. https://doi.org/10.1016/j.chb.2018.05.019

Rozgonjuk D, Saal K, Täht K (2018c) Problematic smartphone use, deep and surface approaches to learning, and social media use in lectures. Int J Environ Res Public Health 15(1):92. https://doi.org/10.3390/ijerph15010092

Rozgonjuk D, Rosenvald V, Janno S, Täht K (2016) Developing a shorter version of the Estonian smartphone addiction proneness scale (E-SAPS18). Cyberpsychology (Brno) 10(4). https://doi.org/10.5817/cp2016-4-4

Rubin AM (2009) The uses-and-gratifications perspective on media effects. In: Bryant J, Oliver MB (eds) Media effects: advances in theory and research, 3rd edn. Routledge, New York, NY pp 165–184

Saeb S, Zhang M, Karr CJ et al (2015) Mobile phone sensor correlates of depressive symptom severity in daily-life behavior: an exploratory study. J Med Internet Res 17(7):e175. https://doi.org/10.2196/jmir.4273

Sariyska R, Rathner E-M, Baumeister H, Montag C (2018) Feasibility of linking molecular genetic markers to real-world social network size tracked on smartphones. Front Neurosci 12. https://doi.org/10.3389/fnins.2018.00945

Sha P, Sariyska R, Riedl R et al (2019) Linking internet communication and smartphone use disorder by taking a closer look at the Facebook and WhatsApp applications. Addict Behav Rep 9:100148. https://doi.org/10.1016/j.abrep.2018.100148

Stachl C, Hilbert S, Au JQ et al (2017) Personality traits predict smartphone usage. Eur J Pers 31(6):701–722. https://doi.org/10.1002/per.2113

Stafford TF, Stafford MR, Schkade LL (2004) Determining uses and gratifications for the internet. Decis Sci 35(2):259–288. https://doi.org/10.1111/j.00117315.2004.02524.x

Sundar SS, Limperos AM (2013) Uses and grats 2.0: new gratifications for new media. J Broadcast Electron Media 57(4):504–525. https://doi.org/10.1080/08838151.2013.845827

van Deursen AJAM, Bolle CL, Hegner SM, Kommers PAM (2015) Modeling habitual and addictive smartphone behavior: The role of smartphone usage types, emotional intelligence, social stress, self-regulation, age, and gender. Comput Hum Behav 45:411–420. https://doi.org/10.1016/j.chb.2014.12.039

Wang JL, Wang HZ, Gaskin J, Wang LH (2015) The role of stress and motivation in problematic smartphone use among college students. Comput Hum Behav 53:181–188. https://doi.org/10.1016/j.chb.2015.07.005

Wilcockson TDW, Ellis DA, Shaw H (2018) Determining typical smartphone usage: what data do we need? Cyberpsychol Behav Soc Netw 21(6):395–398. https://doi.org/10.1089/cyber.2017.0652

Wilmer HH, Sherman LE, Chein JM (2017) Smartphones and cognition: a review of research exploring the links between mobile technology habits and cognitive functioning. Front Psychol 8. https://doi.org/10.3389/fpsyg.2017.00605

Woods HC, Scott H (2016) #Sleepyteens: social media use in adolescence is associated with poor sleep quality, anxiety, depression and low self-esteem. J Adolesc 51:41–49. https://doi.org/10.1016/j.adolescence.2016.05.008

Yarkoni T (2012) Psychoinformatics: new horizons at the interface of the psychological and comput-
    ing sciences. Curr Dir Psychol Sci 21(6):391–397. https://doi.org/10.1177/0963721412457362
Young KS (1996) Psychology of computer use: XL. addictive use of the internet: a case that breaks
    the stereotype. Psychol Rep 79(3):899–902. https://doi.org/10.2466/pr0.1996.79.3.899
Young KS, Rogers RC (1998) The relationship between depression and internet addiction.
    Cyberpsychol Behav 1(1):25–28. https://doi.org/10.1089/cpb.1998.1.25
Zhitomirsky-Geffet M, Blau M (2016) Cross-generational analysis of predictive factors of addictive
    behavior in smartphone usage. Comput Hum Behav 64:682–693. https://doi.org/10.1016/j.chb.
    2016.07.061

Chapter 12
Connecting Domains—Ecological Momentary Assessment in a Mobile Sensing Framework

Thomas Kubiak and Joshua M. Smyth

Abstract Ecological Momentary Assessment (EMA) is the state-of-the-art methodology to capture an individual's experiences (e.g., feelings, thoughts, behaviors) in daily life in an ecologically valid way. In this chapter, we outline the prominent role of EMA within a broader mobile sensing framework connecting domains of data acquisition from a range of sensing sources. We particularly highlight the advantages of context-aware assessment strategies that link the assessment of experiences to specific sensing events or patterns. Finally, we discuss strategies that go beyond assessment to implement innovative Ecological Momentary Interventions in real-life.

12.1 Ecological Momentary Assessment

Ecological Momentary Assessment (EMA) is a state-of-the-art assessment approach that aims at capturing momentary self-reports in naturalistic settings employing "electronic diary" style methodologies. In this chapter, we will briefly highlight the role of EMA as a method for investigating core research questions in behavioral science in the context of mobile sensing. EMA is an essential tool that can be used for linking mobile sensing data, behavioral and physiological data, and environmental data, to an individual's real-life experience that lies at the core of most research questions in the behavioral science. Starting with a brief characterization of EMA among other real-life methodologies and its key features of (quasi) real-time, real-life, and high frequency measurements, we will provide examples of research demonstrating how EMA may be linked to other sources of mobile sensing data in fruitful ways. Finally, we will discuss how the use of EMA may be extended to interventions in a broader mobile sensing framework.

T. Kubiak (✉)
Johannes Gutenberg University Mainz, Mainz, Germany
e-mail: kubiak@uni-mainz.de

J. M. Smyth
The Pennsylvania State University, State College, USA
e-mail: smyth@psu.edu

© Springer Nature Switzerland AG 2019
H. Baumeister and C. Montag (eds.), *Digital Phenotyping and Mobile Sensing*,
Studies in Neuroscience, Psychology and Behavioral Economics,
https://doi.org/10.1007/978-3-030-31620-4_12

12.2   Capturing Experiences in Real-Life

EMA (and related approaches, such as Experience Sampling) is generally construed as comprising a range of self-report approaches that aim to repeatedly capturing momentary experiences in everyday life, providing an opportunity to model intra-individual dynamic processes (Smyth et al. 2017; Stone et al. 2007). Although EMA has traditionally focused on self-reports of internal states, behavior, and context (Kubiak and Stone 2012), as will be discussed below, there is growing application of EMA methods supplemented by other data streams (e.g., wearable sensors, performance tasks). Currently, EMA is typically implemented on smartphones or other devices that are readily carried or worn in everyday life (Kubiak and Stone 2012; Kubiak and Krog 2012). In a typical EMA design, individuals are prompted (e.g., their device makes a sound or vibrates), usually several times throughout the day, to complete self-reports on their momentary experiences; these reports are delivered via short surveys, diary entries, text messages, and other modes.

EMA is a core method within the broader research framework of Ambulatory Assessment (Kubiak and Stone 2012). Ambulatory Assessment is an 'umbrella' concept comprising methodologies that share the common aim of capturing phenomena in situ (Fahrenberg 1996; Fahrenberg et al. 2007). In addition to the self-report data typically captured by EMA, these 'real-life' methodologies also include—but are certainly not limited to—wearable or other sensors that allow for the ambulatory monitoring of physiological signals, actigraphy (e.g., to measure activity and movement), and GPS-based location tracking. Many performance tasks for smartphones, such as ambulatory cognitive testing (e.g., Sliwinski et al. 2018), are also being developed and integrated into these methodologies as well (For additional information on Ambulatory Assessment approaches, see Trull and Ebner-Priemer 2009; Mehl and Conner 2012).

EMA shares key features with other strategies of Ambulatory Assessment with its (near) real-time assessment and its focus on real-life settings outside the laboratory (Trull and Ebner-Priemer 2009), and typically generates rich within-individual data with frequent repeated measurements (often referred to as intensive longitudinal data; e.g., Shiffman et al. 2008; Bolger and Laurenceau 2013). Given this in situ collection style relying on momentary (not lengthy retrospective) recall, EMA data exhibit enhanced ecological validity, minimize recall and reporting biases, and allow for the detection of fine-grained dynamic changes in behavioral process over time in real-life contexts (Shiffman et al. 2008).

12.3   EMA at the Center of a Mobile Sensing Framework

Although EMA can be used to assess self-reported behaviors (such as eating behavior or interpersonal interactions), a unique strength and opportunity for EMA lies in connecting to the *experiential domain* (Conner and Feldman Barrett 2012): EMA allows for capturing an individual's feelings and thoughts in a given moment, such as the assessment of emotions and affect, appraisals of situations, thoughts/cognitions, intentions, and emotion regulation strategies an individual employs (among many other possibilities). Importantly, there are few, if any, other methods for capturing intrapsychic thoughts and feelings *reliably and with specificity*; thus, in any research context or question where these are valuable (or requisite) sources of information, self-report is essential—and EMA provides one valuable method to capture this data. In doing so, we posit that EMA may be considered to be at the core of a mobile sensing framework (see Fig. 12.1): EMA may serve as a central, integrative hub linking mobile sensing data from different sources to the experiential component (in real-life and real-time) and, thus, be important for a range of applications in behavioral science including domains addressed in this volume (e.g., persuasive health technologies, digital markers of cognitive function). Given the rapid technological progress in the field of mobile sensing, one can expect the data sources mentioned in our framework to expand considerably in the future. For example, there are promising developments for the monitoring of biomarkers under real-life conditions, and emerging approaches for speech analysis that may serve as additional data sources for affect and emotions (Mehl et al. 2012).

An example from our own research is the integrative assessment framework developed within the European Determinants of Diet and Physical Activity (DEDIPAC) knowledge hub (Lakerveld et al. 2014). The aim of the DEDIPAC integrative assessment framework is to capture eating behavior, physical activity, and sedentary behaviors concurrently with individual, interpersonal, and environmental correlates and

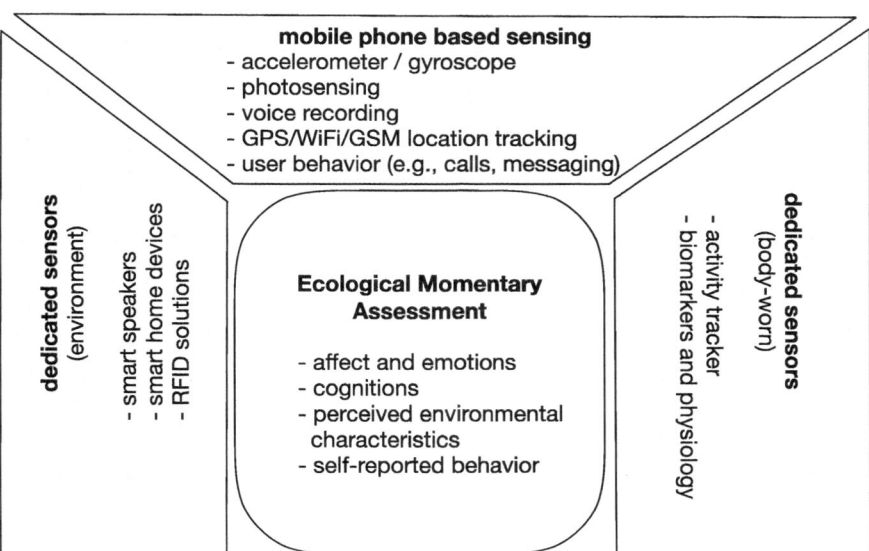

Fig. 12.1 Ecological Momentary Assessment within a mobile sensing framework

determinants within a single assessment tool. In this framework, self-reports on affect and self-regulatory strategies are assessed via smartphone and complemented with data sources from activity monitoring (accelerometry), location tracking based on the Global Positioning System (GPS), and mapping to Geographical Information Systems (GIS; van Laerhoven et al. 2017). In addition, a bar code scanning feature is implemented to capture a range of food products bought or consumed. Initial piloting focused on the consumption of sugar-sweetened beverages and yielded promising data for this system to serve as an integrated data capture tool (Van Laerhoven et al. 2017; Wenzel et al. 2019); many similar approaches exist and such integrative strategies hold great promise for research and practice.

12.4   Sampling Approaches, Including Context-Aware Sampling

EMA typically builds on signal-contingent sampling, event-contingent sampling, interval-contingent (or time-based) sampling, or a combination of these. In *signal-contingent sampling* individuals are signaled ('beeped') by the smartphone to complete a set of standardized questions on their momentary experience (or experiences since the prior signal), usually several times a day in random intervals spaced across the waking hours. Signal-contingent sampling plans are particularly well suited to capture snapshots of experiences and psychological processes that occur throughout the day (e.g., affect), and are most likely to capture "representative" (i.e., typical) moments for an individual. In *event-contingent sampling*, data entry is triggered by the individuals themselves in response to some eliciting stimulus (either internal or external, as specified by the researcher). Event-contingent sampling is best used to capture well-defined, circumscribed episodes that a research participant can identify and respond to. Event-contingent triggers might include participants pressing a button to activate an EMA survey shortly after an event of interest occurs (e.g., drinking a sweetened beverage), or if an earlier EMA survey confirmed the presence of such an event (Shiffman 2014). *Interval-contingent* sampling are based on specific times, such as on the hour each hour, and might be used to characterize temporal processes or to study the impact of temporally entrained stimuli.

Integrating EMA within a broader mobile sensing framework opens up new avenues of sampling an individual's experiences: In *context-aware sampling* (Intille et al. 2003; Intille 2012), any predefined data pattern acquired through any of the sources within the mobile sensing framework may trigger a signal for EMA data entry. For example, an increase in heart rate captured via dedicated sensors may prompt the individual to complete a self-report on his or her perceived stress. Episodes of physical activity assessed via an actigraph device may trigger self-report on a person's feelings afterwards. Using a similar approach, Ebner-Priemer and colleagues (2013) were able to capture bouts of physical activity and concomitant affect that may have easily gone unnoticed if a signal-contingent sampling scheme had been used (that would have relied on sampling probability to capture, or co-occur, with sufficient physical activity bouts to be statistically modeled).

12.5 Ecological Momentary Interventions

The EMA approach is increasingly being extended beyond a mere assessment tool by implementing interventions to be delivered in situ, often via software on the smartphone itself. This approach of Ecological Momentary Intervention (EMI, Heron and Smyth 2010) offers novel opportunities for delivering individually tailored interventions when they are most effective and needed, one example being the just-in-time intervention approach (Smyth and Heron 2016). EMI has been found to be an effective approach for a range of applications, including mindfulness training (Rowland et al. 2016, 2018), smoking cessation (Hébert et al. 2018), and stress-management (Smyth and Heron 2016).

If integrated in a mobile sensing framework, context-aware strategies may substantially enhance interventions: A compelling example in this regard is the study by Gustafson and colleagues (2014) examining the impact of implementing 'warning messages' (based on location tracking) whenever individuals out of alcohol rehabilitation approached a high-risk location like a bar or a liquor vending shop. They demonstrated that the provision of these messages was associated with much improved treatment outcome (e.g., risky drinking days). In a similar vein, comprehensive mobile sensing approaches that integrate an EMA/EMI component have been developed within the European Innovation Partnership on Active and Healthy Ageing (EIP on AHA) that aims to enable older individuals to live independently despite functional and/or cognitive impairments. One example from this work is the development and implementation of systems that produces a mobile sensing based estimation of the risk of falls (one of the most important risks for this sample); this information is able to be collected, processed, visualized and then fed back to at-risk frail older adults and their health care professionals as a means of fall prevention (e.g., De Backere et al. 2017).

12.6 Summary and Conclusions

For many applications in the behavioral and health sciences, EMA can serve as a core component within a mobile sensing framework. Its unique contribution lies in connecting different data sources (across types of data and various sampling densities) to an individual's experiences (the assessment of which can be curated by researchers to topics of interest and relevance). Implementing context-aware strategies and providing individuals interventions during at-risk moments or in risk-elevating contexts based on real-time EMA and/or sensor data opens up tremendously promising avenues for research and care. Given these systems, multi-modal information (from EMA, sensors, and other data sources) can be collected, processed, and used to trigger intervention efforts or provided to patients, caregivers, and/or providers. Well-developed mobile sensing platforms coupled with empirically informed data analysis/processing approaches (i.e., to extract the meaningful information and suggest clinical or other action) have tremendous potential for enhancing the reach, efficiency, and sophistication of monitoring, prevention, and intervention efforts at the individual and public health levels.

References

Bolger N, Laurenceau J-P (2013) Intensive longitudinal methods: an introduction to diary and experience sampling research. Guilford Press

Conner TS, Feldman Barrett L (2012) trends in ambulatory self-report: the role of momentary experience in psychosomatic medicine. Psychosom Med 74(4):327–337. https://doi.org/10.1097/PSY.0b013e3182546f18

De Backere F, den Bergh JV, Coppers S et al (2017) Social-aware event handling within the FallRisk project. Methods Inf Med 56(1):63–73. https://doi.org/10.3414/ME15-02-0010

Ebner-Priemer UW, Koudela S, Mutz G, Kanning M (2013) Interactive multimodal ambulatory monitoring to investigate the association between physical activity and affect. Front Psychol 3. https://doi.org/10.3389/fpsyg.2012.00596

Fahrenberg J (1996) Ambulatory assessment. In: Fahrenberg J, Myrtek M (eds) Ambulatory assessment. Hogrefe & Huber, Seattle

Fahrenberg J, Myrtek M, Pawlik K, Perrez M (2007) Ambulatory assessment—monitoring behavior in daily life settings. Eur J Psychol Assess 23(4):206–213. https://doi.org/10.1027/1015-5759.23.4.206

Gustafson DH, McTavish FM, Chih M-Y et al (2014) A smartphone application to support recovery from alcoholism: a randomized clinical trial. JAMA Psychiatry 71(5):566. https://doi.org/10.1001/jamapsychiatry.2013.4642

Hébert ET, Stevens EM, Frank SG et al (2018) An ecological momentary intervention for smoking cessation: the associations of just-in-time, tailored messages with lapse risk factors. Addict Behav 78:30–35. https://doi.org/10.1016/j.addbeh.2017.10.026

Heron KE, Smyth JM (2010) Ecological momentary interventions: incorporating mobile technology into psychosocial and health behaviour treatments. 15:1–39

Intille SS, Rondoni J, Kukla C et al (2003) A context-aware experience sampling tool. CHI '03 extended abstracts on human factors in computing systems. ACM, New York, NY, USA, pp 972–973

Intille SS (2012) Emerging technology for studying daily life. In: Mehl MR, Conner TS (eds) Handbook of research methods for studying daily life. Guilford, New York, pp 267–284

Kubiak T, Krog K (2012) Computerized sampling of experiences and behaviour. In: Mehl MR, Conner TS (eds) Handbook of research methods for studying daily life. Guilford, New York, pp 124–143

Kubiak T, Stone AA (2012) Ambulatory monitoring of biobehavioral processes in health and disease. Psychosom Med 74(4):325–326. https://doi.org/10.1097/PSY.0b013e31825878da

Lakerveld J, van der Ploeg HP, Kroeze W et al (2014) Towards the integration and development of a cross-European research network and infrastructure: the DEterminants of DIet and Physical ACtivity (DEDIPAC) knowledge Hub. Int J Behav Nutr Phys Act 11(1):143. https://doi.org/10.1186/s12966-014-0143-7

Mehl MR, Robbins ML, große Deters F (2012) Naturalistic observation of health-relevant social processes: the electronically activated recorder (EAR) methodology in psychosomatics. Psychosom Med 74(4):410–417. https://doi.org/10.1097/psy.0b013e3182545470

Mehl MR, Conner TS (eds) (2012) Handbook of research methods for studying daily life. Guilford, New York

Rowland Z, Wenzel M, Kubiak T (2016) The effects of computer-based mindfulness training on self-control and mindfulness within Ambulatorily assessed network systems across Health-related domains in a healthy student population (SMASH): study protocol for a randomized controlled trial. Trials 17(1):570. https://doi.org/10.1186/s13063-016-1707-4

Rowland Z, Wenzel M, Kubiak T (2018) A mind full of happiness: how mindfulness shapes affect dynamics in daily life. Emotion No Pagination Specified-No Pagination Specified. https://doi.org/10.1037/emo0000562

Shiffman S (2014) Conceptualizing analyses of ecological momentary assessment data. Nicotine Tob Res 16(2):S76–S87. https://doi.org/10.1093/ntr/ntt195

Shiffman S, Stone AA, Hufford MR (2008) Ecological momentary assessment. Annu Rev Clin Psychol 4(1):1–32. https://doi.org/10.1146/annurev.clinpsy.3.022806.091415

Sliwinski MJ, Mogle JA, Hyun J et al (2018) Reliability and validity of ambulatory cognitive assessments. Assessment 25(1):14–30. https://doi.org/10.1177/1073191116643164

Smyth JM, Heron KE (2016) Is providing mobile interventions "just-in-time" helpful? an experimental proof of concept study of just-in-time intervention for stress management. In: 2016 IEEE Wireless Health (WH). pp 1–7

Smyth JM, Juth V, Ma J, Sliwinski M (2017) A slice of life: ecologically valid methods for research on social relationships and health across the life span. Soc Pers Psychol Compass 11(10):e12356. https://doi.org/10.1111/spc3.12356

Stone A, Shiffman S, Atienza A, Nebeling L (2007) The science of real-time data capture: self-reports in health research. Oxford University Press

Trull TJ, Ebner-Priemer UW (2009) Using experience sampling methods/ecological momentary assessment (ESM/EMA) in clinical assessment and clinical research: introduction to the special section. Psychol Assess 21(4):457–462. https://doi.org/10.1037/a0017653

Van Laerhoven K, Wenzel M, Geelen A et al (2017) Experiences from a wearable-mobile acquisition system for ambulatory assessment of diet and activity. In: Proceedings of the 4th international workshop on sensor-based activity recognition and interaction. ACM, New York, NY, USA, p 3

Wenzel M, Geelen A, Wolters M, et al (2019) The role of self-control and the presence of enactment models on sugar-sweetened beverage consumption: a pilot study. Front Psychol 10. https://doi.org/10.3389/fpsyg.2019.01511

Chapter 13
Momentary Assessment of Tinnitus—How Smart Mobile Applications Advance Our Understanding of Tinnitus

Winfried Schlee, Robin Kraft, Johannes Schobel, Berthold Langguth, Thomas Probst, Patrick Neff, Manfred Reichert and Rüdiger Pryss

Abstract Tinnitus is a condition associated with a continuous noise in the ears or head and can arise from many different medical disorders. The perception of tinnitus can vary within and between days. In the recent years, Ecological Momentary Assessments of tinnitus have been used to investigate these tinnitus variations during the daily life of the patients. In the last five years, several independent studies have used Ecological Momentary Assessment to assess tinnitus. With this chapter, we want to review the current state of this research. All the EMA studies revealed a considerable variability of tinnitus loudness and tinnitus distress. It has been found that emotional states and emotional dynamics, the subjectively perceived stress level and the time of the day exert influence on the tinnitus variability. In summary, the EMA method revealed a good potential to improve our scientific understanding of tinnitus. Furthermore, it also showed that it can be used to understand the individual differences of tinnitus—and may even be used as a tool for individualized diagnostic and treatment. We conclude, that the results of the EMA studies can lead to improvements of existing research methods in the field of tinnitus.

Tinnitus is a condition associated with a continuous noise in the ears or head and can arise from many different medical disorders. The condition is very common and affects approximately 10–15% of the adult population in the western societies (Nondahl et al. 2012). Severe cases of tinnitus can be accompanied by anxiety, depression, insomnia, and concentration problems all of which can impair quality of life (Kreuzer et al. 2013). Although much progress has been made, tinnitus remains a scientific and clinical enigma. There is little evidence for effective tinnitus treatments with

W. Schlee (✉) · R. Kraft · J. Schobel · B. Langguth · T. Probst · P. Neff · M. Reichert
Department of Psychiatry and Psychotherapy, University of Regensburg, Universitätsstraße 84, 93053 Regensburg, Germany
e-mail: winfried.schlee@gmail.com

R. Pryss
Institute of Clinical Epidemiology and Biometry, University of Würzburg, Josef-Schneider-Str. 2, 97080 Würzburg, Germany

© Springer Nature Switzerland AG 2019
H. Baumeister and C. Montag (eds.), *Digital Phenotyping and Mobile Sensing*,
Studies in Neuroscience, Psychology and Behavioral Economics,
https://doi.org/10.1007/978-3-030-31620-4_13

cognitive behavioral therapy being the best evidence-based treatment option so far (Cima et al. 2014) and no licensed pharmacological therapy. The research on tinnitus and the development of effective treatments is challenging for two reasons: First, there is a large patient heterogeneity among the tinnitus sufferers (Elgoyhen et al. 2015). At least 13 different causes of tinnitus have been identified (Baguley et al. 2013) and the clinical phenotype is largely variable on various dimensions such as aetiology, perceptual characteristics of the sound (i.e. pitch and loudness), time since onset, levels of conscious awareness and perceived distress. All of them potentially influence the individual treatment response of the patient. Second, the majority of patients report that their conscious perception of tinnitus varies within and between days. There are moments with loud and prominent tinnitus perception, but also moments with reduced tinnitus loudness. The reasons for these fluctuations are largely unknown. This not only introduces a methodological challenge for the assessment of tinnitus in basic and clinical research. It also leads to the question about the underlying neurobiological mechanisms of this moment-to-moment fluctuation. A better understanding of these mechanisms might reveal innovative ways for clinical interventions. The systematic assessment of these tinnitus fluctuations, however, would require the measurement of tinnitus with a high sampling rate, ideally during the everyday life of the patient and with a minimum of disturbance to the routine activities. The classical paper-and-pencil questionnaires that are typically used in tinnitus research are hardly suitable for this type of research. In the year 2013, we developed the TrackYourTinnitus App (Schlee et al. 2016) that allows to measure the individual tinnitus fluctuations using the research methodology called Ecological Momentary Assessment (EMA, (Stone and Shiffman 1994). In this chapter we provide a review of almost five years of EMA research on tinnitus and summarize the results of this line of research.

13.1 Ecological Momentary Assessment in Tinnitus

Ecological Momentary Assessment (EMA) is a new research method allowing to systematically collect self-reports of cognition, behavior and emotions in the daily lives of the participants (Kubiak and Smyth 2019). The method is also known by the names "Experienced Sampling" (Czsikszentmihalyi and Larson 1987) or ambulatory assessments, but in the recent years the term "Ecological Momentary Assessment" is more often used. The idea of this method is based on the desire to measure human behavior, cognition and perception in real-world settings rather than laboratory or clinical settings. Retroactive recall of perceptions, cognitions, emotions—and also behavior—can be biased by memory decay and mental reconstruction (see e.g. Fredrickson 2000). Therefore, the EMA methodology favors a prospective measurement to assess the self-report data in the current state. A technical signaling device is used to prompt the participant with a short questionnaire asking questions on the current situation. In the context of tinnitus, EMA is used to collect self-reports about the current perception of the tinnitus and other factors that are closely related to tinnitus such as stress or emotional arousal. While earlier studies on EMA relied on different technical handheld devices, nowadays almost all EMA studies utilize smartphone devices which are owned by a large percentage of the society. In the recent years, the number of studies using smartphone-based or internet-based technologies have largely increased (Kalle et al. 2018). Among them, two studies have also investigated the effects of the long-term use of EMA on the tinnitus distress of the patients: Henry and colleagues reported that the tinnitus distress of 24 study participants did not change significantly during an EMA study with a duration of two weeks (Henry et al. 2012). These effects were again tested on two different groups of participants of the TrackYourTinnitus app (Schlee et al. 2016), using the app for more than one month (n = 66) or less than a month (n = 134). In both groups, there was no significant change in the tinnitus loudness nor tinnitus distress over time. This is an important prerequisite for future EMA studies in the field of tinnitus. First, these results show that the method of repeated question about the tinnitus is not increasing the perceived tinnitus loudness/distress of the study participants. Second, the results suggest that the repeated measurement of tinnitus is not introducing a systematic bias towards an increase or decrease of tinnitus loudness/distress. The methods seems to be appreciated by the patients: in a study by Goldberg and colleagues, 80% of the participants in an EMA study on tinnitus felt that this is a good way to measure tinnitus (Goldberg et al. 2017).

13.2 Tinnitus Fluctuates Within and Between Days

In the recent years, three different studies on tinnitus fluctuations using the ecological assessment method have been published. They all demonstrated relatively large fluctuations of the individual tinnitus perception.

Henry and colleagues reported 2012 a two-weeks study using a Palm Pilot device (Henry et al. 2012). The palm device created four alerts per day and prompted the 24 participants to answer a short version of the Tinnitus Handicap Inventory (THI-S). The authors described a strong variability measured with the THI-S in different time blocks and locations. The highest mean THI-S score with 19.3 was reported for the time period 8 a.m. to 11 a.m. when the participants were travelling, while the lowest mean THI-S score of 13.2 was reported in the evening from 5 pm-8 pm when the participants recorded that they are "somewhere else". Mixed model analysis showed a significant main effect for the location, but not for the time block.

In another study, Wilson and colleagues reported an EMA study on 20 participants during a period of two weeks (Wilson et al. 2015). At four random time points per day, the tinnitus participants received a text message with a hyperlink to an online survey of six questions. The response of the participants to the main question "In the last 5 min, how bothered have you been by your tinnitus?" varied substantially within the participant. A coefficient of variation (CV) was calculated for each participant and was reported in the range between 11.5 and 109.9%. The median CV was 48.8%, indicating a large variation of tinnitus perception over the two weeks study period.

In an analysis of the TrackYourTinnitus database (Probst et al. 2017a), Probst reported similar variability measures for the tinnitus loudness and the tinnitus distress. Tinnitus distress and loudness were both measured on a visual analog scale in the range between 0 and 1. In a sample of 306 users, the mean intra-individual variability of tinnitus distress was reported with 0.18, and the variability of the tinnitus loudness 0.17. In another analysis, Probst showed significant within-day variations of tinnitus with an increase of tinnitus loudness and distress during the night and early morning (Probst et al. 2017a).

In summary, all EMA approaches on tinnitus demonstrated large within- and between-day fluctuations of tinnitus. The amount of fluctuations varies between the individual participants. Examples of the tinnitus variability and the individual differences of it are given in all three EMA studies (Henry et al. 2012; Wilson et al. 2015; Schlee et al. 2016). These results demonstrate the feasibility of measuring the tinnitus fluctuation in the everyday life of the patients and open up a new line of research that allows to systematically investigate the influencing factors and the underlying neurobiological mechanisms. Studies on neurobiological mechanisms of these fluctuations are rare. In one study, it could be shown that the resting-state alpha activity in the auditory cortex of tinnitus sufferers fluctuates within minutes (Schlee et al. 2014). However, a relationship between these fluctuations of neuronal activity and the fluctuations of the tinnitus perception has not been shown yet. A better scientific understanding of the tinnitus variability is needed and can help to improve tinnitus treatment in the future. Some work on influencing factors for tinnitus fluctuations using EMA methodology will be discussed in the following chapter.

13.3   Which Factors Influence the Tinnitus Perception?

The TrackYourTinnitus project is an ongoing EMA study collecting data from tinnitus patients since 2013. The app is freely available in the app stores for iOS and Android devices in English, German and Dutch. The questionnaire of the app not only asks about the perception of tinnitus loudness and tinnitus distress. It also assesses information about the emotional arousal, and emotional valence using the self-assessment manikins (Bradley and Lang 1994), as well as the subjectively perceived stress level and concentration at the same time point. This data was used for several publications to further investigate the factors influencing the tinnitus perception.

Emotional states as mediators between tinnitus loudness and tinnitus distress. In an analysis on 658 users of the TrackYourTinnitus app, Probst et al. (2016a) discovered that emotional states partially explain the association between tinnitus loudness and tinnitus distress. Tinnitus loudness describes the individually perceived loudness of the tinnitus sound, while tinnitus distress describes the psychological annoyance that is associated with the tinnitus. In general, there is a linear correlation between tinnitus loudness and tinnitus distress. Hiller and Goebel reported a significant correlation of $r = 0.45$ between them (Hiller and Goebel 2007). Please note, a correlation estimate of r = 0.45 means that only 20.3% of the variance of tinnitus distress can be explained by tinnitus loudness. The remaining 79.7% of the variance need to be explained by other influencing factors. These other influencing factors are currently unknown. Possible factors are, among others, personality characteristics, the influence of comorbidities or emotional states.

Probst and colleagues hypothesized that the emotional states of the patient mediate the relationship between tinnitus loudness and tinnitus distress. The emotional states were characterized by two measures: emotional arousal and emotional valence. Emotional arousal can be understood as the physiological arousal associated with the emotional state. Higher values of emotional valence indicate positive emotional states, while smaller values represent negative emotions. In accordance with earlier studies, the results revealed a significant positive relationship between tinnitus loudness and distress. Additionally, it was found that emotional arousal and emotional valence are both significant mediators for the relationship between tinnitus loudness and tinnitus distress. An increase of tinnitus loudness also affects the emotional state leading to higher arousal and more negative emotions. This state of negative emotional arousal leads to increased tinnitus distress. Accordingly, more positive emotions are associated with lower tinnitus distress.

In Fig. 13.1, we show the path diagram summarizing these relationships and provide the numerical estimators of the model. Included in this figure is also the perceived stress level, which is an additional mediator reported by the same paper.

Stress as a mediator between tinnitus loudness and tinnitus distress. Furthermore, in the same publication, Probst reported that stress explains a significant proportion of the relationship between tinnitus loudness and distress (Probst et al. 2016b). There

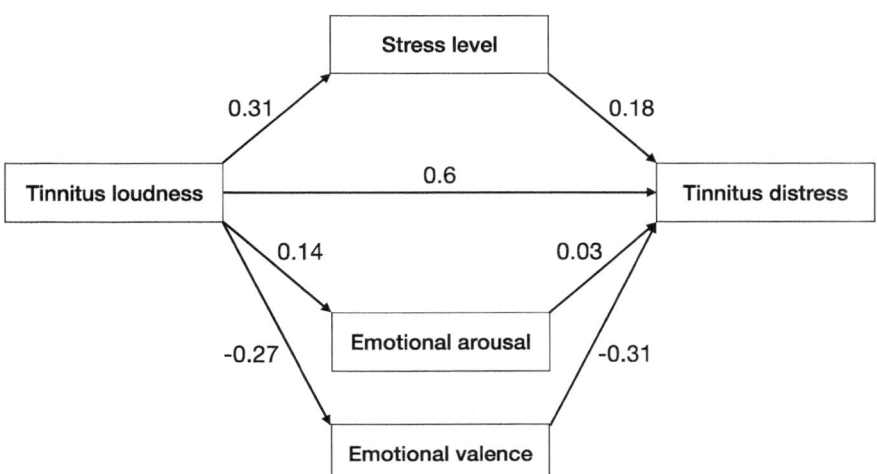

Fig. 13.1 Path diagram summarizing the multiple mediation model explaining the association between tinnitus loudness, tinnitus distress, stress level, emotional arousal and valence. Figure redrawn from Probst et al. (2016a)

is a positive influence of the tinnitus loudness on the general stress level, indicating that a tinnitus loudness increases the general stress level (see the path diagram in Fig. 13.1, path estimate 0.31). The influence of the stress level was with 0.18 also positive, indicating that higher stress levels lead to higher tinnitus distress. Importantly, a statistical comparison of the mediator effects of stress and emotional valence revealed no significant difference. This indicates that both factors, stress and emotional valence, contribute similarly to the process between tinnitus loudness and tinnitus distress. Based on this data we would expect that clinical interventions reducing the stress level and enhancing positive emotions would lead to a reduction of tinnitus distress.

Emotional dynamics are associated with increases of tinnitus loudness, tinnitus distress and the relationship between them. According to Kuppens and colleagues, short-term dynamical changes of the emotional state within a person can be described by the following two measures (Kuppens 2015): Intra-individual variability of affect intensity (pulse) and intra-individual variability of affect quality (spin). Valence and arousal ratings were used for these measures of emotion dynamics. The intra-individual variability of the emotional intensity is called the "pulse". It characterizes the variability of the emotional arousal over time. A person with high pulse is therefore characterized by quick changes between high and low emotional arousal. The pulse is calculated by the intra-individual variability of arousal measures in a longitudinal study. The intra-individual variability of the emotional quality is called "spin". A person that can quickly change from sad to happy emotions would be characterized as a person with a high spin. The spin is calculated by the intra-individual variance of the valence measures in a longitudinal study.

Probst and colleagues analyzed the relationship between the emotional dynamics and tinnitus on 306 users of the TrackYourTinnitus application (Probst et al. 2016c). They found significant positive correlations between tinnitus distress, measured with the Mini-Tinnitus Questionnaire (Mini-TQ), and the pulse ($r = 0.19$, $p = 0.001$). Also, the correlation between spin and tinnitus distress was significant ($r = 0.12$, $p = 0.035$). Patients with higher emotional dynamics are more often found with stronger tinnitus distress. Because the Mini-TQ was only used at the beginning of the study, the causal relationship between tinnitus and emotional dynamics could not clarified. In an additional analysis, multilevel modelling revealed that the relationship between tinnitus loudness and distress is stronger in patients with high emotional dynamics. In other words: patients with high emotional dynamics are more distressed if the tinnitus loudness increases, while patients with low emotional dynamics are less likely to show this reaction. Additionally, the authors report that high spin—but not pulse—is associated with increases of tinnitus loudness over time ($p < 0.01$). Based on these analyses, it can be hypothesized that clinical interventions improving emotional stability of the patients can be beneficial for patients with high emotional dynamics.

Tinnitus loudness and distress is higher during the night and early morning. In another analysis of the TrackYourTinnitus database, Probst and colleagues reported that the tinnitus perception depends on the time of day (Probst et al. 2017a). They analyzed the data of 350 tinnitus patients that have filled out enough assessments to allow a systematic analysis of within-day tinnitus variability. A total of 17'209 assessments were included. Using multilevel modeling with random effects, they analyzed the tinnitus perception in six different time periods: early morning (4 a.m. to 8 a.m.), late morning (8 a.m. to 12 p.m.), afternoon (12 p.m. to 4 p.m.), early evening (4 p.m. to 8 p.m.), late evening (8 p.m. to 12. a.m.), and night (12 a.m. to 4 a.m.). They found that the tinnitus loudness and distress are significantly higher during the night and early morning hours. Further research will be needed to understand the neurobiological mechanism underlying this circadian tinnitus rhythm.

In summary, EMA studies are able to measure systematic patterns of intra-individual tinnitus variability by repeated sampling of short questionnaires. The emotional states, the stress level, emotional dynamics and the time of the day have been identified as possible factors influencing the moment-to-moment variability of the tinnitus perception. However, the predictive value of these influencing factors—calculated over a large patient sample—is low, and differences between the individual patients are high. This leads to the idea that a long-term assessment of the tinnitus variability together with potential influencing factors can be used to identify for each individual tinnitus patient those factors with an impact on the subjective tinnitus perception. If the result of this analysis is fed back to the individual patient, she or he can potentially learn about situations that lead to an increase or decrease of tinnitus. With this knowledge about the own tinnitus, the patient would be in a position to control the tinnitus by changing the behavior. This concept was implemented in a mobile feedback service that was presented by Pryss and colleagues (2017). Future research will be needed to empirically test if this intervention leads to an improvement of tinnitus symptoms. If the claims hold true, this concept can be applied as an individualized treatment for chronic tinnitus patients.

13.4   Limitations of the EMA Studies in Tinnitus

There are several limitations of the EMA studies in tinnitus, which mainly affect the generalizability of the results. In 2017, Probst and colleagues compared the sample of the TrackYourTinnitus database with the users of the TinnitusTalk internet forum, and the patients representing at the University Clinic Regensburg (Probst et al. 2017b). The three patient groups differed significantly from each other with respect to age, gender and tinnitus duration. As shown in Fig. 13.2, the most prominent differences are: Participants between the age of 25 and 44 are overrepresented in the TrackYourTinnitus study, while older participants with an age over 65 years are less likely to use the TrackYourTinnitus app. The percentage of women using the modern technologies of the internet forum TinnitusTalk and the smartphone app TrackYourTinnitus is larger than in the sample of patients seeking help in the tinnitus center Regensburg. The percentage of tinnitus sufferers with acute tinnitus (duration less than three months) in the TrackYourTinnitus sample is higher compared to Tinnitus Center Regensburg. Patients with a tinnitus duration longer than 20 years

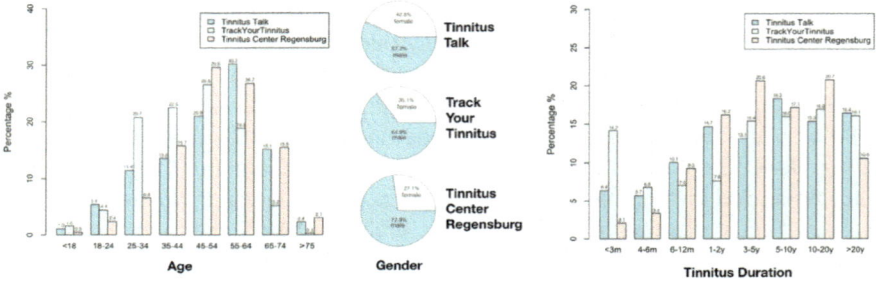

Fig. 13.2 Comparison between the samples of the EMA study TrackYourTinnitus, the online patient forum TinnitusTalk and the university-based tinnitus center Regensburg, published in Probst et al. (2017a, b)

less likely seek help in the tinnitus center, but rather use the TrackYourTinnitus app or the TinnitusTalk forum. These differences in the study samples need to be kept in mind when interpreting the results of EMA studies on tinnitus. Several interpretations of these differences are discussed in the paper (Probst et al. 2017b).

Another sampling bias may be introduced by the operating system that is addressed with the smartphone app. There are studies on the differences between the operating systems and how they influence the user's choice for the operating system and the cell phone; even minor personality differences have been found (Lim et al. 2014; Götz et al. 2017). With the TrackYourTinnitus app being available for iOS and Android, Pryss and colleagues investigated the differences between the iOS and Android users and found significant group differences in the age and tinnitus duration (Pryss et al. 2018a, b): The Android users were on average 1.5 years older than iOS users ($t(1443) = -2.1$, $p = 0.035$). Also, the Android users also reported a longer tinnitus duration with a mean of 11.2 years, while the iOS users reported on average only 7.2 years ($t(1214) = -6.72$, $p < 0.001$). No differences were found for gender, family history or causes of tinnitus. While the age differences between the user groups was only small, a remarkable group difference between the users of the different smart phone systems was found in tinnitus duration. Studies that are addressing only one operating system will need to take this difference into account when interpreting the results.

Furthermore, in 2015, Pryss and colleagues analyzed the number of users participating in the TrackYourTinnitus (Pryss et al. 2015a, b). At this time, 1'718 users had downloaded the app, but only 822 have registered for the TrackYourTinnitus study. Among them, only 150 users had used the app more than ten times. These 150 users contributed to approx. 90% of the collected data. This analysis shows that the researchers lost a large percentage of participants in the process from downloading the app to the regular usage. This loss of users may introduce a bias to the sample and also reduces the efficacy of the study. Additional work will be needed to understand the reasons for this high attrition rate and find ways to avoid it. Promising research is already on the way with analyses concentrating on the incentive management of EMA studies in the medical domain (Agrawal et al. 2018). The goal of this research, also known as persuasive design research (Baumeister et al. 2019) is to identify features in eHealth smartphone apps that are enhancing the motivation of the users to participate for a long time. This knowledge can be important for both, smartphone apps delivering a clinical service to the patients as well as for smartphone apps in basic research.

13.5 Lessons Learned for Clinical Assessment of Tinnitus

With the ongoing TrackYourTinnitus study, several lessons for clinical assessment of tinnitus have been learned:

Assessment of tinnitus in clinical trials. The perception of tinnitus is characterized by within- and between-day variability that can bias the clinical assessment of tinnitus. So far, it has been shown that the emotional states and emotional dynamics, stress level, and the time of the day exert influence on the tinnitus perception. This is a danger for the reliability of the outcome measures in clinical trials. Since clinical trials often rely on a single measurement time point before and after the clinical intervention, the control of these measurement points is of importance. With the circadian fluctuation of the tinnitus (Probst et al. 2017a), the assessment should always be done at the same time of the day. Furthermore, since the tinnitus assessment can be influenced by the stress level and the emotions of the patients, efforts should be made to ensure comparable stress level and emotional states at all time points of tinnitus measurement. Additionally, based on the lessons learned from the EMA studies on tinnitus, we would suggest to collect tinnitus measurement at multiple time points before and after the clinical intervention.

Retrospective assessment of tinnitus. In the assessment of the clinical characteristics of tinnitus, some questionnaires include retrospective reporting of the tinnitus patients. Such a retrospective assessment can be biased by failure to recall the events correctly from memory. This recall bias was investigated in two studies using the TrackYourTinnitus app.

In one study, Pryss and colleagues (Pryss et al. 2018a, b) concentrated on the variation of the tinnitus loudness. At the beginning of the study, participants were asked retrospectively if they have noticed variations in the tinnitus loudness in the past. Based on this answer, the participants were divided into a group experiencing the tinnitus loudness as varying and a group experiencing the tinnitus loudness as non-varying. Both groups used the TrackYourTinnitus app for at least 10 days to assess prospectively the variability of tinnitus loudness. The day-to-day variability of tinnitus loudness was calculated for all participants and compared across the two groups. There was no significant difference between the two groups ($t(258) = 0.19$, $p = 0.85$). The same test was repeated with patients using the app for more than 25 days; there was again no significant group difference ($t(126) = 0.96$, $p = 0.34$). If the patients would be able to recall the tinnitus loudness variability correctly from memory, we should have seen a significant difference between the groups. Both results were far from statistical significance. This demonstrates that the majority of the patients are not able to correctly recall the variability of their tinnitus loudness in the past.

In another analysis, the participants were asked retrospectively about the influence of stress on their tinnitus perception before starting the TrackYourTinnitus app (Pryss et al. 2018a, b). The participants were able to rate if stress worsens their tinnitus, improves their tinnitus or has no influence. Using the prospectively assessed data on the tinnitus loudness, tinnitus distress and stress levels, multilevel models were calculated for all three groups. The results showed that higher stress levels are associated with higher tinnitus loudness and higher tinnitus distress ratings (all $p < 0.001$). Even in the participant group that reported retrospectively that stress improves their tinnitus, the analysis of prospective data indicated that increased stress leads to tinnitus worsening. In summary, both studies demonstrate a large discrepancy between the retrospective reporting of the tinnitus patients and the analysis of the prospective assessment of tinnitus.

13.6   Summary and Future Perspectives

Several independent studies have used Ecological Momentary Assessment to assess tinnitus under daily living conditions. They all revealed a considerable variability of tinnitus loudness and tinnitus distress. It has been found that emotional states and emotional dynamics, the subjectively perceived stress level and the time of the day exert influence on this variability. In these studies, the EMA method revealed a good potential to improve our scientific understanding of tinnitus and also showed that it can be used to understand the individual differences of tinnitus—and may even be used as a tool for individualized diagnostic and treatment. Furthermore, the results of the EMA studies can lead to improvements of existing research methods in the field of tinnitus.

With the rapid development of modern technology and methods in data mining, many advancements can be expected for the near future. To give two examples: Schickler and his colleagues presented 2016 a pre-study using smart watches for the EMA research (Schickler et al. 2016). They experimented with different implementations for answering questionnaire items on the limited screen space of the watch. Another example is a paper by Muniandi and colleagues applying the subspace discovery algorithm PROCLUS on the TrackYourTinnitus database to search for similarities of individual patients in the tinnitus evolution over time (Muniandi et al. 2018). Clustering of the temporal evolution of tinnitus patients offers the possibility to predict the future development of the tinnitus loudness and tinnitus distress based on the cluster membership of the patient.

In its relatively short history—and with only a small number of researchers working on it—EMA research on tinnitus has already contributed well to a better understanding of tinnitus. More discoveries and developments are already on the horizon and we hope that more researchers will start using this promising method.

References

Agrawal K, Mehdi M, Reichert M et al (2018) Towards incentive management mechanisms in the context of crowdsensing technologies based on TrackYourTinnitus insights. Procedia Comput Sci 134:145–152. https://doi.org/10.1016/j.procs.2018.07.155

Baguley D, McFerran D, Hall D (2013) Tinnitus. Lancet 382(9904):1600–1607. https://doi.org/10.1016/S0140-6736(13)60142-7

Baumeister H et al (2019) Persuasive e-health design for behavior change. In: Baumeister H, Montag C (eds) Mobile sensing and digital phenotyping in psychoinformatics. Springer, Berlin, pp x–x

Bradley MM, Lang PJ (1994) Measuring emotion: the self-assessment manikin and the semantic differential. J Behav Ther Exp Psychiatry 25(1):49–59. https://doi.org/10.1016/0005-7916(94)90063-9

Cima RFF, Andersson G, Schmidt CJ, Henry JA (2014) Cognitive-behavioral treatments for tinnitus: a review of the literature. J Am Acad Audiol 25(1):29–61. https://doi.org/10.3766/jaaa.25.1.4

Czsikszentmihalyi M, Larson R (1987) Validity and reliability of the experience sample method. J Nerv Ment Dis 175(9):526–536

Elgoyhen AB, Langguth B, De Ridder D, Vanneste S (2015) Tinnitus: perspectives from human neuroimaging. Nat Rev Neurosci 16(10):632–642. https://doi.org/10.1038/nrn4003

Fredrickson BL (2000) Extracting meaning from past affective experiences: the importance of peaks, ends, and specific emotions. Cogn Emot 14(4):577–606. https://doi.org/10.1080/026999300402808

Goldberg RL, Piccirillo ML, Nicklaus J et al (2017) Evaluation of ecological momentary assessment for tinnitus severity. JAMA Otolaryngol Head Neck Surg 143(7):700–706. https://doi.org/10.1001/jamaoto.2017.0020

Götz FM, Stieger S, Reips U-D (2017) Users of the main smartphone operating systems (iOS, Android) differ only little in personality. PLoS ONE 12(5):e0176921–18. https://doi.org/10.1371/journal.pone.0176921

Henry JA, Galvez G, Turbin MB et al (2012) Pilot study to evaluate ecological momentary assessment of tinnitus. Ear Hear 32(2):179–290. https://doi.org/10.1097/AUD.0b013e31822f6740

Hiller W, Goebel G (2007) When tinnitus loudness and annoyance are discrepant: audiological characteristics and psychological profile. Audiol Neurotol 12(6):391–400. https://doi.org/10.1159/000106482

Kalle S, Schlee W, Pryss RC et al (2018) Review of smart services for tinnitus self-help, diagnostics and treatments. Front Neurosci 12:162–168. https://doi.org/10.3389/fnins.2018.00541

Kreuzer PM, Vielsmeier V, Langguth B (2013) Chronic tinnitus: an interdisciplinary challenge. Dtsch Arztebl Int 110(16):278–284. https://doi.org/10.3238/arztebl.2013.0278

Kubiak T, Smyth JM (2019) Connecting domains—ecological momentary assessment in a mobile sensing framework. In: Baumeister H, Montag C (eds) Mobile sensing and digital phenotyping in psychoinformatics. Springer, Berlin, pp x–x

Kuppens P (2015) It's about time: a special section on affect dynamics. Emot Rev 7(4):297–300. https://doi.org/10.1177/1754073915590947

Lim SL, Bentley PJ, Kanakam N et al (2014) Investigating country differences in mobile app user behavior and challenges for software engineering. IIEEE Trans Softw Eng 41(1):40–64. https://doi.org/10.1109/TSE.2014.2360674

Muniandi LP, Schlee W, Pryss R et al (2018) Finding tinnitus patients with similar evolution of their ecological momentary assessments. In: 2018 IEEE 31st international symposium on computer-based medical systems (CBMS), pp 112–117. https://doi.org/10.1109/cbms.2018.00027

Nondahl DM, Cruickshanks KJ, Huang G-H et al (2012) Generational differences in the reporting of tinnitus. Ear Hear 33(5):640–644. https://doi.org/10.1097/AUD.0b013e31825069e8

Probst T, Pryss R, Langguth B, Schlee W (2016a) Emotional states as mediators between tinnitus loudness and tinnitus distress in daily life: results from the "TrackYourTinnitus" application. Nature Publishing Group, pp 1–8. https://doi.org/10.1038/srep20382

Probst T, Pryss R, Langguth B, Schlee W (2016b) Emotional states as mediators between tinnitus loudness and tinnitus distress in daily life: results from the "TrackYourTinnitus" application, pp 1–8. https://doi.org/10.1038/srep20382

Probst T, Pryss R, Langguth B, Schlee W (2016c) Emotion dynamics and tinnitus: daily life data from the "TrackYourTinnitus" application. Nature Publishing Group, pp 1–9. https://doi.org/10.1038/srep31166

Probst T, Pryss RC, Langguth B et al (2017a) Does tinnitus depend on time-of-day? An ecological momentary assessment study with the "TrackYourTinnitus" application. Front Ag Neurosci 9:1–9. https://doi.org/10.3389/fnagi.2017.00253

Probst T, Pryss RC, Langguth B et al (2017b) Outpatient tinnitus clinic, self-help web platform, or mobile application to recruit tinnitus study samples? Front Ag Neurosci 9:49–7. https://doi.org/10.3389/fnagi.2017.00113

Schickler M, Pryss R, CBMS MR et al (2016) Using wearables in the context of chronic disorders: Results of a pre-study. ieeexploreieeeor. https://doi.org/10.1109/cbms.2016.40

Pryss R, Reichert M, Herrmann J (2015a) Mobile crowd sensing in clinical and psychological trials–a case study

Pryss R, Reichert M, Langguth B (2015b) Mobile crowd sensing services for tinnitus assessment, therapy, and research

Pryss R, Schlee W, Langguth B, Reichert M (2017) Mobile crowdsensing services for tinnitus assessment and patient feedback

Pryss R, Probst T, Schlee W et al (2018a) Prospective crowdsensing versus retrospective ratings of tinnitus variability and tinnitus–stress associations based on the TrackYourTinnitus mobile platform. Int J Data Sci Anal 43:1–12. https://doi.org/10.1007/s41060-018-0111-4

Pryss R, Reichert M, Schlee W et al (2018b) Differences between Android and iOS users of the TrackYourTinnitus Mobile Crowdsensing mHealth platform. In: 2018 IEEE 31st international symposium on computer-based medical systems (CBMS), pp 411–416

Schlee W, Schecklmann M, Lehner A et al (2014) Reduced variability of auditory alpha activity in chronic tinnitus. Neural Plast 2014. https://doi.org/10.1155/2014/436146

Schlee W, Pryss RC, Probst T et al (2016) Measuring the moment-to-moment variability of tinnitus: the TrackYourTinnitus smart phone app. Front Ag Neurosci 8:215–8. https://doi.org/10.3389/fnagi.2016.00294

Stone AA, Shiffman S (1994) Ecological momentary assessment (EMA) in behavioral medicine. Ann Behav Med 16(3):199–202. https://doi.org/10.1093/abm/16.3.199

Wilson MB, Kallogjeri D, Joplin CN et al (2015) Ecological momentary assessment of tinnitus using smartphone technology: a pilot Study. Otolaryngol Head Neck Surg 152(5):897–903. https://doi.org/10.1177/0194599815569692

Chapter 14
Mobile Crowdsensing in Healthcare Scenarios: Taxonomy, Conceptual Pillars, Smart Mobile Crowdsensing Services

Rüdiger Pryss

Abstract Recently, new paradigms like crowdsensing emerged in the context of mobile technologies that promise to support researchers in life sciences and the healthcare domain in a new way. For example, by the use of smartphones, valuable data can be quickly gathered in everyday life and then easily compared to other crowd users, especially when taking environmental factors or sensor data additionally into account. In the context of chronic diseases, mobile technology can particularly help to empower patients in coping with their individual health situation more properly. However, to utilize the achievements of mobile technology in the aforementioned contexts is still challenging. Following this, the work at hand discusses two important and relevant aspects for mobile crowdsensing in healthcare scenarios. First, the status quo of mobile crowdsensing technologies and their relevant perspectives on healthcare scenarios are discussed. Second, salient aspects are presented, which can help researchers to conceptualize mobile crowdsensing to a more generic software toolbox that is able to utilize data gathered with smartphones and their built-in sensors in everyday life. The overall toolbox goal is the support of researchers to conduct studies or analyzes on this new and less understood kind of data source. On top of this, patients shall be empowered to demystify their individual health condition more properly when using the toolbox, especially by exploiting the wisdom of the crowd.

14.1 Introduction

mHealth, which means to support medical and health questions/issues/aspects with mobile technology, becomes increasingly important in supporting patients with chronic diseases in their everyday life. Smartphones can be used to remind patients to take their medication, provide them with context information, and foremostly, contribute to the patients' self-help as patients can monitor their condition by performing

R. Pryss (✉)
Institute of Databases and Information Systems, Ulm University, James-Franck-Ring, 89081 Ulm, Germany
e-mail: ruediger.pryss@uni-ulm.de

© Springer Nature Switzerland AG 2019   221
H. Baumeister and C. Montag (eds.), *Digital Phenotyping and Mobile Sensing*,
Studies in Neuroscience, Psychology and Behavioral Economics,
https://doi.org/10.1007/978-3-030-31620-4_14

measurements, acquire insights on when the disease feels worst, and decide in a more informed way when they should ask a physician for help. Along this trend, clinicians as well as researchers try to exploit the advantages of mobile technology. As mHealth applications can gather data in everyday life and are able to easily integrate sensors (e.g., the GPS sensor to measure the walking speed) in order to collect more valuable data (e.g., by adding contextual information), *mHealth related opportunities* are more and more utilized (Kubiak and Smyth 2019; Messner et al. 2019; Pryss et al. 2019). In addition, compared to traditional cohort recruitment methods, the opportunities of smartphones may lead to the fact that large amounts of data can be collected in a rather short time (Rozgonjuk et al. 2019; Sariyska and Montag 2019; Vaid and Harari 2019). Moreover, patients often behave differently in a clinical setting compared to their daily life behavior and, hence, mHealth applications can collect data in an environment that reflects the daily behavior more accurately. Although technical solutions emerged that deal with such a data collection setting, such as using mobile crowdsensing, the opportunities, challenges, and risks are still less understood. For example, one opportunity is given by easily comparing retrospective health ratings of patients with their prospective assessments (Pryss et al. 2017a, b). Following such findings, especially in the context of chronic health conditions, opportunities arise that patients may be empowered to better demystify their health condition by the use of mHealth solutions. Mobile crowdsensing thereby refers to a recent paradigm, where individuals collectively gather and share data with their own smart mobile devices and it is mainly characterized by two recent trends. First, mobile crowdsensing became a popular technology through the so-called bring your own device principle. Private smartphones can be easily used for meaningful sensing purposes as almost all privately owned devices have powerful computational capabilities. Second, users are more and more interested to utilize these powerful capabilities to earn money or share data for a more common interest. Technically, mobile crowdsensing mainly focuses on the assessment of large-scale phenomena by using community sensing paradigms such as (1) participatory (i.e., actively provided sensor data such as pictures) or (2) opportunistic sensing approaches (i.e. passively provided sensor data such as GPS traffic tracking) (Ganti et al. 2011).

In this context, consider the following scenario for the chronical disease *migraine*: Petra Pain wakes up late on a Saturday morning after a hard and stressful week. Since she didn't get a lot of sleep and she couldn't attend her sports class on Monday and Thursday, she decides to do some running exercises after breakfast. Unfortunately, she cannot even enjoy her breakfast neither do sports since a severe migraine attack destroys her plans. Until Sunday night, she only will be able to stay in bed, feeling dizzy and suffering from the symptoms of her migraine. From a friend she learns about a mobile crowdsensing mHealth application, which helps her to structure her everyday life according to her personal needs and gives her a specific prevention and care program based on her data. She is given help (explanation), specific exercises, and contact to other patients and specialists through the mobile crowdsensing application. When and how the help is provided depends on her situation; i.e., data is collected about her body state, surrounding situation, and routines.

To exploit mHealth as needed by a migraine patient like Petra Pain, the following questions emerge for a technical mHealth crowdsensing solution:

Q1: How to collect, accommodate, combine, and store efficiently the data entered into the smartphone, either by the patients themselves or through sensors?
Q2: How to analyze those data, taking into account that some patients do not want to share their data, while others are willing to have their data analyzed for their own benefit, for the benefit of others, or the benefit of clinical research?
Q3: How to turn those methods into smart mobile crowdsensing services, to be used both by the patients themselves and by the clinical researchers, dealing with the needs, interests, and background knowledge of both target groups?

Smart mobile crowdsensing services can particularly contribute to patient self-empowerment, to personalized medicine, and to research on the evolution of a chronic disease. For example, the active performance of Ecological Momentary Assessments (Kubiak and Smyth 2019; Schlee et al. 2019) to capture the daily and individual moment-to-moment variability of patients and the sensing of the ambient environment can help the patients gain insights on how their symptom incidents manifest themselves and evolve, and ultimately assist them in demystifying their disease. Demystification is promoted by using the smartphone recordings as basis for the patient-physician interaction: instead of describing, for example, a migraine incident the way the patient remembers it long after the incident's onset, the description will be based on a combination of objective (sensor-based) data and patient diary entries from the moments of the incident. From this information, the physicians can adjust the treatment, but can also acquire understanding on different migraine profiles.

To unravel these opportunities, smart mobile crowdsensing services should aim at a data-driven collection procedure that can be used by (1) researchers as well as (2) patients and health care providers. Thereby, the data sources and evaluation possibilities should be utilized arising through the use of mobile devices in everyday life for disease management and questions related to life sciences. To enable this, a proper mobile crowdsensing toolbox should be built on three fundamental considerations:

First, the toolbox that shall be developed must be based on the requirements and needs of patients, health care providers and researchers. In this context, a data model must be derived that serves as a basis for a new data pool that stores data gathered by smartphones in everyday life. Following this, it must be investigated how existing data analysis methods can be applied to this data pool or whether new data analysis methods become necessary.

Second, the toolbox must comprise smart mobile crowdsensing services, which are based on real-life data gathered for patients suffering from chronic diseases such as migraine. Those patient groups are particularly interesting for clinical research since the health state of these persons is highly dependent on external factors of the personal environment and on internal factors of the body that can be measured by changes of several clinical indicators (via mobile technology and sensors). By using mobile sensor technology and high-end data analysis, a broad spectrum of situation-based data (including the moment-to-moment variability) can be collected and used as a basis for the development of preventive or curative health care services.

Third, as mobile applications are technically dependent on many factors (Schickler et al. 2015; Schobel et al. 2017), the smart mobile crowdsensing services should be realized as a sustainably specialized IT solution in the context of chronic conditions, patient self-empowerment, life sciences, and the opportunities arising with the proliferation of mobile technology. Consequently, if the functionality is bound to several smart mobile crowdsensing services instead of a particular mobile health application, gathered data can be compared more easily to other types of data sources (e.g., clinical-driven data collection).

The present chapter provides an overview on this important topic, subdivided into six sections organized as follows. In the next section, current research on mobile crowdsensing is discussed. Then, section *Taxonomy* discusses mobile crowdsensing in healthcare scenarios compared to existing approaches. In the sections *Conceptual Pillars of Smart Mobile Crowdsensing Services for Healthcare Scenarios* and *Development of Smart Mobile Crowdsensing Services for Healthcare Scenarios*, important technical foundations are presented. Finally, sections *Discussion and Summary and Outlook*, summarize findings to establish these services in healthcare scenarios and provide an outlook on future work.

14.2   Related Work

Existing research that is relevant in the context of this work constitutes mobile crowdsensing in the medical domain as well as works that deal with mobile crowdsensing enabling Ecological Momentary Assessments (EMA; also known as ambulatory assessment and experience sampling) methods (Kubiak and Smyth 2019; Schlee et al. 2019; Trull and Ebner-Priemer 2013). Mobile crowdsensing is an emerging research topic in various application domains (Luo et al. 2017; Shu et al. 2017). In the healthcare domain, however, this research direction has been has been less picked up so far. The fact that the healthcare domain is less considered by contemporary crowdsensing approaches might be explainable by legal and data privacy issues (Christin et al. 2011). Nevertheless, mobile crowdsensing offers promising perspectives for the medical domain (Ganti et al. 2011), as it exhibits unique features for gathering valuable patient data in the large scale (Demirbas et al. 2010). In particular, mobile crowdsensing allows for the effective, context-aware gathering (Ma et al. 2014) of daily-life patient data (Probst et al. 2016), which, in turn, may shift clinical research to a new level. Furthermore, some recent works exist that that deal with generic crowdsensing approaches to enable human-subject studies (Xiong et al. 2016).

Furthermore, apart from mobile crowdsensing solutions, mobile applications have been presented that incorporate EMA-measurements with valuable healthcare results. In particular, EMA approaches are used to capture various aspects such as pain or feelings in daily life. Although most mental health symptoms are subjective experiences and, thus, most EMA approaches use self-reports to capture these symptoms, some symptoms are behavioral (e.g., avoidance in anxiety disorders) or physiological (e.g., increase of heart rate in anxiety disorders). Especially in this context, mobile applications offer powerful opportunities to measure behavioral or physiological data in daily life (Ebner-Priemer and Kubiak 2007; Montag et al. 2019; Sariyska et al. 2018). Altogether, EMA-driven approaches provide unprecedented opportunities to study mental health symptoms and life science factors under ecologically valid conditions (Myin-Germeys et al. 2009), even though the integration of EMA-related possibilities into mobile crowdsensing solutions is still in its infancy, especially in the medical/healthcare domain. However, as mobile crowdsensing and EMA have already revealed valuable insights for tinnitus (Kubiak and Smyth 2019; Probst et al. 2016; Schlee et al. 2019) and provided help in the context of pregnancy (Ruf-Leuschner et al. 2016), the investigation of mobile technology for life sciences and the medical/healthcare domain is promising. In particular, new data pools, new evaluation methods, and algorithms as well as related tool developments are promising outcomes when investigating this technology in an in-depth manner. Based on this, the work at hand proposes smart mobile crowdsensing services for healthcare scenarios.

14.3 Taxonomy

Prior to considerations how smart mobile crowdsensing services can be realized for healthcare scenarios, another important aspect must be introduced, i.e., the way how mobile crowdsensing has been hitherto characterized in literature and existing technical solutions. Currently, existing works put mobile crowdsensing into one of the following two main categories. Either a participatory (C1) or an opportunistic sensing (C2) approach is pursued (Ganti et al. 2011; Ma et al. 2014). Besides, several existing works do no put their crowdsensing approaches into C1 or C2 and rather describe realized technical infrastructures and their benefits (C3) (Xiong et al. 2016). A brief introduction into C1–C3 will be given in the following and on top of this it is shown that mobile crowdsensing in healthcare scenarios cannot be put directly into C1–C3. Consider therefore Fig. 14.1, in which a new, so-called healthcare-driven mobile crowdsensing, is proposed.

The three existing approaches C1–C3, in turn, have the following characteristics:

Category 1: Participatory Sensing (C1): Sensing tasks (e.g., measure sound level at a certain location) are advertised through a crowdsensing platform and crowd users that are bound to this platform deliberately opt for accomplishing these tasks. Main goal of the participatory sensing is to find the most suitable crowd user for a sensing task in a reasonable period of time. Crowd users either get an incentive to take over such tasks (e.g., by getting money) or want to serve the society by accomplish sensing

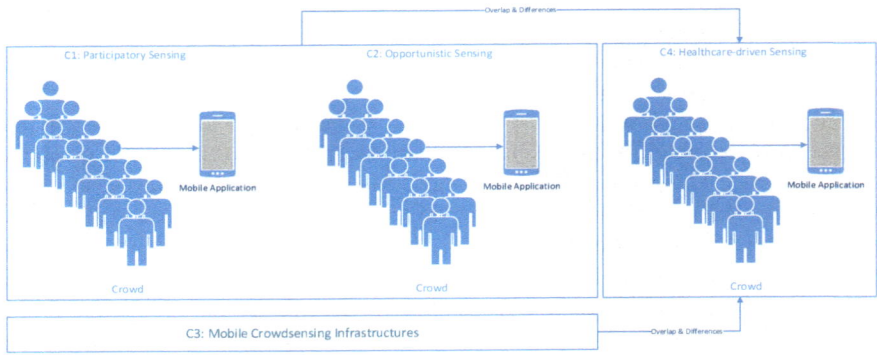

Fig. 14.1 Mobile crowdsensing taxonomy including healthcare scenarios

tasks for a common interest (e.g., by gathering data that can be used to create a noise map for hearing-impaired people).

Category 2: Opportunistic Sensing (C2): Opposed to the participatory sensing, here, crowd users do not deliberately take over sensing tasks, they rather install a crowd-sensing client on their smartphone, which collects data in an automatic and user-unconscious fashion. Beside this difference to C1, opportunistic sensing follows the same goals than C1.

Category 3: Mobile Crowdsensing Infrastructures (C3): There exist several approaches that do not explicitly distinguish whether their realized infrastructure follows the participatory or opportunistic sensing paradigm, they rather describe how a crowdsensing infrastructure or parts of it can be efficiently developed. For example, in the work of (Xiong et al. 2016) it is described how an infrastructure may look like to support human-subject studies. Another example is the sharing of GPS-based traffic information to better cope with congestions or other vehicle-related aspects (Wan et al. 2016).

To justify that a category that is called healthcare-driven mobile crowdsensing (C4: a new paradigm for crowdsensing in health care) should be separately considered, it is presented that the basic goal of the existing approaches C1–C3 differs significantly compared to C4. To be more specific, approaches that follow C1–C3 mainly address the following goal: Sensing tasks exist and the question must be answered, how a suitable crowd user can be found and selected to accomplish the task in the best possible way (in most cases this means timely or inexpensively). As a direct effect of this goal, it is not important who actually accomplishes a sensing task, it is rather important that the crowd is utilized in the best possible way for upcoming sensing tasks. As an example, it is crucial to find the closest crowd user for a task that says to sense the temperature at a certain location. Following this, a fundamental research question of approaches following C1–C3 is to investigate and develop recruitment methods for a sensing task (e.g. Karaliopoulos et al. 2015). Therefore, a proper user motivation or an efficient incentive management are major research targets in this context. Another definition would be that the sensing tasks can be seen as a trade currency, which is utilized by the crowd users as well as the crowdsensing platform vendors to pursue one of the two aforementioned goals.

In healthcare mobile crowdsensing scenarios, in turn, the direction of how sensing tasks are related to executing crowd users is very different. Here, the main goal is that crowd users are the important target and the two main questions arise, (1) how sensing tasks are applied to the crowd users and (2) which data aspects are correlated and actually evaluated. In other words, in healthcare-driven mobile crowd-sensing scenarios, the initial considerations do not start with the sensing task, they start with the crowd user and it is then determined, how a user accomplishes sensing tasks. As another fundamental difference of C4 compared to C1–C3, in most cases, a crowd user shall continuously accomplish the same sensing task to investigate within-day and day-by-day variations of the same sensing task. This shall help to (1) monitor an individual crowd user over time and to (2) enable comparisons with other crowd users that have a similar evolution. For approaches that follow C1–C3, in turn, crowd users opt for different sensing tasks, without pursuing the goal to continuously accomplish the same sensing task. Based on the fundamental difference, i.e., whether the sensing task or the crowd user is the starting point, the addressed research questions differ respectively. For example, in scenarios related to C4, motivation means to find solutions that crowd users do not leave the platform, while in approaches following C1–C3, motivation means to find solutions to increase platform usage. Finally note that several aspects between the different paradigms, although C1–C3 and C4 differ, overlap. For example, proper infrastructures are required for C1–C4. Furthermore, security aspects play an important role in all categories as well. Having in mind that there is a profound difference in using mobile crowdsensing for healthcare scenarios, the considerations for respective smart services can be better aligned to the pursued goals in this context.

14.4   Conceptual Pillars

The presented concept for smart mobile crowdsensing services in healthcare scenarios is given by the following existing and ongoing projects: TrackYourTinnitus (Tinnitus) (Pryss et al. 2017b), KINDEX (Pregnancy) (Ruf-Leuschner et al. 2016), https://www.trackyourhearing.org (Hearing Loss), https://www.trackyourstress.org (Stress), as well as the involvement in studies on diabetes patients in Bulgaria and Spain (EU Grant Agreement Number 761307). Based on this groundwork, the proposed smart mobile crowdsensing are built on the following conceptual pillars: (1) The development of crowdsensing collection procedures for healthcare scenarios, (2) the knowledge about data sets in this context, and the (3) behavior patterns of patients and users in such a collection setting as well as the needs of researchers. Regarding Pillar (1), consider Fig. 14.2, it reflects the crowdsensing collection procedure being revealed in the aforementioned projects. This procedure essentially pursues three goals: First, data shall be collected on a daily basis (cf. Figs. 14.2, 4). However, a

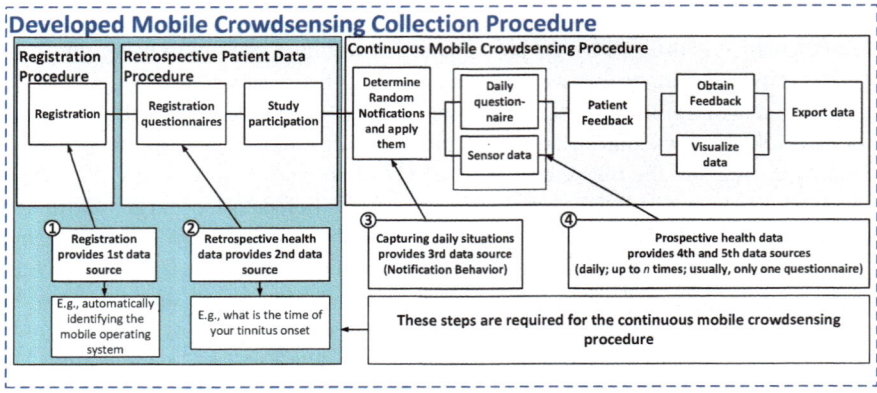

Fig. 14.2 Developed mobile crowdsensing collection procedure

user shall not foresee the times he or she is asked to sense data (cf. Figure 14.2, 3). This is ensured by asking the users in various daily life situations. Second, the collected data shall enable new kinds of data analytics like juxtaposing real-time assessments and retrospective reports (cf. Fig. 14.2, 2; Pryss et al. 2017a, 2018a), or evaluating the different assessment times of a day (Probst et al. 2017). Third, gathered data shall be used to provide feedback to the mobile crowd users. Although the conducted projects revealed that a structured collection procedure as shown in Fig. 14.2 is indispensable, its systematic use is still neglected and less understood. However, a systematic and structured use would enable a more generic utilization as well as allow for more flexible collection procedures in this context, thus potentially leading to more valuable data and analysis opportunities.

As can be also obtained from Fig. 14.2, such a mobile crowdsensing collection procedure reveals different possibilities to establish valuable types of data sources. For example, a data source that reflects retrospective data (i.e., a retrospective assessment of a patient) and one that reflects prospective data (i.e., a prospective assessment of a patient when being asked in a daily situation or on future outcomes).

Taking the different types of data sources into account, regarding Pillar (2), the used data set structure in this context is important for the development of smart mobile crowdsensing services. Therefore, consider Fig. 14.3, it shows an example for such a data set structure used in the aforementioned projects and based on the collection procedure shown in Fig. 14.2. When mobile crowdsensing data is collected along such data set structure, new data pools can be established, which, in turn, may reveal valuable medical insights of a crowd user over time. Regarding the development of new data pools in this context, consider the following realistic example: For a migraine patient, several data sets can be gathered utilizing the procedure shown in Fig. 14.2 on a daily level. Let us assume that 12 gathered data sets each day are realistic and not too burdensome for a migraine patient. Then, when multiplying this with 7 days a week and 52 weeks a year, eventually, for one migraine patient, possibly 4,368 data sets become possible. If one thousand crowd users are applying

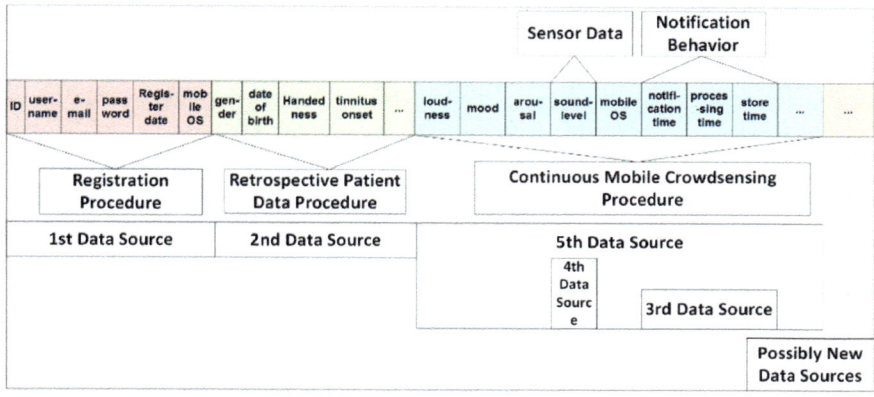

Fig. 14.3 TrackYourTinnitus crowdsensing data set structure

this procedure in the context of their migraine situation, it can be assumed that in one year with only 1,000 users, 4,368,000 potential data sets (i.e., each including all the information shown in Fig. 14.3) become possible. However, not only the amount of data is an important advantage and a data management challenge at the same time, many other evaluations become possible, e.g.:

- Juxtaposing retrospective assessments of patients with their daily assessments.
- Evaluating user behavior of different mobile operating systems compared to their provided data.
- Comparing users with similar and different sociodemographic (e.g., gender, age, education, nationality, ethnicity) and clinical characteristics (e.g. disease status, chronicity, disease progress, comorbidities) to provide benchmark and prediction models.
- Adding sensor data to the gathered patient data and compare subjective assessments with more *objective* assessments.
- By changing the procedure of gathering data it can be observed how patients may change their behavior accordingly.
- Gathering longitudinal medical data in a more efficient way, especially without the potential bias of experimenters compared to traditional clinical trials.

Comparing this structured data amongst all users (for those who share personal anonymous data sets with others) or in a longitudinal study for single users (for those who only track their own personal data sets) allows the development of preventive and care services that are custom-fit for individual users based on the power of the crowd. Finally, Pillar (3) constitutes behavior patterns of users in this context as well as the needs of researchers. For example, when the first data is collected for a migraine patient, it might be the case that patients lose interest to provide data or use the technical solution in general. In this case, an incentive management (e.g., by the use of gamification aspects) becomes necessary for the smart mobile crowdsensing services (e.g. Agrawal et al. 2018).

14.5 Development Phases

In the previous section, conceptual pillars for the development of smart mobile crowd-sensing services in healthcare scenarios were shown. In this section, the development phases that have been identified to realize such services in practice are presented. Prior to this, it is shortly sketched what technical components should be always assumed for a mobile crowdsensing platform in healthcare scenarios. In the mobile crowdsensing projects that have been realized as the basis for the work at hand, 6 fundamental technical components have been identified as a proper basis to enable a useful mobile crowdsensing platform for healthcare settings. First of all, an (1) Android as well as an iOS mobile app must be developed to gather data. Principally, also platform-independent approaches for realizing the apps are a conceivable solution. However, as sensor measurements are performed, which need mobile operating system calls, native apps should be preferred. Second, these apps must be connected to a (2) RESTful API (Pryss et al. 2018b) that governs the communication with (3) a central relational or NoSQL database to which the data is stored to. Fourth, (4) a web application must be developed, which can be used by the crowd users as well as researchers and healthcare professionals to analyze and visualize the collected data. In addition, the website is used for further administrative needs (e.g., user management). Fifth, (5) a sensor component must be realized, which handles the collection of sensor data. Finally, (6) a feedback component must be developed that enables healthcare professionals and researchers to send feedback to a mobile crowd user. To get a better insight into such components, for the TrackYourTinnitus project, they are described in (Pryss et al. 2015). Note that these components are the basis to offer smart mobile crowdsensing services based on the conceptual pillars shown in the previous section. In order to realize the technical components practically, the following 6 development phases have been elaborated, to be accomplished in the shown order:

Phase 1—Requirements and Needs: In this phase, the relevant roles (e.g., patients, researchers, doctors), their requirements (e.g., support of Android and/or iOS smartphones), and their needs (e.g., notification algorithms) must be identified and assessed.

Phase 2—Crowdsensing Collection Procedure: In this phase, the mobile crowdsensing collection procedure must be defined with the help of researchers as well as the healthcare professionals (e.g., medical doctors or psychologists).

Phase 3—Data Model: A data model (i.e., the data set structure) that drives all phases of the mobile crowdsensing collection procedure must be defined. In addition, a data quality model must be defined over the data model.

Phase 4—Data Analysis Methods: All functions to be used for the later data analysis must be defined. For example, exploiting patient similarity could be a research question that shall be dealt with. Therefore, it must be defined whether existing methods can be used or new methods become necessary.

Phase 5—App User Journey and Synchronization Procedure: In this phase, based on the outcomes of Phases 2–4, two procedures must be defined. **First**, the so-called user journey for the app usage must be determined. That means, the dialogue structure (i.e., the provided user views and their logical interdependencies) for using the mobile apps based on the specified mobile crowdsensing collection procedure must be defined. For the TrackYourTinnitus project, the user journey can be found in Agrawal et al. (2018). To this end, the user journey translates the mobile crowdsensing collection procedure into a procedure that users have to follow when using the mobile crowdsensing apps. **Second**, the synchronization procedure must be defined, i.e., the strategy at which points in time a mobile crowdsensing app synchronizes its collected data with the central relational or NoSQL database through the RESTful API. In addition, it must be defined whether the collected data is only stored into the central database or also locally on the mobile crowdsensing apps. Finally, it must be defined how data is cached on the mobile crowdsensing apps. Caching becomes necessary if an app has no internet connection for a longer period of time. For the TrackYourTinnitus project, the synchronization strategy can be found at https://www.trackyourtinnitus. org/process.pdf. In addition, the synchronization procedure is a crucial point for ad hoc analyzes that shall be directly performed on the smart mobile devices of crowd users. For example, if a crowd user wants to compare his assessments over the last week with other crowd users, it must be determined which data should be considered and is located where.

Phase 6—System Evaluation: In this phase, an evaluation service must be developed, which is the fundamental pillar of a service quality framework. The latter shall provide features to be able to continuously monitor the smart mobile crowdsensing services over time.

If the considerations of the previous section are taken into account during the presented 6 development phases, then, the 6 technical components discussed in the beginning of this section can be realized in a way that they can actually provide smart mobile crowdsensing services for healthcare scenarios in a flexible and powerful way. Exemplarily, for the TrackYourTinnitus project, a selected service that takes all these considerations into account—namely, the patient feedback service—can be found in (Pryss et al. 2017b).

Finally, the crucial points of the three presented technical sections are summarized, which are the basis to enable helpful mobile crowdsensing services for patients like the fictive case of Petra Pain as illustrated in the Introduction.

First of all, it was elaborated that the experiences of the performed practical implementations of several crowdsensing platforms have revealed that the offered crowdsensing features should be bound to smart mobile crowdsensing services. Second, it was shown that mobile crowdsensing in healthcare scenarios is different in its nature to non-healthcare scenarios. Third, the revealed conceptual pillars for the smart mobile crowdsensing services were presented. They take the nature of healthcare aspects particularly into account. Fourth, development phases and technical components that are necessary to practically realize these services were shown. For the conducted mobile crowdsensing projects, these aspects were key success factors to maintain and evolve a mobile crowdsensing platform for healthcare scenarios. A recently proposed mobile crowdsensing infrastructure that offers sophisticated mobile crowdsensing services for a healthcare scenario can be found in (Kraft et al. 2019).

14.6 Summary and Outlook

The work at hand discussed mobile crowdsensing technology in the context of chronic diseases in particular and healthcare scenarios in general. To this end, three findings that were discussed along the conducted projects are of particular importance. On one hand, mobile crowdsensing can provide substantial data sources that may lead to new insights for a medical condition or be the basis for patients and health care providers to understand patients' individual health condition and disease courses in a better way. On the other hand, from a more technical perspective, two more findings are important. First, mobile crowdsensing, when used in healthcare scenarios, poses different characteristics to non-healthcare scenarios. This refers particularly to the relationship between sensing tasks and crowd users. To be more precise, in healthcare scenarios, sensing tasks do not exist a priori like for the well-known participatory and opportunistic crowdsensing paradigms. Rather, the sensing tasks are decided by the crowd users themselves. In addition, in healthcare scenarios, results of the sensing tasks of crowd users are compared over time, which is, again, not important for the aforementioned well-known sensing approaches. Second, if mobile crowdsensing shall be used as a generic toolbox in healthcare scenarios, many considerations are necessary, which have been less addressed so far. In particular, the presented considerations on the mobile crowdsensing collection procedure are decisive aspects. Altogether, it can be argued that the opportunities surpass the drawbacks and therefore mobile crowdsensing can play an important role in healthcare scenarios. Future efforts of the work at hand mainly address more powerful configuration opportunities for the smart mobile crowdsensing services on one hand. On the other, incentive management features with the goal to prevent crowd users to early leave a mobile crowdsensing platform will be a main research focus. Successfully implemented, combining ecological momentary procedures and smart sensing possibilities for individuals with the power of crowd-knowledge (i.e. cross individual sensing), will ultimately improve insights into diseases and their individual courses, allowing for just in time interventions tailored to the specific needs of individual patients.

References

Agrawal K, Mehdi M, Reichert M et al (2018) Towards incentive management mechanisms in the context of crowdsensing technologies based on TrackYourTinnitus insights. In: The 15th international conference on mobile systems and pervasive computing, Gran Canaria, Spain, 13–15 August 2018. Procedia Computer Science, Elsevier Science, pp 145–152

Christin D, Reinhardt A, Kanhere SS, Hollick M (2011) A survey on privacy in mobile participatory sensing applications. J Syst Softw 84(11):1928–1946. https://doi.org/10.1016/j.jss.2011.06.073

Demirbas M, Ali Bayir M, Akcora CG et al (2010) Crowd-sourced sensing and collaboration using twitter. In: 2010 IEEE international symposium on "A World of Wireless, Mobile and Multimedia Networks" (WoWMoM), Montreal, QC, Canada, 14–17 June 2010. IEEE, pp 1–9

Ebner-Priemer UW, Kubiak T (2007) Psychological and psychophysiological ambulatory monitoring. Eur J Psychol Assess 23(4):214–226. https://doi.org/10.1027/1015-5759.23.4.214

Ganti R, Ye F, Lei H (2011) Mobile crowdsensing: current state and future challenges. IEEE Commun Mag 49(11):32–39. https://doi.org/10.1109/MCOM.2011.6069707

Karaliopoulos M, Telelis O, Koutsopoulos I (2015) User recruitment for mobile crowdsensing over opportunistic networks. In: 2015 IEEE conference on computer communications (INFOCOM), Kowloon, Hong Kong, 26 April–1 May 2015. IEEE, pp 2254–2262

Kraft R, Birk F, Reichert M et al (2019) Design and implementation of a scalable crowdsensing platform for geospatial data of tinnitus patients. In: 32nd IEEE CBMS international symposium on computer-based medical systems (CBMS 2019), Cordoba, Spanien, 5–7 June 2019. IEEE

Kubiak T, Smyth JM (2019) Connecting domains—ecological momentary assessment in a mobile sensing framework. In: Montag C, Baumeister H (eds) Mobile sensing and digital phenotyping: new developments in psychoinformatics. Springer, Berlin, pp xx–xx

Luo T, Kanhere SS, Huang J et al (2017) Sustainable incentives for mobile crowdsensing: auctions, lotteries, and trust and reputation systems. IEEE Commun Mag 55(3):68–74. https://doi.org/10.1109/MCOM.2017.1600746CM

Ma H, Zhao D, Yuan P (2014) Opportunities in mobile crowd sensing. IEEE Commun Mag 52(8):29–35. https://doi.org/10.1109/MCOM.2014.6871666

Messner E-M, Probst T, O'Rourke T et al (2019) mHealth applications: potentials, limitations, current quality and future directions. In: Montag C, Baumeister H (eds) Mobile sensing and digital phenotyping: new developments in psychoinformatics. Springer, Berlin, pp xx–xx

Montag C, Baumeister H, Kannen C et al (2019) Concept, possibilities and pilot-testing of a new smartphone application for the social and life sciences to study human behavior including validation data from personality psychology. J 2(2):102–115. https://doi.org/10.3390/j2020008

Myin-Germeys I, Oorschot M, Collip D et al (2009) Experience sampling research in psychopathology: opening the black box of daily life. Psychol Med 39(9):1533–1547. https://doi.org/10.1017/S0033291708004947

Probst T, Pryss R, Langguth B, Schlee W (2016) Emotional states as mediators between tinnitus loudness and tinnitus distress in daily life: Results from the "TrackYourTinnitus" application. Sci Rep 6(1):20382. https://doi.org/10.1038/srep20382

Probst T, Pryss RC, Langguth B et al (2017) Does tinnitus depend on time-of-day? An ecological momentary assessment study with the "TrackYourTinnitus" application. Front Aging Neurosci 9:253. https://doi.org/10.3389/fnagi.2017.00253

Pryss R, Reichert M, Langguth B, Schlee W (2015) Mobile crowd sensing services for tinnitus assessment, therapy, and research. In: 2015 IEEE international conference on mobile services, New York City, NY, USA, 27 June–2 July 2015. IEEE, pp 352–359

Pryss R, Probst T, Schlee W et al (2017a) Mobile crowdsensing for the juxtaposition of realtime assessments and retrospective reporting for neuropsychiatric symptoms. In: 2017 IEEE 30th international symposium on computer-based medical systems (CBMS), Thessaloniki, Greece, 22–24 June 2017. IEEE, pp 642–647

Pryss R, Schlee W, Langguth B, Reichert M (2017b) Mobile crowdsensing services for tinnitus
    assessment and patient feedback. In: 2017 IEEE international conference on AI & mobile services
    (AIMS), Honolulu, HI, USA, 25–30 June 2017. IEEE, pp 22–29
Pryss R, Probst T, Schlee W et al (2018a) Prospective crowdsensing versus retrospective ratings
    of tinnitus variability and tinnitus–stress associations based on the TrackYourTinnitus mobile
    platform. Int J Data Sci Anal: 1–12. https://doi.org/10.1007/s41060-018-0111-4
Pryss R, Schobel J, Reichert M (2018b) Requirements for a flexible and generic API enabling
    mobile crowdsensing mHealth applications. In: 2018 4th international workshop on requirements
    engineering for self-adaptive, collaborative, and cyber physical systems (RESACS), Banff, AB,
    Canada, 20 August 2018. IEEE, pp 24–31
Pryss R, Kraft R, Baumeister H et al (2019) Using Chatbots to support medical and psychological
    treatment procedures. In: Montag C, Baumeister H (eds) Mobile sensing and digital phenotyping:
    new developments in psychoinformatics. Springer, Berlin, pp xx–xx
Rozgonjuk D, Elhai JD, Hall BJ (2019) Studying psychopathology in relation to smartphone use.
    In: Montag C, Baumeister H (eds) Mobile sensing and digital phenotyping: new developments
    in psychoinformatics. Springer, Berlin, pp xx–xx
Ruf-Leuschner M, Brunnemann N, Schauer M et al (2016) The KINDEX-App—an instrument
    for assessment and immediate analysis of psychosocial risk factors in pregnant women in
    daily practice by gynecologists, midwives and in gynecological hospitals. Verhaltenstherapie
    26(3):171–181. https://doi.org/10.1159/000448455
Sariyska R, Montag C (2019) Smartphone supported psychodiagnostics in the assessment of per-
    sonality and physical activity. In: Montag C, Baumeister H (eds) Mobile sensing and digital
    phenotyping: new developments in psychoinformatics. Springer, Berlin, pp xx–xx
Sariyska R, Rathner E-M, Baumeister H, Montag C (2018) Feasibility of linking molecular genetic
    markers to real-world social network size tracked on smartphones. Front Neurosci 12:945. https://
    doi.org/10.3389/fnins.2018.00945
Schickler M, Reichert M, Pryss R et al (2015) Entwicklung mobiler Apps: Konzepte, Anwendungs-
    bausteine und Werkzeuge im Business und E-Health. Springer, Berlin, Heidelber
Schlee W, Kraft R, Schobel J et al (2019) Momentary assessment of tinnitus—how smart mobile
    applications advance our understanding of tinnitus. In: Montag C, Baumeister H (eds) Mobile
    sensing and digital phenotyping: new developments in psychoinformatics. Springer, Berlin, pp
    xx–xx
Schobel J, Pryss R, Schlee W et al (2017) Development of mobile data collection applications
    by domain experts: experimental results from a usability study. In: Dubois E, Pohl K (eds)
    Advanced information systems engineering, CAiSE 2017. Lecture notes in computer science.
    Springer International Publishing, Cham, pp 60–75
Shu L, Chen Y, Huo Z et al (2017) When mobile crowd sensing meets traditional industry. IEEE
    Access 5:15300–15307. https://doi.org/10.1109/ACCESS.2017.2657820
Trull TJ, Ebner-Priemer U (2013) Ambulatory assessment. Annu Rev Clin Psychol 9(1):151–176.
    https://doi.org/10.1146/annurev-clinpsy-050212-185510
Vaid SS, Harari GM (2019) Smartphones in personal informatics: Self-tracking with mobile sensing
    for behavior change. In: Montag C, Baumeister H (eds) Mobile sensing and digital phenotyping:
    new developments in psychoinformatics. Springer, Berlin, pp xx–xx
Wan J, Liu J, Shao Z et al (2016) Mobile crowd sensing for traffic prediction in internet of vehicles.
    Sensors 16(1):88. https://doi.org/10.3390/s16010088
Xiong H, Huang Y, Barnes LE, Gerber MS (2016) Sensus: a cross-platform, general-purpose system
    for mobile crowdsensing in human-subject studies. In: Proceedings of the 2016 ACM international
    joint conference on pervasive and ubiquitous computing—UbiComp '16, Heidelberg, Germany,
    12–16 September 2016. ACM Press, pp 415–426

Chapter 15
mHealth Applications: Potentials, Limitations, Current Quality and Future Directions

Eva-Maria Messner, Thomas Probst, Teresa O'Rourke, Stoyan Stoyanov and Harald Baumeister

Abstract Due to the constant use of smartphones in daily life, mHealth apps might bear great potential for the use in health care support. In this chapter the potentials, limitations, current quality and future directions of mHealth apps will be discussed. First, we describe potential benefits like quicker facilitation of information, patient empowerment and inclusion of undersupplied population groups. Furthermore, the use of mHealth apps for diverse somatic and mental health conditions will be discussed. Beyond, the chapter provides the reader with a short overview on the efficacy of mHealth apps for different indications: Exemplary, we provide evidence for the efficacy of mHealth apps in the realm of asthmatic disease, depression and anxiety disorder. Despite the availability of mHealth solutions, the acceptance of among health care providers is still moderate to low. This represents a substantial problem, as health care providers are important gate keepers for intervention uptake. In this context we describe methods to foster acceptance. Furthermore, we address potential risks of mHealth app use including low responsiveness towards critical situations (e.g. self-harm) or the difficulty for users to assess the quality of the app's content. Here we refer to standardized instruments to assess app quality. With respect to the massive amount of sensitive data already being collected through such mHealth apps, we also reflect on the latest current legal situation in Europe and the United States.

E.-M. Messner (✉) · H. Baumeister
Clinical Psychology and Psychotherapy, Ulm University, Ulm, Germany
e-mail: eva-maria.messner@uni-ulm.de

T. Probst · T. O'Rourke
Psychotherapy and Biopsychosocial Health, Danube University Krems, Krems an der Donau, Austria

S. Stoyanov
Centre for Children's Health Research, Institute of Health and Biomedical Innovation and School of Psychology and Counselling, Queensland University of Technology, Brisbane, Australia

© Springer Nature Switzerland AG 2019
H. Baumeister and C. Montag (eds.), *Digital Phenotyping and Mobile Sensing*,
Studies in Neuroscience, Psychology and Behavioral Economics,
https://doi.org/10.1007/978-3-030-31620-4_15

15.1   Current Use of Smartphones

Smartphones are an integral part of daily life. They are widely used all over the world with around 65% owning a smartphone in Europe and America and 33% worldwide (Donner 2008). Especially in developing countries the number of smartphone owners is rising, resulting in major social and economic changes (Marcolino et al. 2018). Given the ubiquitous availability of smartphones, the health care sector might profit from their large distribution. In this context, especially the developing countries have hope, that mHealth apps can complement their routine care while reducing costs (Beratarrechea et al. 2014; Donner 2008; Gurman et al. 2012). In general younger individuals are more likely to own a smartphone than older, leading to diverse digital behaviour across generations (Albrecht 2016). A study from Germany demonstrated that the average person between 15 and 35 years carries the smartphone around all day and spends an average of 162 min using it, thus making smartphones a great opportunity to track behaviour in real life (Montag et al. 2015) particularly in the younger generation at present.

In the context of the present work, it is of interest, that currently 325.000 mHealth apps are available in the app stores (Research2guidance 2016; Statista 2019). The explosion in the number of available mobile health apps is mainly due to their economic profitability: The turnover has quintupled to $23 billion since 2013. The (scientific) assessment of available apps, their potential and limitations as well as app quality and data security issues for end-users and practitioners is therefore as challenging as urgently needed in this rapidly growing secondary healthcare market.

15.2   Current Use of mHealth Apps

Around 60% of smartphone users have at least one mobile Health app (mHealth app) installed and around 76% of individuals report interest in using their smartphone to monitor and improve mental health (Proudfoot et al. 2010). In the last four years the number of downloads of mHealth apps has doubled, indicating a rising acceptance of mHealth app use (Research2guidance 2016). Users of mHealth apps report to use the mHealth app for recording of body and fitness data (27%), educational purposes related to health (20%), to assist in behaviour/lifestyle modification (11%) and to improve health management regarding medication and vaccination (2%). Krebs and Duncan (2015) found that in America 58% have downloaded a mHealth app and used it on a daily basis. In their sample mHealth app users were younger, had higher income and a higher educational status. The most common categories of app use were fitness and nutrition.

There are no gender differences in the frequency of mobile health app use (Albrecht 2016). People between the age of 18 and 29 use mobile health apps the most, people between the age of 30 and 59 use mHealth apps moderately, whilst persons above the age of 60 rarely use mHealth apps at the moment (Albrecht 2016).

Since elderly people are prone to chronic disease, they have a higher need of support which might not always be covered comprehensively by health care professionals, resulting in the potential use of mHealth apps to empower elderly patient's self-help abilities. Hence, the training of technology literacy particularly for the elderly could be promising to complement traditional care systems (Singh et al. 2016). Moreover, the age threshold of mHealth app use most probably will further increase with the next generation of elderly who by then will already show substantial mHealth literacy. Hence, digital natives will probably more easily embrace this new (self-help) intervention approaches.

15.3 Potential Benefits of mHealth Apps

MHealth apps have the potential to transform the way health services are delivered by quicker facilitation of health information (Marcolino et al. 2018). Future health care systems will rely on synchronized care processes shared by diverse care givers, resulting in the need of embedded broad information systems. Furthermore, patient empowerment (e.g. seeing the patient as a major driver of documentation, healing and health behaviour) will be emphasized in the future (Nasi et al. 2015). Those developments could be assisted through mHealth technology. Moreover, there is great potential for new app technologies, because population groups currently being undersupplied might profit from their use. Among these are people from rural areas, people from developing countries, children and adolescents, elderly people, disabled people and minorities (Donner 2008; Marcolino et al. 2018; Singh et al. 2016). Mobile health apps could be used independently of geographical situation, language abilities and further barriers to treatment. They can be used as assistance in clinical diagnostics, for the assessment of disease course, in blended therapy settings, as self-help tool while waiting for routine care or in relapse prevention (Baumeister et al. 2017; Ebert et al. 2018; Rathner et al. 2018a). Assistance in diagnostics and measurements of change can be achieved through active user input or passive behavioural tracking (Rathner et al. 2018b; Sariyska et al. 2018). Furthermore, they could complement standard care by giving less stigmatized access to therapeutic content or medical treatment (Bloomfield et al. 2014; Torous et al. 2014).

Moreover, there is evidence from systematic reviews that mHealth apps can help to improve treatment adherence overall and especially in chronic or stigmatized disease management (Beratarrechea et al. 2014; Bloomfield et al. 2014; Hamine et al. 2015; Marcolino et al. 2018). Such improvements in adherence are rooted in the embedding of interventions into everyday life by using real-time situation triggered reminders, pushes and notifications (Marcolino et al. 2018). In addition, mobile health apps offer the opportunity for real-time interventions (e.g. triggering a breathing exercise in case of acute need) and for automated data input (e.g. passive tracking of smartphone usage behaviour, movement data, sleeping times) as well as manual data input (e.g. mental state, homework) (Kubiak and Smyth 2019; Rabbi et al. 2019).

Furthermore, people with chronic diseases represent a significant target group. People suffering from chronic conditions can benefit in a great extent from lifestyle changes with regard to the prospective disease process. Moreover, it might be cost-effective if individuals with high treatment needs and corresponding high treatment costs would use (blended) internet-and mobile based interventions (IMIs) (Bendig et al. 2018; Singh et al. 2016). In this sense IMIs also might help to empower patients to a self-determined health management (Bendig et al. 2018). The use of mobile health apps verifiably leads to higher levels of autonomy and increase perceived self-efficacy. Mobile health apps have the potential to assist people with and without clinical diagnosis to promote desired behaviours (Bakker et al. 2016).

The use and the prescription of mHealth interventions might be of particular interest, as patients believe that these apps could improve quality of care through better communication (Krebs and Duncan 2015). Moreover, from a patient's perspective, other domains of mHealth apps are of relevance. Individuals with mental health needs report, that they think content (90.8%), ease of use (89.6%), cost (79.2%), encryption (74.2%), interactive features (73.7%), customization (70.9%), privacy policy (70.5%), direct and indirect research evidence (68.1%), simple language (60.7%), user ratings (59.4%) and user reviews (58.7%) are important in mHealth adoption. Furthermore, burdened individuals highlight, that they value ease of use (27%), visual appeal (18.2%), simple language (17.4%) and content (14.4%) in mHealth apps they already use (Schueller et al. 2018).

15.4 Evidence on the Efficacy of mHealth Apps

Worldwide mHealth apps are mainly used for patient communication, monitoring, education, for disease management, to facilitate health services, to improve clinical diagnostics, to foster treatment adherence and for the management of chronic diseases (Devi et al. 2015; Gurman et al. 2012). Although they are widely used and there is a common belief that mHealth apps can improve the quality of care, lead to a reduction in costs and can be adapted on large scale, strong evidence on the effectiveness and cost-effectiveness is still lacking in most areas (Ebert et al. 2017). Note that this lack of evidence is specific for mHealth, while there is strong evidence for the effectiveness and cost-effectiveness of Internet-based (therapeutically guided) self-help interventions (Bendig et al. 2018; Domhardt et al. 2018; Ebert et al. 2018; Paganini et al. 2018).

In a systematic review of systematic reviews, Marcolino et al. (2018) came to the conclusion that mHealth apps are helpful for individuals who suffer from asthmatic disease in regard to a better symptom control, decrease in medication, increased treatment adherence and a reduction in hospitalizations when compared to a treatment as usual control group. Patients in cardiac rehabilitation profit from mHealth app use with respect to their exercise capacity as well as in reduction of blood pressure and body mass index. Patients with congestive heart failure reported fewer symptoms and their relative risk of death or hospitalisation decreased, while their quality of life increased when using mHealth interventions. Individuals suffering from diabetes profited in various bio-parameters such as HbA_{1c}, cholesterol and microalbuminuria. The treatment adherence and viral load of people suffering from HIV could be improved by using mHealth apps. Regarding lifestyle changes including promotion of physical activity, smoking cessation and safe sexual behaviour mHealth seems promising, too. However, the latter assumption relies on an expert opinion rather than on empirical evidence (Marcolino et al. 2018).

Firth et al. (2017a) conducted a meta-analysis on randomized controlled trials (RCTs) of depression mobile apps. They showed that depressive symptoms decrease significantly when using a mHealth app when compared to active (e.g. treatment as usual) or inactive (e.g. waitlist control) control groups. MHealth Apps focusing on improvement of general mental health were superior to those focusing on cognitive training. These results indicate that mHealth apps are a promising self-help tool to manage depression. In another meta-analysis and systematic review Firth et al. (2017b) found that mHealth apps are also beneficial for individuals suffering from anxiety disorders. Taken together these findings build a solid body of evidence for the efficacy of mHealth apps in specific disorders and point towards possible efficacy in others. The mechanisms of change while using mHealth interventions are still unknown (Domhardt et al. 2019) and might stem from factors in the individual, characteristics of the mHealth app and general attitudes of the socioeconomic system.

15.5   Acceptance of mHealth Apps in Health Care Providers

Despite the positive findings on the effectiveness of mHealth apps there is moderate to low acceptance towards the use of mHealth apps among health care providers (Krebs and Duncan 2015). This represent a highly relevant problem since end users (here patients) can be influenced by expert opinions (East and Havard 2015). The acceptance of mHealth apps depends on the characteristics of the intervention as well as internal factors related to the patient (e.g. knowledge, competences and attitudes) and external factors (e.g. policy, health care, technical and institutional resources, attitudes of the social system) (Liu et al. 2014; Phillips et al. 2015). Health care providers are mainly concerned about data security and low responsiveness towards critical situations (e.g. reactivity if a person is self-harming or a hazard towards others). Scepticism about privacy especially occurs when health care providers do not feel adequately informed or have little technology literacy (Gagnon et al. 2016). The highest acceptance among health care professionals for the use of e- and mHealth interventions can be registered in the areas of prevention and relapse prevention, rehabilitation, self-help and psychoeducation (Surmann et al. 2017).

Teenagers and young adults form the most relevant target group according to practitioners (Hennemann et al. 2016). With regard to therapeutic indication, the acceptance for the use of new technologies in practitioners is the highest among their use in the treatment of depression and anxiety disorders (Liu et al. 2014). Furthermore, age and familiarity with technologies have a significant influence on the attitude and acceptance towards digital health offers (Baumeister et al. 2015; Lin et al. 2018). Practitioners with a higher assurance in handling technologies expect a higher benefit of mHealth apps (Gagnon et al. 2016).

15.6 Acceptance of mHealth Apps in Patients

There is a high acceptance of low-threshold mobile health applications for recording disease course (tracking) among outpatients. Overall 61% would like to receive text messages from their health care providers, 73% would like to access general health care information via the phone, 70% would like to download and use an app to track their mental health condition daily. People over the age of 60 show less interest in using mHealth apps (Torous et al. 2014). In individuals who suffer from depression, anxiety or stress the interest in using mHealth apps was higher when symptoms were present compared to symptom free episodes (Proudfoot et al. 2010). In younger individuals, like university students, the acceptance of the use of apps in health behaviour change is high. Self-reported app use in students is influenced by worries about security, required effort and the immediate effect on mood when using the mHealth app (Dennison et al. 2013). In the general population lack of interest, cost, concern about data collection and data storage are barriers in the adoption of mHealth apps (Krebs and Duncan 2015). In individuals with mental health needs the barriers to uptake and adoption of mHealth apps are doubts about effectiveness (31.4%), finding the right app (27.3%), costs (13.7%), lack of interest (11.1%), concerns about privacy and data security (10.7%), lack of time to use the app (6.6%), lack of space on the personal phone (6%) and usability issues (5%) (Schueller et al. 2018).

The attitude of patients suffering from conditions such as diabetes, depression or pain towards the use of mHealth could be influenced by acceptance facilitation interventions (e.g. a video). Acceptance facilitating interventions subsume any methods that are likely to influence an individual's opinion (e.g. video clips, workshops, advertisements, articles, etc.). In people seeking pain-management the information video changed their attitude towards the use of mHealth significantly. In chronic pain patients this overall effect was not present. Specific subgroups such as younger and female patients as well as patients with a higher actual burden, could be influenced by the video. As a consequence, acceptance-promoting interventions (e.g. videos, brochures, advanced education, journal articles and training) unfold their potential when targeted to the group (Baumeister et al. 2014, 2015; Ebert et al. 2015; Liu et al. 2014).

Half of the individuals who use a mHealth app stop using it after a while. Reasons to stop using mHealth apps are high data entry burden, loss of interest and hidden costs (Krebs and Duncan 2015). Thus, highlighting the importance of passive tracking and the inclusion of persuasive technology to maintain user adherence (Baumeister et al. 2019).

15.7 Risks Associated with mHealth App Use

The CHARISMHA study (Albrecht 2016), a study dealing with the topic of the use of mobile technologies in health-related areas from various perspectives, points to the following risks for mHealth app usage: lack of functionality, dissemination of false information, misdiagnosis, mistreatment and unknown unwanted side effects. An important concern regarding mHealth app use, is the lack of reactivity of algorithms in case of emergency (e.g. self-endangerment or hazards of others). Singh et al. (2016) showed that only 23% of mobile health apps responded adequately to dangerous user input (e.g.: suicidal ideations). This illustrates the enormous need of improvement in terms of responsiveness of mHealth apps in potentially dangerous situations.

Furthermore, it is hard for patients to detect a helpful app and to assess the quality of the content by themselves (Krebs and Duncan 2015). Individuals mainly retrieve information about mHealth apps in social media (45.1%), through personal search (e.g.: in the app stores, via Google or in web forums like Reedit (42.7%) or with the help of a family member or friend (36.9%). Formal sources of information such as health care providers were only used by 24.6% of the individuals (Schueller et al. 2018).

To enable a safe use of mHealth apps, regarding the quality of the therapeutic content, it is necessary to define internationally accepted quality criteria and to distribute this knowledge on a broad scale in an easy understandable way (e.g. quality seals, databases of expert ratings, etc.). In regard to informed health care decisions and patient empowerment, it is essential for patients to have access to standardised ratings of mHealth apps, particularly since the quality of user star ratings is questionable (Terhorst et al. 2018). In regard to safe app use three criteria should be evaluated (1) quality of the therapeutic content, (2) functionality and (3) data safety and protection (Neary and Schueller 2018). While (1) can be reached through open access standardized psychometrically valid expert or user ratings, (2) can be guaranteed through the implementation of a medical device law (e.g. IEC82304 in Europe) and (3) is addressed by data protection regulations such as explicated for example by the European Union.

One current problem is, that many of the above mentioned scientifically examined mHealth apps do not appear in the app stores. In contrast, the majority of available mHealth apps are not tested for effectiveness. To assess the quality of available mHealth apps, our research group evaluated in Germany available mobile health apps for treatment of depression and anxiety disorders as well as for supporting physical activity with a standardized internationally accepted procedure, the mobile applications rating scale (MARS; Stoyanov et al. 2015). To collect the available mobile health apps we developed a web crawler, a program that circumvents the filter settings of the app stores (iTunes, Windows and google play). The ratings of the experts were verified in a further step by an editor and after that published open access on the page www.mhad.science, so that end-users, practitioners and service providers are empowered to take informed health decisions. Overall, the results are eye-opening: the expert rating of all German-speaking depression apps revealed that only 25% of these apps correspond to national guidelines for the treatment of depression. Only 10% of the rated depression apps could be recommended at least based on minimal standards for blended-care or self-help (Terhorst et al. 2018). In terms of mobile health apps that are offered for support and treatment of anxiety disorders in the European app stores, we found similar results. Efficacy studies were present in less than 1% of the rated anxiety apps and none of the depression apps. Anxiety apps developed by universities or non-governmental non-profit organizations (NGO's) showed a higher overall quality. Behaviour therapy alignment as well as the number of offered exercises of the mHealth apps resulted in an elevated quality rating. There was no difference with regard to app quality in relation to the app store and the price of the mHealth app. Analogous to previous studies there was no connection between the star ratings of the app stores (usually given by the app users) and standardized expert ratings (Messner et al., in prep.). One possible explanation of this missing correlation is the lack of knowledge among end users about the applied therapeutic methodology/background as well as the tendency of end-users to judge usability and design instead of effectiveness. In summary, it can be stated that available mHealth apps in Western countries such as Germany with usually high demands regarding patient safety are far below their potential possibilities concerning quality and safety of therapeutic content.

15.8 Standardized Instruments to Assess App Quality

Diverse quality assessment approaches in the evaluation of mHealth apps tend to have reached relative consensus. Current mHealth app quality assessment tools and recommendations tend to focus on a set of criteria such as evidence, information quality, security, engagement and ease of use (often linked to user experience—UX), functionality and visual design. Widely-used tools include the Mobile App Rating Scale (MARS; Stoyanov et al. 2015), the American Psychiatric Association—recommended App Evaluation Model (American Psychiatric Association 2019), and the ASPECTS guide (Torous et al. 2016). Such instruments allow health organizations and information hubs (i.e. Psyberguide.org; KidsHelpline.com.au; Reachout. com; mhad.science) to evaluate and recommend high-quality apps to their clients with increased confidence. Nevertheless, while strongly related to each other, *quality* does not equal *efficacy*. *Effectiveness* and *efficacy* evaluation of mHealth apps remains a time-consuming, complicated and costly process but nevertheless a hot topic in digital health research (Zanaboni et al. 2018). Until international efficacy evaluation and regulation standards are developed, quality ratings may be used for guidance only, but cannot guarantee safety and effectiveness of the intervention.

Another instrument for the evaluation of mHealth programs is Enlight (Baumel et al. 2017), a multidimensional, criteria-based set of scales that assesses the quality of mHealth interventions independently of clinical aim (health related behavior or mental health) or delivery medium (mobile app-based interventions or web-based interventions). It includes two novel concepts that have not been assessed by previous scales: therapeutic persuasiveness and therapeutic alliance. Therapeutic persuasiveness refers to persuasive design, which directly influences users' behavior and a program's therapeutic potential (Webb et al. 2010). Therapeutic alliance is one of the most robust measures for predicting psychotherapy success (Klein et al. 2003) and was included in the quality assessment section, as it has been shown that therapeutic alliances between users and online intervention programs do exist (Bickmore et al. 2005). The instrument consists of a quality assessment section and a checklist section. The quality assessment is comprised of the six core dimensions usability, visual design, user engagement, content, therapeutic persuasiveness, and therapeutic alliance. In contrast, the checklists are based on criteria that do not directly impact the user's experience and include credibility, privacy explanation, basic security, and evidence-based program ranking. The checklist scores are calculated by summing up the scores in each of the categorical items, apart from the evidence-based program ranking, as it is based on a five-point scale.

15.9 Current Legal Situation of mHealth Apps

The so far not adopted regulation of data safety in mHealth apps raises concerns about mHealth app quality and safety of use (Powell et al. 2014). The advances in soft- and hardware development made it possible to collect a vast amount of behavioural and medical data at low cost. This data is very sensitive by nature and an adequate protection is crucial for the safe use of mHealth applications (Kargl et al. 2019; Papageorgiou et al. 2018). The majority of available mobile health apps do not have an understandable privacy policy, so the end users do not know what data will be recorded, how the data will be transferred, where the data be stored and with whom the data will be shared (Messner et al., in prep.; Terhorst et al. 2018). In general privacy policies lack intelligibility (Papageorgiou et al. 2018). This situation will hopefully change with the implementation of the General Data Protection Regulation of the European Union (GDPR) and the statement of the US Food and Drug administration (FDA). Due to their novelty there is no research on the impact of those regulations on app quality so far.

The new GDPR states a fundamental right of data security for individuals. Every organization has to disclose their practices of data processing, data sharing, and data storage. In case of violation of the GDPR companies can be fined with 4% of annual revenue or up to 20 million€. Users are more seen as owners of their own data, leading to the theory of a basic right to access and delete one's own data as well as to receive an understandable declaration of consent. The informed consent has to be actively accepted per click. In addition, software which is used for administration, maintenance or improvement of the health of individuals succumbs to the "Medical Devices Law" under IEC82304 since May 2017 (International Organisation for Standardization 2016). Thus, increased demands apply on the functionality of mobile health apps. Both legislative changes have the potential to contribute to the improvement of functionality and data security of mHealth apps in the future.

The current situation in the US is more liberal. The US Food and Drug administration (FDA) grades mHealth apps into categories regarding to their potential harm on the user: apps that are medical devices which require regulatory oversight, apps that are medical devices which will not be regulated and apps that are no medical device. Regarding to BinDhim et al. (2015) most mHealth apps will not be classified as regulated medical device and therefore receive no formal regulation, resulting in a need of further quality assessment.

15.10   Summary

Mobile health apps offer great potential with regard to the complementation of routine care, especially in the developing world and for specific subgroups (e.g.: elderly, minorities, rural regions, etc.). The majority of individuals own a smartphone and already use at least one mHealth app. The acceptance of mHealth apps in health care providers is average to low. Concerns regarding the use of apps in health care are mainly about data and user safety as well as effectiveness. There is preliminary evidence for the effectiveness of mHealth app usage. Despite that fact, for the implementation of mHealth apps into routine care social legislation and institutional framework, regulations regarding data safety and an adequate procedure for quality assessment are missing. There are first efforts to increase transparency of app quality through standardized expert and user ratings.

mHealth is a complex domain which requires diverse expertise of health professionals, researchers, designers, developers, legal advisors, and importantly the end-users themselves. It is, therefore, understandable that the development and implementation of best practices and regulations is a slow process. After all, Internet data privacy regulation took decades and a lot of ethical breaches before reaching its current state in 2018. Yet, it is important to adapt and become more agile in our policy-making practices, as the 21st century is yet to expand the new horizons of artificial intelligence and robotics well beyond app use. We should carefully consider our role and contribution towards building the foundations of scientifically, societal and ethically-sound approaches to digital health, to ensure the safe and effective integration of new technologies into healthcare.

References

Albrecht U-V (2016) Chancen und Risiken von Gesundheits-Apps (CHARISMHA). Universitäts-bibliothek der Technischen Universität Braunschweig
American Psychiatric Association (2019) App evaluation model. https://www.psychiatry.org/psychiatrists/practice/mental-health-apps/app-evaluation-model. Accessed 02 Aug 2019
Bakker D, Kazantzis N, Rickwood D, Rickard N (2016) Mental health smartphone apps: review and evidence-based recommendations for future developments. JMIR Ment Health 3(1):e7. https://doi.org/10.2196/mental.4984

Baumeister H, Nowoczin L, Lin J et al (2014) Impact of an acceptance facilitating intervention on diabetes patients' acceptance of internet-based interventions for depression: a randomized controlled trial. Diabetes Res Clin Pract 105(1):30–39. https://doi.org/10.1016/j.diabres.2014.04.031

Baumeister H, Seifferth H, Lin J et al (2015) Impact of an acceptance facilitating intervention on patients' acceptance of internet-based pain interventions: a randomized controlled trial. Clin J Pain 31(6):528–535. https://doi.org/10.1097/AJP.0000000000000118

Baumeister H, Lin J, Ebert DD (2017) Internet- und mobilebasierte Ansätze: Psychosoziale Diagnostik und Behandlung in der medizinischen Rehabilitation. Bundesgesundheitsblatt Gesundheitsforschung Gesundheitsschutz 60(4):436–444. https://doi.org/10.1007/s00103-017-2518-9

Baumeister H, Pryss R, Baumel A, Messner E-M (2019) Persuasive e-health design for behavior change. In: Montag C, Baumeister H (eds) Mobile sensing and digital phenotyping: new developments in psychoinformatics. Springer, Berlin

Baumel A, Faber K, Mathur N et al (2017) Enlight: a comprehensive quality and therapeutic potential evaluation tool for mobile and web-based eHealth interventions. J Med Internet Res 19(3):e82. https://doi.org/10.2196/jmir.7270

Bendig E, Bauereiß N, Ebert DD et al (2018) Internet-based interventions in chronic somatic disease. Dtsch Arztebl Int 115(40). https://doi.org/10.3238/arztebl.2018.0659

Beratarrechea A, Lee AG, Willner JM et al (2014) The impact of mobile health interventions on chronic disease outcomes in developing countries: a systematic review. Telemed J E Health 20(1):75–82. https://doi.org/10.1089/tmj.2012.0328

Bickmore T, Gruber A, Picard R (2005) Establishing the computer–patient working alliance in automated health behavior change interventions. Patient Educ Couns 59(1):21–30. https://doi.org/10.1016/j.pec.2004.09.008

BinDhim NF, Shaman AM, Trevena L et al (2015) Depression screening via a smartphone app: cross-country user characteristics and feasibility. J Am Med Inform Assoc 22(1):29–34. https://doi.org/10.1136/amiajnl-2014-002840

Bloomfield GS, Vedanthan R, Vasudevan L et al (2014) Mobile health for non-communicable diseases in sub-saharan africa: a systematic review of the literature and strategic framework for research. Glob Health 10(1):1–9. https://doi.org/10.1186/1744-8603-10-49

Dennison L, Morrison L, Conway G, Yardley L (2013) Opportunities and challenges for smartphone applications in supporting health behavior change: qualitative study. J Med Internet Res 15(4):e86. https://doi.org/10.2196/jmir.2583

Devi BR, Syed-Abdul S, Kumar A et al (2015) mHealth: an updated systematic review with a focus on HIV/AIDS and tuberculosis long term management using mobile phones. Comput Methods Programs Biomed 122(2):257–265. https://doi.org/10.1016/j.cmpb.2015.08.003

Domhardt M, Steubl L, Baumeister H (2018) Internet- and mobile-based interventions for mental and somatic conditions in children and adolescents: a systematic review of meta-analyses. Z Kinder Jugendpsychiatr Psychother: 1–14. https://doi.org/10.1024/1422-4917/a000625

Domhardt M, Geßlein H, von Rezori RE, Baumeister H (2019) Internet- and mobile-based interventions for anxiety disorders: a meta-analytic review of intervention components. Depress Anxiety 36(3):213–224. https://doi.org/10.1002/da.22860

Donner J (2008) Research approaches to mobile use in the developing world: a review of the literature. Inf Soc 24(3):140–159. https://doi.org/10.1080/01972240802019970

East ML, Havard BC (2015) Mental health mobile apps: from infusion to diffusion in the mental health social system. JMIR Ment Health 2(1):e10. https://doi.org/10.2196/mental.3954

Ebert DD, Berking M, Cuijpers P et al (2015) Increasing the acceptance of internet-based mental health interventions in primary care patients with depressive symptoms. A randomized controlled trial. J Affect Disord 176:9–17. https://doi.org/10.1016/j.jad.2015.01.056

Ebert DD, Cuijpers P, Muñoz RF, Baumeister H (2017) Prevention of mental health disorders using internet- and mobile-based interventions: a narrative review and recommendations for future research. Front Psychiatry 8:116. https://doi.org/10.3389/fpsyt.2017.00116

Ebert DD, Van Daele T, Nordgreen T et al (2018) Internet- and mobile-based psychological interventions: applications, efficacy, and potential for improving mental health. Eur Psychol 23(2):167–187. https://doi.org/10.1027/1016-9040/a000318

Firth J, Torous J, Nicholas J et al (2017a) The efficacy of smartphone-based mental health interventions for depressive symptoms: a meta-analysis of randomized controlled trials. World Psychiatry 16(3):287–298. https://doi.org/10.1002/wps.20472

Firth J, Torous J, Nicholas J et al (2017b) Can smartphone mental health interventions reduce symptoms of anxiety? A meta-analysis of randomized controlled trials. J Affect Disord 218:15–22. https://doi.org/10.1016/j.jad.2017.04.046

Gagnon M-P, Ngangue P, Payne-Gagnon J, Desmartis M (2016) m-Health adoption by healthcare professionals: a systematic review. J Am Med Inform Assoc 23(1):212–220. https://doi.org/10.1093/jamia/ocv052

Gurman TA, Rubin SE, Roess AA (2012) Effectiveness of mHealth behavior change communication interventions in developing countries: a systematic review of the literature. J Health Commun 17(sup1):82–104. https://doi.org/10.1080/10810730.2011.649160

Hamine S, Gerth-Guyette E, Faulx D et al (2015) Impact of mHealth chronic disease management on treatment adherence and patient outcomes: a systematic review. J Med Internet Res 17(2):e52. https://doi.org/10.2196/jmir.3951

Hennemann S, Rudolph FM, Waldeck E et al (2016) Online-Gesundheitsprogramme in der stationären Rehabilitation: Akzeptanz und Bedarf bei Mitarbeitern und Rehabilitanden. In: Deutsche Rentenversicherung Bund (ed) 25. Rehabilitationswissenschaftliches Kolloquium, 109th edn. Deutsche Rentenversicherung Bund, Berlin, pp 141–143

International Organisation for Standardization (2016) IEC 82304-1: Health software—part 1 general requirements for product safety. https://www.iso.org/standard/59543.html. Accessed 02 Aug 2019

Kargl F, Van der Heijden RW, Erb B, Bösch C (2019) Privacy in mobile sensing. In: Montag C, Baumeister H (eds) Mobile sensing and digital phenotyping: new developments in psychoinformatics. Springer, Berlin

Klein DN, Schwartz JE, Santiago NJ et al (2003) Therapeutic alliance in depression treatment: controlling for prior change and patient characteristics. J Consult Clin Psychol 71(6):997–1006. https://doi.org/10.1037/0022-006X.71.6.997

Krebs P, Duncan DT (2015) Health app use among US mobile phone owners: a national survey. JMIR mHealth uHealth 3(4):e101. https://doi.org/10.2196/mhealth.4924

Kubiak T, Smyth JM (2019) Connecting domains—ecological momentary assessment in a mobile sensing framework. In: Montag C, Baumeister H (eds) Mobile sensing and digital phenotyping: new developments in psychoinformatics. Springer, Berlin

Lin J, Faust B, Ebert DD et al (2018) A web-based acceptance-facilitating intervention for identifying patients' acceptance, uptake, and adherence of internet- and mobile-based pain interventions: randomized controlled trial. J Med Internet Res 20(8):e244. https://doi.org/10.2196/jmir.9925

Liu L, Miguel Cruz A, Rios Rincon A et al (2014) What factors determine therapists' acceptance of new technologies for rehabilitation—a study using the unified theory of acceptance and use of technology (UTAUT). Disabil Rehabil 37(5):447–455. https://doi.org/10.3109/09638288.2014.923529

Marcolino MS, Oliveira JAQ, D'Agostino M et al (2018) The impact of mHealth interventions: systematic review of systematic reviews. JMIR mHealth uHealth 6(1):e23. https://doi.org/10.2196/mhealth.8873

Messner E-M, Terhorst Y, Baumeister H (in prep.) When the fear kicks in. A systematic review and evaluation of apps that tackle anxiety. J Anxiety Disord

Montag C, Błaszkiewicz K, Sariyska R et al (2015) Smartphone usage in the 21st century: who is active on WhatsApp? BMC Res Notes 8(1):331. https://doi.org/10.1186/s13104-015-1280-z

Nasi G, Cucciniello M, Guerrazzi C (2015) The role of mobile technologies in health care processes: the case of cancer supportive care. J Med Internet Res 17(2):e26. https://doi.org/10.2196/jmir.3757

Neary M, Schueller SM (2018) State of the field of mental health apps. Cogn Behav Pract 25(4):531–537. https://doi.org/10.1016/j.cbpra.2018.01.002

Paganini S, Teigelkötter W, Buntrock C, Baumeister H (2018) Economic evaluations of internet- and mobile-based interventions for the treatment and prevention of depression: a systematic review. J Affect Disord 225:733–755. https://doi.org/10.1016/j.jad.2017.07.018

Papageorgiou A, Strigkos M, Politou E et al (2018) Security and privacy analysis of mobile health applications: the alarming state of practice. IEEE Access 6:9390–9403. https://doi.org/10.1109/ACCESS.2018.2799522

Phillips CJ, Marshall AP, Chaves NJ et al (2015) Experiences of using the theoretical domains framework across diverse clinical environments: a qualitative study. J Multidiscip Healthc 8:139–146. https://doi.org/10.2147/JMDH.S78458

Powell AC, Landman AB, Bates DW (2014) In search of a few good apps. JAMA 311(18):1851–1852. https://doi.org/10.1001/jama.2014.2564

Proudfoot J, Parker G, Hadzi Pavlovic D et al (2010) Community attitudes to the appropriation of mobile phones for monitoring and managing depression, anxiety, and stress. J Med Internet Res 12(5):e64. https://doi.org/10.2196/jmir.1475

Rabbi M, Klasnja P, Choudhury T et al (2019) Optimizing mHealth interventions with a bandit. In: Montag C, Baumeister H (eds) Mobile sensing and digital phenotyping: new developments in psychoinformatics. Springer, Berlin

Rathner E-M, Djamali J, Terhorst Y et al (2018a) How did you like 2017? Detection of language markers of depression and narcissism in personal narratives. In: Proceedings Interspeech 2018. ISCA, pp 3388–3392

Rathner E-M, Terhorst Y, Cummins N et al (2018b) State of mind: classification through self-reported affect and word use in speech. In: Proceedings Interspeech 2018. ISCA, pp 267–271

Research2guidance (2016) mHealth App Developer Economics 2016

Sariyska R, Rathner E-M, Baumeister H, Montag C (2018) Feasibility of linking molecular genetic markers to real-world social network size tracked on smartphones. Front Neurosci 12:945. https://doi.org/10.3389/fnins.2018.00945

Schueller SM, Neary M, O'Loughlin K, Adkins EC (2018) Discovery of and interest in health apps among those with mental health needs: survey and focus group study. J Med Internet Res 20(6):e10141. https://doi.org/10.2196/10141

Singh K, Drouin K, Newmark LP et al (2016) Many mobile health apps target high-need, high-cost populations, but gaps remain. Health Aff 35(12):2310–2318. https://doi.org/10.1377/hlthaff.2016.0578

Statista (2019) Number of mHealth apps available in the Apple App Store from 1st quarter 2015 to 2nd quarter 2019 [Graph]. In Statista. https://www.statista.com/statistics/779910/health-apps-available-ios-worldwide/. Accessed 12 Aug 2019

Stoyanov SR, Hides L, Kavanagh DJ et al (2015) Mobile app rating scale: a new tool for assessing the quality of health mobile apps. JMIR mHealth uHealth 3(1):e27. https://doi.org/10.2196/mhealth.3422

Surmann M, Bock EM, Krey E et al (2017) Einstellungen gegenüber eHealth-Angeboten in Psychiatrie und Psychotherapie: Eine Pilotumfrage auf dem DGPPN-Kongress 2014. Nervenarzt 88(9):1036–1043. https://doi.org/10.1007/s00115-016-0208-8

Terhorst Y, Rathner E-M, Baumeister H, Sander L (2018) "Hilfe aus dem App-Store?": Eine systematische Übersichtsarbeit und Evaluation von Apps zur Anwendung bei Depressionen. Verhaltenstherapie 28(2):101–112. https://doi.org/10.1159/000481692

Torous J, Friedman R, Keshavan M (2014) Smartphone ownership and interest in mobile applications to monitor symptoms of mental health conditions. JMIR mHealth uHealth 2(1):e2. https://doi.org/10.2196/mhealth.2994

Torous JB, Chan SR, Yellowlees PM, Boland R (2016) To use or not? Evaluating ASPECTS of smartphone apps and mobile technology for clinical care in psychiatry. J Clin Psychiatry 77(6):e734–e738. https://doi.org/10.4088/JCP.15com10619

Webb TL, Joseph J, Yardley L, Michie S (2010) Using the internet to promote health behavior change: a systematic review and meta-analysis of the impact of theoretical basis, use of behavior change techniques, and mode of delivery on efficacy. J Med Internet Res 12(1):e4. https://doi.org/10.2196/jmir.1376

Zanaboni P, Ngangue P, Mbemba GIC et al (2018) Methods to evaluate the effects of internet-based digital health interventions for citizens: systematic review of reviews. J Med Internet Res 20(6):e10202. https://doi.org/10.2196/10202

Chapter 16
Using Chatbots to Support Medical and Psychological Treatment Procedures: Challenges, Opportunities, Technologies, Reference Architecture

Rüdiger Pryss, Robin Kraft, Harald Baumeister, Jens Winkler, Thomas Probst, Manfred Reichert, Berthold Langguth, Myra Spiliopoulou and Winfried Schlee

Abstract The advent of chatbots may influence many treatment procedures in the medical and psychological fields. In particular, chatbots may be useful in many situations before and after medical procedures when patients are back at home. For example, while being in the preparation phase of a colonoscopy, a chatbot might answer patient questions more quickly than a doctor. Moreover, it is more and more discussed whether chatbots may be the first entry point for (urgent) medical questions instead of the consultation of a medical expert, as there exist already well-established algorithms for some of these situations. For example, if a new medical symptom occurs, a chatbot might serve as the first "expert" to relieve a patient's condition. Note that the latter situation to use chatbots is mainly driven by the trend that patients often have to wait too long for appointments with a proper medical expert due to capacity problems of many healthcare systems. While the usage of supporting "at home actions" of patients with chatbot technologies is typically welcomed by medical experts, the use of this technology to "replace" them in their core competence, namely diagnosis and therapy, is generally seen highly critical. Apart from the

R. Pryss (✉) · R. Kraft · J. Winkler · M. Reichert
Institute of Databases and Information Systems, Ulm University, James-Franck-Ring, 89081 Ulm, Germany
e-mail: ruediger.pryss@uni-ulm.de

H. Baumeister
Department of Clinical Psychology and Psychotherapy, Ulm University, Ulm, Germany

T. Probst
Department for Psychotherapy and Biopsychosocial Health, Danube University Krems, Krems an der Donau, Austria

B. Langguth · W. Schlee
Clinic and Policlinic for Psychiatry and Psychotherapy, University of Regensburg, Regensburg, Germany

M. Spiliopoulou
Department of Technical and Business Information Systems, Otto-von-Guericke-University Magdeburg, Magdeburg, Germany

© Springer Nature Switzerland AG 2019
H. Baumeister and C. Montag (eds.), *Digital Phenotyping and Mobile Sensing*,
Studies in Neuroscience, Psychology and Behavioral Economics,
https://doi.org/10.1007/978-3-030-31620-4_16

249

domain side, it must be carefully considered what currently available chatbot technologies can do or cannot do. Moreover, it has also to be considered, how existing technologies can be established in highly critical medical and interdisciplinary fields with possible emergency situations (e.g., if a chatbot gets the message of a patient that indicates to commit suicide), involving ethical questions as well as questions of responsibility and accountability. Therefore, this work raises aspects that might be the basis for medical as well as technical experts to better work together for proper chatbot solutions. Thereby, the work at hand proposes an architecture that should serve as a reference for various medical and psychological scenarios. When using suitable technical solutions, we argue that chances emerge, which mitigate upcoming challenges significantly.

16.1   Introduction

Although the idea of chatbots, i.e., a smart software application that is able to simulate a conversion with human users, especially for the medical and psychological fields, is not a recent one (Weizenbaum 1966), the development for its widespread use in the aforementioned contexts is interestingly a recent trend, which is mainly driven by three facts. First, the computational capabilities have been dramatically increased during the last decades. For example, research achievements in the field of artificial intelligence have fanned the use of chatbot technology in the medical context (e.g. Crutzen et al. 2011). Second, the way how present healthcare systems are operated compel many patients to consider new treatment opportunities. For example, for many patients with mental health problems, access to adequate healthcare offers is limited (Ebert et al. 2018). These common waiting time problems across countries and medical conditions might have crucial drawbacks (Siciliani et al. 2014). Evidence based chatbots may help to overcome this global limitation of timely access to health care offers. Third, in many treatment procedures, patients have to accomplish tasks when being at home. While performing these tasks, patients often feel uncomfortable or they have questions how to accomplish their tasks properly. For example, in the preparation phase of a colonoscopy, patients have often many questions and concerns. However, it is not always possible to quickly talk to a medical expert, what might affect the performance of the colonoscopy. In such settings, medical experts are open for new technologies like chatbots. Although the trend is plausible with respect to the presented reasons, the following three questions emerged in practical projects we are involved in when technical solution shall be implemented:

- If chatbot technology shall be used, can it be offered off-the-shelf?
- Is the technology really already mature enough for clinical use, especially when off-the-shelf solutions are used?
- How much interdisciplinary work is needed to establish chatbot technology for scenarios of different application domains?

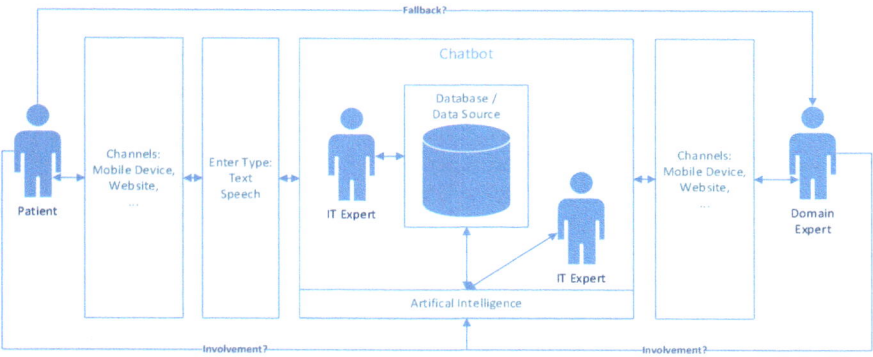

Fig. 16.1 Interplay of decisive aspects of a chatbot system

To answer these questions, we identified an interplay of decisive aspects between patients, domain and IT experts, which must be carefully considered (cf. Fig. 16.1).

Thereby, the most important aspects we identified with respect to the overall interaction procedure are as follows:

(1) Which channels shall be provided to a patient to interact with the chatbot? To interact with a mobile device, a website, or both channels has implications for a technical solution. If, for example, the chatbot needs some responsive time for an answer, then, in the case of a mobile device, a text message can be sent, which, in turn, is more difficult to achieve for a stationary PC setting.

(2) Which ways are allowed for a patient to enter data? There is a huge difference if a patient can only enter textual data compared to speech data or a combination of both. In addition, it must be specified, which languages shall be supported?

(3) Do already clinical databases, standard algorithms, or guidelines exist on which the chatbot answers can be based on?

(4) What determines if the answer of a chatbot is reliable? Are there any metrics and what reliability is acceptable for the different medical settings?

(5) In addition to the latter aspect, shall the system support the opportunity that a domain expert is contacted if the chatbot cannot answer in a reliable manner?

(6) How will the domain experts be involved to establish, maintain, and evolve a data source?

(7) Which channels must be provided for domain experts to interact with patients on one hand? On the other, which channels are required to interact with the chatbot system?

(8) How can patients be involved to efficiently evolve a data source?

Furthermore, the most important technical aspects are as follows:

(9)   Are there existing any standard technologies or off-the-shelf systems that fit to the requirements of the medical and psychological fields in a flexible way?

(10) What kind of IT experts are actually needed or should be involved?

(11)   What means artificial intelligence in this context and can it be measured?
(12)   Are there existing (standard) procedures how to establish, maintain, and evolve a data source the chatbot is based on?

Due to space limitations, the work at hand cannot answer all questions in detail. However, three main findings are provided, which are a basis to further tackle the questions. First, we will present the core aspects for establishing a powerful data source the chatbot is working with. In this context, the way a chatbot provides the *intelligence* has to be intertwined with the data source in a way that the chatbot is able to automatically learn and improve itself over time. Second, despite the many challenges that must be tackled, it can be reasonably argued that the opportunities to use chatbots surpass the challenges. However, existing technology is in a premature state that needs fundamental enhancements and considerations. Third, we present a reference architecture to realize chatbot systems for the medical context.

The remainder of this work involves six sections organized as follows. In the next section, current research on chatbots is discussed. Then, the Sect. 16.3 *Challenges* and Sect. 16.4 *Opportunities* present selected pros and cons, which are particularly crucial to deal with. The Sect. 16.5 *Reference Architecture* presents a generic architecture, which we figured out as being suitable for various practical settings. Section 16.6 *Discussion*, in turn, summarizes our findings to establish chatbots in the medical and psychological fields. Finally, Sect. 16.7 *Summary and Outlook* concludes this paper and gives an outlook on future work.

16.2   Related Work

The idea of dialogues between humans and machines is as old as evidenced by the development of the Touring-test in 1950, i.e., a test to reveal whether a machine is able to provide an intelligent behaviour comparable to the behavior of a human. In the sixties and seventies of the last century, the first chatbot implementations for medical and psychological domains have been proposed, e.g., ELIZA (Weizenbaum 1966), which simulates a psychotherapist, or PARRY (Colby 1973), which simulates a paranoid patient. A more recent chatbot implementation, in turn, is ALICE (Wallace 2003), which extends the concept of ELIZA and is based on knowledge of patterns and response templates stored in Artificial Intelligence Markup Language (AIML) files. All of the aforementioned implementations are purely based on *pattern matching* on predefined keywords and templates. Notably, this type of chatbot implementations are currently predominant.

More recent chatbots like Jabberwacky (Carpenter 2011) and Cleverbot (Carpenter 2018) are able to automatically learn new responses from the interaction with the user and save them to their database. In general, in recent years, numerous chatbot platforms emerged, which enable developers and businesses to setup their own customized chatbots with minimal efforts. Examples for these platforms are

Pandorabots, which is based on AIML (Pandorabots 2018), as well as IBM's *Watson Assistant*, and Google's *Dialogflow*, which let developers build dialogs through the use of web interfaces by configuring *intents* (the user's anticipated goal) and *entities* (terms or objects that provide context for intents) (Watson Assistant 2018; Dialogflow 2018). However, existing approaches lack holistic considerations how to use the technology in various settings. Furthermore, research on chatbot implementations for the medical and psychological context includes solutions for medical consultations (Lokman et al. 2009; Comendador et al. 2015; Brixey et al. 2017), diagnosis (Morales-Rodríguez et al. 2010; Mujeeb et al. 2017; Divya et al. 2018), primary care (Ni et al. 2017), psycho-social counseling (Oh et al. 2017), or (medical) education (Heller et al. 2005; Kazi et al. 2012). A recent scoping review provides evidence on chatbots in clinical psychology and psychotherapy (Bendig et al. 2019). The majority of these implementations are based on AIML or another implementation technique based on pattern matching. An example of a chatbot application that is supposed to help dealing with depression is given by Woebot[1] (Fitzpatrick et al. 2017). The latter applies principles of cognitive behavioral therapy (CBT) to guide patients with their mental health problems.

Technically, suitable generic architectures or other generic technical solutions are less proposed so far (Yan et al. 2016; Reshmi and Balakrishnan 2016), which draw general questions on how to establish chatbot systems in various settings. In addition, guidelines are missing that can be used when starting with new projects. However, more and more approaches show that the need for new findings is great (Cahn 2017). Finally, although platforms like *Pandorabots* exist, configurable off-the-shelf solutions cannot be easily purchased. Thereby, mainly the development of a suitable data source is the reason why powerful solutions cannot be realized in a quick and easy manner. Hence, this work aims to present general considerations on chatbots for medical and psychological treatment procedures.

16.3 Challenges

In this section, first of all, the technical challenges are summarized to understand the general operation procedure of a chatbot system. Based on this, general considerations on the used knowledge base are presented.

Generally, the technical challenges that emerge when designing a chatbot can be categorized into two main areas: *understanding what the user says (or wants)* and *giving an appropriate response*. The first category includes Natural Language Processing (NLP) functionalities like parsing, filtering, normalizing, and segmenting the user input, as well as applying pattern matching methods. Additionally, advanced chatbots should be context aware, consider the conversation history with the user, and perform a sentiment analysis on the user's input. The second category is primarily dependent on a *knowledge base*, which should contain all information that is

---

[1] https://woebot.io/.

needed to provide a reasonable response to the user's queries in a specific application context. Best case, this knowledge base is extendable by manual input, integration of existing databases and ontologies, as well as machine learning mechanisms in order to continuously improve the response quality (Abdul-Kader and Woods 2015). If the chatbot is supposed to simulate a human conversation partner, another challenge constitutes to make the user believe that he or she is actually communicating with a human being in contrast to a machine; in other words, an ideal chatbot should pass the Turing Test (Turing 2009). To this end, the chatbot might incorporate personalized and contextual responses, varying responses, sensible default responses, canned responses, typing errors and simulation of keystrokes, a model of personal history, and non sequiturs (Abdul-Kader and Woods 2015).

With respect to the technical challenges, a decisive step is to understand how the aforementioned knowledge base can be established. In this context, the following questions must be dealt with:

(1) Can existing knowledge bases be used? In addition, are they specifically appropriate for the target setting or do they need adaptions? If adaptions are needed, the question emerges, whether adaptions are technically supported?
(2) If existing knowledge bases are planned to be used, who owns the knowledge base rights?
(3) How can existing knowledge bases be compared with respect to their quality?
(4) If different knowledge bases exist, do they use the same or a similar vocabulary to configure and maintain data entries? In general, the vocabulary, including the technical configuration and maintenance are crucial aspects for the communication between domain and IT experts.
(5) If no knowledge bases exist or the existing ones are not appropriate, how can a new knowledge base be established from scratch?

Practical projects have shown that the comprehension in what way such knowledge bases can be created from scratch is crucial. Thereby, three general options exist: The first option (I) is to create the knowledge base only manually. This means, IT experts provide an option to domain experts to manually enter data to the knowledge base. For example, IBM's *Watson Assistant* and Google's *Dialogflow* provide such functionality. If the knowledge base is created this way, the overall creation takes time. Note that we figured out in practical settings to establish a reasonable knowledge base from scratch with the IBM's *Watson Assistant* is still very challenging and time-consuming. If existing knowledge bases providing data for diagnostic or therapeutic management or treatment guidelines can be used or bought, then the time period to create the knowledge base can be decreased. The second option (II) is based on the idea to crawl data of existing data sources, which already provide valuable data. In most cases, such data sources are not intended to be used for chatbots. For example, Internet fora can be used to establish a knowledge base for chatbots. However, such approach is technically challenging. In (Dandage et al. 2018), such an approach is proposed to evaluate fora threads to help Tinnitus patients in using the fora in question more efficiently. Note that the workflow described in Dandage et al. (2018) can be generally considered to establish knowledge bases for chatbots from scratch.

Although such an approach is technically challenging, three main advantages arise: First, the *wisdom of the crowd* can be exploited. Second, knowledge bases can be established much faster. Third, the workflow to crawl data can be accomplished again and again to improve the overall data quality of the knowledge base over time. The third option (III) is to combine options (I) and (II). Based on the obtained practical experiences, the third option is most promising, particularly, as currently only few knowledge bases exist, which can be easily used off-the-shelf. Therefore, the discussed options to establish a knowledge base are important to consider.

16.4   Opportunities

Chatbots constitute extensive opportunities in the medical and psychological context. They could be used for diagnostic assessment, general counseling, counseling for disease management, counseling before and after diagnostic and therapeutic interventions, disease prevention, treatment and therapy accompaniment, follow-up treatment, (self-) diagnosis, and—ultimately—for offering treatment, e.g., psychotherapy, replacing physicians' and psychotherapists' tasks to a certain extent. Another opportunity is given by the availability and scalability of chatbots. A chatbot can be available 24 h a day, 7 days a week, at any place with an internet connection, for thousands of concurrent users and can be operated at minimal costs. Providing a comparable availability with trained human experts would be economically unfeasible. The location-independent availability is of particular benefit for immobile users, users in rural areas, or in remote areas with poor medical infrastructure (e.g., workers on oil rigs, ship's crews, war refugees, soldiers etc.). Finally, chatbots would be ideally suited to offer services to people in foreign environments, as they can communicate with the chatbot in their language, whereas the use of the medical infrastructure around them is hampered by the language barrier.

16.5   Reference Architecture

We propose a reference architecture for a general chatbot system in the medical and psychological context as shown in Fig. 16.2. Rather than focusing on basic chatbot functionalities like NLP and pattern matching, we wanted to take advantage of existing chatbot services by building a comprehensive system on top of them. On one hand, this enables a more flexible system regarding overall configuration efforts. On the other, a higher precision of responses can be expected. Furthermore, we can take advantage of the evolvements of the integrated services. Moreover, we assume that modern chatbot systems are mainly utilized by the use of smart mobile devices. More specifically, the proposed system consists of six major components: a central backend, a *Conversation API*, a database, a web interface for the service provider, and a mobile application for the user as well as the domain expert.

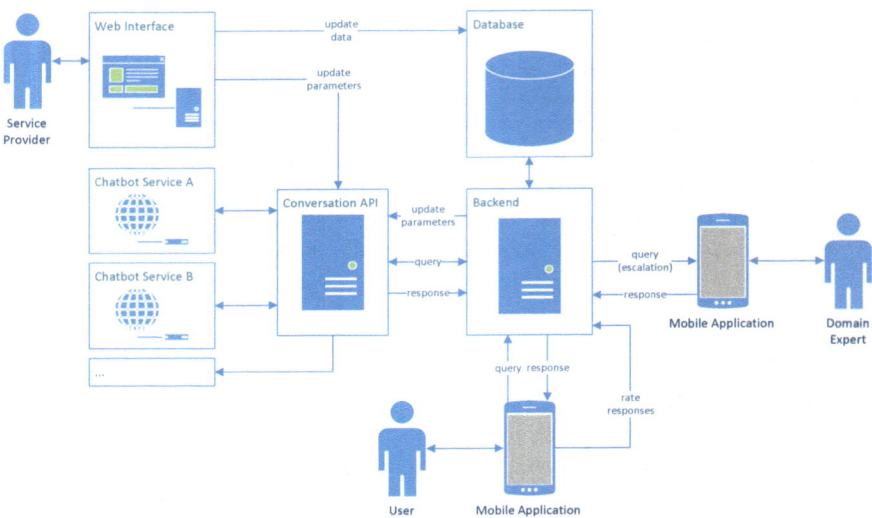

Fig. 16.2 Reference architecture for a chatbot system

The Conversation API operates as an abstract layer for third-party chatbot services. In our reference implementation, we integrated IBM's *Watson Assistant*[2] and Google's Dialogflow.[3] The API provides a generic interface for *intents* and *entities* as well as artefacts for the chatbot interaction. This way, (a) the different clients do not need to know about the underlying specifics of the third-party chatbot services, (b) replacing a third-party service is facilitated, and (c) the system can use another service as fallback mechanism if one service fails. The database stores user accounts, chat histories of users, messages of domain experts, response ratings, and other meta data. The web interface, in turn, provides access to the database for the service provider. Additionally, it communicates with the Conversation API in order to create and edit chatbot dialog templates. Two customized mobile applications are used for the communication between the user and the chatbot as well as the required communication between the user and a domain expert in case of an escalation. The latter reflects the scenario in which the chatbot is not able to provide a reliable response. Such escalation is a powerful way to improve the overall system quality—and therefore its potential acceptance—for patients as well as domain experts. Patients benefit as they receive answers to all their questions, while the domain experts learn what requests cannot be answered by the chatbot. This, in turn, enables the chatbot system to evolve by incorporating the response of the domain experts. More technically, a typical interaction between the user and the system can look like as follows:

(1) The user enters a text *query* (e.g., a question or problem statement) into the mobile application.

---

[2]https://www.ibm.com/watson/ai-assistant/.
[3]https://dialogflow.com/.

(2) The query is sent to the backend, which, in turn, forwards the query to the Conversation API

(3) The Conversation API forwards the query to one or multiple chatbot services and returns the result to the backend. Thereby, one of the two following options apply:

    a. The chatbot service is able to respond to the query; the response is therefore sent back to the user (i.e., to the mobile application)

    b. The chatbot service is unable to respond to the query **or** a specific keyword has been detected (e.g., "suicide suspicion") → escalation is required

       i. the query is forwarded to a group of domain experts; a generic response is sent to the user[4]

      ii. a domain expert responds to the query and inputs the chatbot parameters into the system (e.g., intents and entities for IBM Watson)

     iii. the parameters of the chatbot services are updated through the Conversation API in order to be able respond to the query automatically the next time it is detected

     iv. the response is sent back to the user.

(4) The mobile application displays the response to the user

(5) (Optional) The user might rate the response. Response ratings are stored in the database and the chatbot service parameters are updated accordingly, either automatically by the system or manually by an expert at a later time. Such feature can again improve the overall system quality and acceptance significantly. One could even integrate a Turing-test –like feature to evaluate whether the chatbot answers are distinguishable from human domain experts' answers.

Three aspects are fundamental when considering the reference architecture. First, a generic interface is used to integrate different chatbot services. This way, the flexibility and quality of the overall system can be improved. Second, an escalation service is used to handle requests that cannot be performed automatically. Third, a rating system is used that enables patients to rate the quality of responses. Although this architecture provides no answer for a creating or using proper knowledge base, it provides mechanisms to evolve and maintain existing knowledge bases. In addition, it enables the creation of an independent knowledge base, which can combine existing ones.

16.6 Discussion

This paper raised questions that emerged in practical projects, which have the goal to design and realize a chatbot system that is able to support medical and psychological

---

[4]e.g., "Your request has been forwarded to one of our experts. We will try to answer it as soon as possible."

treatment procedures. Due to space limitations, not all questions can be answered in detail and, hence, the most important findings are summarized. Currently, only few knowledge bases exist, which can be used off-the-shelf. Therefore, questions to establish such knowledge base must be considered. In addition, the alternatives to build a knowledge base from scratch might be helpful to evaluate a final solution. In this context, five issues are of paramount importance. First, the quality of the knowledge base is key to the quality of the entire system. Since the efficiency and acceptance of the system is mainly dependent on exact responses to the user, much efforts should be spent to reliably build the knowledge base. Second, the development of chatbot systems requires a highly interacting multidisciplinary team consisting of IT experts, mathematicians, statisticians, domain experts, communication experts/linguists, and software engineers. For example, IT experts must comprehensible explain the vocabulary of existing systems to domain experts, otherwise misunderstandings may decrease the overall response quality. Third, an escalation service is indispensable to compensate knowledge gaps of the system and to enable the knowledge base to evolve over time. Fourth, a rating system should be also a mandatory service of the system. This way, users' feedbacks can be utilized to improve the system quality. Fifth, at a very early stage, it should be carefully evaluated whether the chatbot system shall accept text input, speech input, or a combination of both. In addition, it should be evaluated which languages must be supported. The presented reference architecture as well as the raised questions may be a valuable starting point to tackle these five important issues to realize a powerful chatbot system that is able to support medical and psychological treatment procedures.

16.7   Summary and Outlook

The paper provided insights on practical projects with the goal to add beneficial IT support to medical and psychological treatment procedures. The obtained insights indicate that the most decisive aspects emerge already at the beginning of any chatbot project. More specifically, a set of questions that must be dealt with was presented and on top of this, the ways how a knowledge base can be established. Moreover, a reference architecture was conceived to realize a powerful and reliable chatbot system. The way in which the reference architecture is proposed is particularly beneficial with respect to a continuous system evolvement. Although the presently existing challenges surpass the opportunities, in future, it is very likely that chatbot systems will play an increasingly important role to support medical and psychological treatment procedures in daily life.

References

Abdul-Kader SA, Woods JC (2015) Survey on chatbot design techniques in speech conversation systems. Int J Adv Comput Sci Appl 6(7). https://doi.org/10.14569/ijacsa.2015.060712

Bendig E, Erb B, Schulze-Thuesing L, Baumeister H (2019) Next generation: chatbots in clinical psychology and psychotherapy to foster mental health—a scoping review. Verhaltenstherapie: 1–15. https://doi.org/10.1159/000499492

Brixey J, Hoegen R, Lan W et al (2017) SHIHbot: a Facebook chatbot for sexual health information on HIV/AIDS. In: Proceedings of the 18th annual SIGdial meeting on discourse and dialogue, Saarbrücken, Germany, August 2017

Cahn J (2017) CHATBOT: architecture, design, and development. University of Pennsylvania, School of Engineering and Applied Science, Department of Computer and Information Science

Carpenter R (2011) Jabberwacky. http://www.jabberwacky.com/. Accessed 19 Sept 2018

Carpenter R (2018) Cleverbot. http://www.cleverbot.com/. Accessed 19 Sept 2018

Colby KM (1973) Simulations of belief systems. Comput Models Thought Lang: 251–286

Comendador BEV, Francisco BMB, Medenilla JS et al (2015) Pharmabot: a pediatric generic medicine consultant chatbot. J Autom Control Eng 3:137–140. https://doi.org/10.12720/joace.3.2.137-140

Crutzen R et al (2011) An artificially intelligent chat agent that answers adolescents' questions related to sex, drugs, and alcohol: an exploratory study. J Adolesc Health 48(5):514–519. https://doi.org/10.1016/j.jadohealth.2010.09.002

Dandage S et al (2018) Patient empowerment through summarization of discussion threads on treatments in a patient self-help forum. In: Maglaveras N, Chouvarda I, de Carvalho P (eds) Precision medicine powered by pHealth and connected health ICBHI 2017 IFMBE proceedings, vol 66. Springer, Singapore

Dialogflow (2018) In: Dialogflow. https://dialogflow.com/. Accessed 27 Sept 2018

Divya S, Indumathi V, Ishwarya S et al (2018) A self-diagnosis medical chatbot using artificial intelligence. J Web Dev Web Des 3

Ebert DD et al (2018) Internet and mobile-based psychological interventions: applications, efficacy and potential for improving mental health. A report of the EFPA E-Health taskforce. Eur Psychol 23(3):269. https://doi.org/10.1027/1016-9040/a000318

Fitzpatrick KK, Darcy A, Vierhile M (2017) Delivering cognitive behavior therapy to young adults with symptoms of depression and anxiety using a fully automated conversational agent (Woebot): a randomized controlled trial. JMIR Ment Health 4(2):e19. https://doi.org/10.2196/mental.7785

Heller B, Proctor M, Mah D et al (2005) Freudbot: an investigation of chatbot technology in distance education. In: EdMedia: world conference on educational media and technology, Association for the Advancement of Computing in Education (AACE), Athabasca University, Montreal, Canada, 27 June 2005

Kazi H, Chowdhry BS, Memon Z (2012) MedChatBot: an UMLS based chatbot for medical students. Int J Comput Appl 55:1–5. https://doi.org/10.5120/8844-2886

Lokman AS, Zain JM, Komputer FS, Perisian K (2009) Designing a chatbot for diabetic patients. In: International conference on software engineering & computer systems (ICSECS '09), University Malaysia Pahang, Kuantan, Pahang, October 2009

Morales-Rodríguez ML, González BJJ, Florencia Juárez R et al (2010) Emotional conversational agents in clinical psychology and psychiatry. In: Sidorov G, Hernández Aguirre A, Reyes García CA (eds) Advances in artificial intelligence, vol 780. Springer, Berlin, Heidelberg, pp 458–466

Mujeeb S, Hafeez M, Arshad T (2017) Aquabot: a diagnostic chatbot for achluophobia and autism. Int J Adv Comput Sci Appl 8. https://doi.org/10.14569/ijacsa.2017.080930

Ni L, Lu C, Liu N, Liu J (2017) MANDY: towards a smart primary care chatbot application. In: Chen J, Theeramunkong T, Supnithi T, Tang X (eds) Knowledge and systems sciences. Communications in Computer and Information Science, vol 780. Springer, Singapore, pp 38–52

Oh K-J, Lee D, Ko B, Choi H-J (2017) A chatbot for psychiatric counseling in mental healthcare service based on emotional dialogue analysis and sentence generation. In: 2017 18th IEEE international conference on mobile data management (MDM), Daejeon, South Korea, 29 May-01 June 2017

Pandorabots (2018) In: Pandorabots Inc. https://home.pandorabots.com/home.html. Accessed 27 Sept 2018

Reshmi S, Balakrishnan K (2016) Implementation of an inquisitive chatbot for database supported knowledge bases. sādhanā, 41(10):1173–1178. https://doi.org/10.1007/s12046-016-0544-1

Siciliani L, Moran V, Borowitz M (2014) Measuring and comparing health care waiting times in OECD countries. Health Policy 118(3):292–303. https://doi.org/10.1016/j.healthpol.2014.08.011

Turing AM (2009) Computing machinery and intelligence. In: Epstein R, Roberts G, Beber G (eds) Parsing the turing test. Springer, Dordrecht, pp 23–65

Wallace R (2003) The elements of AIML style. Alice AI Found 139

Watson Assistant (2018) In: IBM Watson Assistant. https://www.ibm.com/watson/ai-assistant/. Accessed 27 Sept 2018

Weizenbaum J (1966) ELIZA—a computer program for the study of natural language communication between man and machine. Commun ACM 9(1):36–45

Yan M, Castro P, Cheng P, Ishakian V (2016) Building a chatbot with serverless computing. In: Proceedings of the 1st international workshop on mashups of things and APIs Article No. 5, Trento, Italy, 12–16 December 2016

Chapter 17
Persuasive E-Health Design for Behavior Change

Harald Baumeister, Robin Kraft, Amit Baumel, Rüdiger Pryss and Eva-Maria Messner

Abstract At a time, in which people are more and more suffering from lifestyle-related diseases such as cardiovascular diseases, diabetes, or obesity, changing health behavior and preserving a healthy lifestyle are salient factors of any public health effort. Hence, research on predictors and pathways of health behavior change is increasingly important. Following this, new ways of implementing *behavior change interventions* become possible based on internet technologies, allowing for technological approaches fostering behavior change. Such union of media informatics and psychology is denoted as *persuasive design* and refers to all technological intervention components, which help people to take, regularly use and re-take (after relapses into unwanted behavior) interventions. Along this trend, the present chapter introduces (1) theories of health behavior change and summarizes (2) present persuasive design approaches, thereby ending with (3) future directions in the field.

Keywords Persuasive design · Health behavior change · E-health · Lifestyle interventions

17.1 Introduction

A common target of e-health interventions is behavior change towards an increased health-related behavior. This might refer to e.g. less alcohol and nicotine consumption, increased physical activity, less stressful lifestyle or work-life-balance, safer sexual behavior, medication adherence, or a more positive treatment motivation in

H. Baumeister (✉) · R. Kraft · E.-M. Messner
Department of Clinical Psychology and Psychotherapy, Institute of Psychology and Education, Ulm University, Ulm, Germany
e-mail: harald.baumeister@uni-ulm.de

R. Pryss
Institute of Databases and Information Systems, Ulm University, Ulm, Germany

A. Baumel
Department of Community Mental Health, University of Haifa, Haifa, Israel

© Springer Nature Switzerland AG 2019
H. Baumeister and C. Montag (eds.), *Digital Phenotyping and Mobile Sensing*,
Studies in Neuroscience, Psychology and Behavioral Economics,
https://doi.org/10.1007/978-3-030-31620-4_17

general. The latter includes the affinity towards the uptake of indicated, evidence-based health care measures (Baumeister et al. 2008; Renneberg and Hammelstein 2006; Schwarzer 2004). Two research areas have been recently combined to investigate possibilities to increase the likelihood of behavior change: (1) The field of health psychology-research provides a longstanding expertise on theories and interventions that relates to motivation and more generally health behavior change (Oinas-Kukkonen and Harjumaa 2009; Riley et al. 2011); and (2) the field of (media) informatics, which has developed and examined a multitude of technological features that can foster motivating strategies. These two research fields combined, introduced as *persuasive design*, might enable scholars to address a common and fundamental challenge in the field of evidence-based health care by dealing with the lack of sufficiently motivated patients who cannot be motivated in sufficient numbers in on-site face-to-face settings (Van Ballegooijen et al. 2014; Wangberg et al. 2008). Persuasive design thus allows for more sophisticated and welcomed Internet- and mobile-based health interventions (IMI) by overcoming two of the major challenges in this context: a general low uptake rate and high attrition rates (Ludden et al. 2015; Riley et al. 2011). Along the described challenges, the chapter at hand provides a summary of (1) treatment motivation and behavior change theories, (2) technological approaches to support behavior change and (3) the integration of both fields to leverage the potential of e-health behavior change interventions.

17.2   Treatment Motivation and Behavior Change

Motivation (latin: motus = motion) refers to a theoretical construct that defines the direction and intensity of a behavior. Motivation is a key predictor (a) towards behavior and (b) maintenance of this behavior, mediated by one's own volition to realize intentions. Thereby, volition refers to the means one chooses to realize the intended behavior and the efforts as well as commitments one is willing to invest (Ryan et al. 2011).

Over the last decades, a multitude of health behavior change theories have been established aiming to facilitate the understanding about the reasons why people do (not) live and behave in a beneficial way to their health, especially when considering the present health risks associated with a risky lifestyle or a dysfunctional behavior. Therefore, in the following, the current state of theories on health behavior change is summarized based on Baumeister et al. (2008), providing the basis for developing and implementing technological and digital persuasive solution for facilitating intended behavior.

17.2.1 Health Behavior Change Models

Only a few decades ago, health behavior change knowledge of professionals was rather simple and straightforward. For example, a physician suggests something and the patient adheres to it, such as "smoking can kill you", with the expectation that this risk indication will actually stop people from smoking. It became quickly obvious that such approaches, associated with a paternalistic communication style that ignores patients´ attitudes and motivations, do not work (Schwarzer 2004). Thus, while risk perception, as described as a core predictor of behavior in the health belief model (Becker 1947), is still a valid and integral part of most models, two further core predictors of one's intention have been established in several other models such as the theory of planned behavior (TPB; Ajzen 1985), the social-cognitive theory (SCT; Bandura 2001), the transtheoretical model (TTM; Prochaska and Velicer 1997), the health action process approach (HAPA; Schwarzer 2008), or the technology-related Unified Theory of Acceptance and Use of Technology (UTAUT; Venkatesh et al. 2003): Predictor (1) self-efficacy and Predictor (2) outcome expectancy. Note that these constructs, in addition with risk perception are the three predictors of intention as defined in the HAPA model (Schwarzer 2008). Furthermore, they can be found in a similar way, but often denoted differently in many other models (see Table 17.1).

Table 17.1 Overlapping constructs of health behavior change models (modified after Renneberg and Hammelstein 2006; Baumeister et al. 2008)

Models	Social-cognitive factors of health behaviour change				
	Self-efficacy	Outcome expectancy	Risk perception	Goal setting	Planning
HBM	–	✓[c]	✓	–	–
TPB	✓[a]	✓	–	✓	–
SCT	✓	✓	✓[b]	✓	–
TTM	✓	✓[d]	✓	–	–
UTAUT	✓[e]	✓	–	–	–
HAPA	✓	✓	✓	✓	✓

[a]Perceived behavior control,
[b]Implicitly inclosed,
[c]Pros minus cons,
[d]Decisional balance,
[e]Performance expectancy
HAPA = Health Action Process Approach; HBM = Health Believe Model; TPB = Theory of Planned Behavior; SCT = Social-Cognitive Theory; TTM = TransTheoretical Model; UTAUT = Unified Theory of Acceptance and Use of Technology

17.2.1.1 Risk Perception

Risk perception as a predictor of motivation (most often operationalized as behavior intention) combines the perceived severity of risks (e.g., diseases following alcohol consumption) and the perceived vulnerability of a person (Baumeister et al. 2008). Thereby, risk perception is viewed as a necessary, but not sufficient condition for behavior change. While one might not think of changing anything in case of a lack of risk perception, it is known that when solely communicating the risks associated with a specific behavior such as smoking, alcohol consumption, or risky sexual behavior does not change the respective behavior at large (Ferrer and Klein 2015; Schwarzer 2004). Therefore, the transtheoretical model (Prochaska and Velicer 1997) additionally specifies that for a risk perception leading to intention and actual behavior change, a person must consider the value of a risk negatively for one's own life. Cognitions such as "what do I care if my life is short but lived to the fullest" exemplify the gap between general risk perception and health behavior actions. In other words, humans are always motivated, but maybe not towards the directions health care professionals and caring third parties expect them to be. Thus, next to negatively valued risk perception, one particularly needs to believe that the intended behavior change can be achieved (self-efficacy) and it results in a favorable outcome (= positive outcome expectancy) (Baumeister et al. 2008; Hardcastle et al. 2015; Schwarzer 2004; Sheeran et al. 2016).

17.2.1.2 Self-efficacy

Self-efficacy refers to the subjective certainty of being capable to master new or challenging situations due to one's own competency (Schwarzer 2004). The construct was introduced in the social-cognitive-theory (SCT) (Bandura 2001) and is viewed as a core predictor of health behavior change (Hardcastle et al. 2015; Schwarzer 2004; Sheeran et al. 2016). Thereby, a distinction between generic and context specific self-efficacy has been suggested (Schwarzer 2004). While generic self-efficacy describes a global assessment of one's own confidence of being capable to solve new or challenging tasks, context-specific self-efficacy refers to the expectation of being able to handle a specific situation (e.g., quit smoking, start or maintain physical activity). Current health behavior change models further differentiate self-efficacy by regarding the phases of the health behavior change process, starting with *motivation-related self-efficacy* (confidence of being able to achieve the goal), followed by *volition-related self-efficacy* (see below) (Baumeister et al. 2008).

17.2.1.3 Outcome Expectancy

Different to the construct of self-efficacy, which is used similarly across the different theories, outcome expectancy occurs in most models, but is labeled differently with also varying connotations (Schwarzer 2004; Renneberg and Hammelstein

2006). At least, implicitly the models define outcome expectancy as a subjective cost-benefit assessment by regarding the expected outcomes of behavior changes. In some of these models, such as the SCT (Bandura 2001), outcome expectancy already includes the construct *social norm*, while others, such as the TPB (Ajzen 1985), and the UTAUT (Venkatesh et al. 2003), define outcome expectancy as a separate predictor. Thereby, social norm has been theorized as working through both a social pressure to act (e.g. spouse kindly suggesting to lose weight) and an anticipated reinforcement by meaningful others (e.g., anticipated compliment given to the improved body shape) (Schwarzer 2004). While the latter refers to outcome expectancy (anticipated approval), the first rather can be explained as operant conditioning (negative reinforcement due to the expected discontinuation of the social pressure once the behavior has been changed).

A similar controversy exists regarding the perceived costs of an action. For example, if one thinks of reducing alcohol consumption, a person's expected negative consequence might be abstinence symptoms (= outcome expectancy). Probably, this person would anticipate at the same time that lowering alcohol consumption would be accompanied by substantial emotional stress (= perceived costs of the action). These aspects do not exactly match with the term outcome expectancy and might better be operationalized as action-related expectancies. In the process of an action, such expectancies might refer to cognitions prior (e.g. expected opportunity costs like "when I go jogging twice a week on top of everything else I can't watch my favored TV series anymore), during ("I will be quite exhausted and fun is something else"), and after the action ("I will be in such a good shape").

17.2.1.4 Intention

Intention is the construct that describes the case that a person decides to change a respective behavior, given a present risk perception and associated psychological strain, sufficient task specific self-efficacy, and outcome expectancy. For a long time, intention has been the postulated core predictor of health behavior change (Knoll et al. 2005). However, as we all experienced failures in regard to New Year's resolutions, the difference between intention and actual behavior change becomes obvious. This is introduced as the *intention-behavior-gap* phenomenon (Conner 2008; Sheeran and Webb 2016; Sutton 2008) and has led current health behavior change models, such as the HAPA (Schwarzer 2004), to introduce a volitional phase, following on one's intention to change a behavior.

17.2.1.5 Volitional Factors of Health Behavior Change

The term volition refers to a process that focuses on the actual realization of a behavioral intention. In the social sciences, volition is seen as a construct that is linked to the philosophical discussion on a free will, which limits the possibility of empirically examining the process behind the *intention-behavior-gap,* with a still ongoing controversial discussion whether volition can be validly assessed (Zhu 2004). However, in the fields of health psychology and motivation research, volition has become an inherent part of modern health behavior change theories (Heckhausen 2007; Renneberg and Hammelstein 2006; Schwarzer 2004).

The HAPA model for example describes a three stepped volitional phase, consisting of a pre-actional, actional and post-actional phase (Schwarzer 2004; Zhang et al. 2019). In the *pre-actional phase,* intentions are transformed into more specific plans about when, where, and how the intended behavior shall take place ("action planning", Sniehotta et al. 2005; "implementation intentions", Gollwitzer (1999). Additionally, one should anticipate possible barriers and challenges to successfully conduct the intended behavior, such as situational temptations (e.g., going in a pub with friends while trying to stay abstinent; "coping planning", Sniehotta et al. 2005). The *actional phase* is characterized by conducting the intended behavior and maintaining it over time (e.g. gym visits twice a week for the next year). Research shows that even freely chosen health behavior actions conducted as part of an experiment already lasted 66 days (median) to become an automatism (Lally et al. 2010). Hence, a core challenge in this actional phase is to protect the intended behavior against alternatives, sometimes tempting motives and aims until the behavior has become part of one's daily life. Finally, the action is evaluated in the *post-actional phase* and becomes reinforced according to the concept of operant conditioning, which is theorized to impact the reoccurrence of a behavior in dependence of the evaluation and reinforcement.

Again, self-efficacy has been postulated to be a core factor in this volitional phase, with a sub-categorization into *"action self-efficacy"*, referring to one's believe of being able to conduct the behavior, *"coping self-efficacy"*, referring to ones believe of being able to protect the planned behavior against other plans and temptations, and *"recovery self-efficacy"*, referring to the believe in one's own ability to recover from setbacks instead of showing disengagement (Schwarzer 2008).

17.2.1.6 Person- and Personality Characteristics Associated with Health Behavior Change

Inter-individual differences regarding the health behavior baseline as well as differences in the ability of health behavior change are well documented (Kaprio et al. 2002). Most prominently, gender has been examined extensively. For a long period, women were seen as less prone to drinking, smoking, and unhealthy diet, but less physically active compared to men; a view that might have become more complex (McDade-Montez et al. 2007). Regarding age, most risk behaviors decrease with increasing age, while physical activity becomes less likely (McDade-Montez et al. 2007). Most importantly in this context is that gender or age are not causal for health behavior, but different bio-psycho-social factors make a specific behavior more likely in one population compared to another. This becomes obvious when looking at the simplification of homosexuality being the core risk factor for HIV in the 1980s.

Not homosexuality, but unprotected sexual behavior was always the causal risk factor, which has more frequently been practiced by homosexual men (Hammelstein et al. 2006). Focusing on homosexuality, instead of unprotected sexual intercourse in risk communication and prevention strategies, might therefore be the reason for heterosexual intercourse having become more frequently been associated with HIV than homosexual intercourse in the following years (Hammelstein 2006). Next to socio-demographic variables, personality traits such as the "big five" *openness*, *conscientiousness*, *extraversion*, *agreeableness* and *neuroticism* have been suggested as relevant moderators of health behavior and behavior change (Bogg and Roberts 2004; McDade-Montez et al. 2007; Roberts et al. 2005). Thereby, *conscientiousness* and *agreeableness* have been associated with positive and *neuroticism* with negative health behavior (McDade-Montez et al. 2007), while results are less conclusive regarding *openness* and *extraversion*. Further personality constructs are frequently discussed, such as *optimism* as way of interpreting information, which might impact ones outcome expectancy (Hammelstein et al. 2006; McDade-Montez et al. 2007; Schwarzer 2004) and *self-directedness* as the ability to regulate and adapt behavior to individually chosen, voluntary goals (Cloninger et al. 1994; Sariyska et al. 2014). Finally, mental disturbances, such as depressive symptoms, are discussed as motivational and volitional barriers towards health behavior change, which one might misinterpret as being non-compliant (Baumeister et al. 2008).

17.3 Persuasive Design: Technological Features to Enhance Health Behavior Change

Digitalization is currently a frequent keyword when it comes to preparing our healthcare services for future challenges, particularly for an aging population with tremendous health care needs in resource-limited healthcare systems (Singh et al. 2016). Several technological solutions for a variety of health conditions, mental disturbances and unfavorable lifestyles have been developed in the last years (Christensen et al. 2009; Day and Sanders 2018; Van Ballegooijen et al. 2014; Wangberg et al. 2008). However, intervention adherence is often one of the core limitations of these otherwise helpful interventions (Baumel et al. 2017; Baumel and Yom-Tov 2018). Persuasive design is one of the constructs that specifically focuses on this human-machine-interaction problem, for which the machine would do the trick if only the human would work like a machine (Kok et al. 2004; Muench and Baumel 2017; Perski et al. 2017). Conflicting motives, a lack of self-efficacy, perceived high costs of the behavior, unfavorable outcomes expectancies, a lack of self-efficacy, as well as missing skills and potential temptations in the volitional phase are the key factors, for which persuasive interventions can make a difference (Venkatesh et al. 2003).

In the last decade, there has been substantial research efforts in the area of persuasive design. Persuasive technologies are defined as interactive systems, which are intentionally designed to influence their users in order to change their attitude and/or behavior (Fogg 1998). These technologies and their design principles can further be categorized in (a) *primary task support*, (b) *computer-human dialogue support*, (c) *system credibility* and (d) *social support* (Hamari et al. 2014; Oinas-Kukkonen and Harjumaa 2009).

Primary task design principles aim to support the user by achieving his primary goal when using the system (e.g., a successful intervention). The design principles in this category include reduction of complex behavior, guiding the user through the system, tailoring and personalization of content, as well as providing functionalities for self-monitoring (e.g., by visualizing and tracking progress), simulations, or (virtual) rehearsals of behavior. Tikka and colleagues (2018), for example, presented a gamification approach to promote rehearsals by repeatedly letting the user play a food categorization game in order to improve their game score. However, their study results indicate that rehearsal may not be enough to result in a positive behavior change (Tikka et al. 2018). Anagnostopoulou and colleagues (2018), as a second example, used personalized persuasive messages in a route planning application in order to motivate users to opt for more environmentally friendly route choices. These context-aware messages implemented self-monitoring (e.g., by providing feedback for past traffic usage) and suggestion (e.g. by suggesting to walk short distances) as persuasive features, and were perceived as useful by the users within a pilot study (Anagnostopoulou et al. 2018). Self-monitoring and mood/behavior-feedback systems have also been implemented frequently in health apps (e.g. Kauer et al. 2012; Montag et al. 2019). However, the component effects of such monitoring features in particular, or *primary task* design principles in general, are still largely unknown.

Computer-human dialogue design principles are supposed to help the users to move towards their goal or target behavior by implementing system feedback (e.g., audio, visual or textual), through direct feedback (i.e., positive and negative reinforcement), rewards (i.e., gamification through credits, points and achievements), reminders and alerts, suggestions and advice, as well as by designing the system in a way that it is appealing to its users and adopts a social role for them (e.g., by incorporating virtual agents). Reddy and colleagues (2018) conducted a feasibility study in this field of persuasive design technology for a phone-based recommendation system with the goal to change energy usage behavior at home, showing that recommendations may influence participant behavior by increasing their contextual awareness. Another field study showed that an animated character can be used as an imaginative trigger to foster healthy smartphone use (Chow 2018). As a third example, Wais-Zechmann and colleagues (2018) used personalized reminders and rewards to assist in meeting physical activity goals for patients with COPD (chronic obstructive pulmonary disease). They investigated the perceived persuasiveness within an online study utilizing storyboards, and concluded that these persuasive strategies are rated above average (Wais-Zechmann et al. 2018). While such automatic prompts and reminders are already part of established and well evaluated Internet- and mobile-based interventions (Bendig et al. 2018; Domhardt et al. 2018; Ebert et al. 2018, 2017), the question on when to prompt and remind users in what way and dosage to achieve the best possible behavior change is an open question not yet well understood (Baumeister et al. 2014; Domhardt et al. 2019; Fry and Neff 2009).

Design principles in the *system credibility* category focus on designing a system that is credible to its users by providing verifiably qualified, truthful, fair and unbiased information, demonstrating experience and competence, having a competent look and feel, and referring to real-world people and respected third-party endorsements. Wais-Zechmann and colleagues (2018) state that information and suggestions coming from an authority (like physicians or acknowledged institutions) are more persuasive for persons with COPD. Several interventions have already been developed and examined that used such persuasive messages referring to authorities (e.g. doctors and experts) in order to improve participants' intervention expectancy and adherence (e.g. Lin et al. 2017a, b; Sander et al. 2017; Spelt et al. 2018). Whether such authority focused approaches are indeed the best way to optimize system credibility, however, is a question for future studies, which should compare the persuasiveness and effectiveness of authority based strategies against other possible approaches such as professional look-and-feel strategies, strategies including a buddy avatar, or strategies using labels and certificates of well-respected organizations.

Finally, design features in the *social support* category describe how to design the system in a way that motivates its users by leveraging social influence through functionalities to observe, compare, and learn from other users as well as facilitating interaction, cooperation, competition, and recognition of successfully achieving behavior change goals, e.g., through the sharing of leaderboards or rankings (Hamari et al. 2014; Naslund et al. 2017; Oinas-Kukkonen and Harjumaa 2009; Orji and Moffatt 2018). Examples for this type of features are: interactive tools like messaging and chats with other users, user groups, social media sharing functions, rankings, the possibility to follow and mentoring functions (Mylonopoulou et al. 2018). Wunsch and colleagues (2015) implemented persuasive strategies in order to encourage biking as low-energy mode of transportation by utilizing recognition (awards based on the number of bike rides), competition (email updates with a leaderboard), cooperation (collective goals), and social comparison (options to compare the number of bike rides with others). They observed an increase in bike sharing for participants receiving the intervention as compared to the control group (Wunsch et al. 2015).

Overall, the field of persuasive design is still in its infancy and the correlation between health-related behavior change and persuasive design enhanced interventions is still unclear. A recent systematic review concludes that in 75% of the included studies persuasive design was superior to standard design in regard to health behavior change, whereas in 17% of the examined studies positive and negative outcomes have been reported. Finally, 8% of the studies reported no effect of persuasive design on the intended health behavior change (Orji and Moffatt 2018). Hamari and colleagues (2014) reported in their literature review 52% of positive, 36% of mixed, and 7% of negative outcomes related to persuasive design approaches in health behavior change.

Altogether, when it comes to sustained behavior change in the real world, the different persuasive design components in the system (e.g., feedback, rewards, support) should correspond properly to create a holistic user experience that helps change human behavior in real life. For example, providing people with feedback and rewards without adapting the program based on a user's progress, or without making sure they understand the expectations and relevance of the intervention before they begin, might fail the creation of the holistic experience that is expected to nurture a behavior change. Another aspect that is key in nurturing such an experience is that the quality of persuasive design is important and not only the question of whether a certain checklist of different components was included within the development process. For example, rewarding a person by offering a badge, if it cannot be presented to a group of people this person cares about and who also understand the meaning of the badge, would probably not achieve that same outcome as the intrinsic reward for doing an activity for oneself that can be mirrored through a compassionate statement.

Trying to answer these gaps, recent research introduced the concept of *therapeutic persuasiveness*, which is the way a program is designed as a whole to encourage users to make positive behavior change in their life. *Therapeutic persuasiveness* includes (1) call to action (e.g., goal setting, prompts), (2) load reduction of activities, (3) real data-driven/adaptive content (monitoring of user state and ongoing adaption of the intervention according to a user's individual progress), (4) ongoing feedback and rewards, and (5) clarity of therapeutic pathway and rational (Baumel et al. 2017). In this sense, *therapeutic persuasiveness* captures the quality of support a user receive from a technological system in his or her own path to achieve the desired goals. Furthermore, therapeutic persuasiveness aims to assess the degree a software is assisting in overcoming emerging difficulties during the behavior change process.

17.4   A Field Moving Forward

Recent technological progress and research trends as well as the prevalence of mobile devices and wearables create promising opportunities in the field of persuasive design. In Table 17.2, the most present persuasive design techniques in the scientific literature are heuristically mapped to the underlying psychological factors that predict health behavior change. While this mapping is based on expert consensus only and should therefore be interpreted as preliminary, it illustrates the broad range of technological approaches for each psychological behavior change dimension. However, common persuasive design techniques address mostly *outcome expectancy* and *self-expectancy*, while *particularly the important predictors of the volitional phase goal setting* and *planning* are less frequently taken into account yet. Therefore, persuasive design approaches that aim to increase health behavior change should include strategies for this volitional phase as well in order to not only facilitating intention, but leading to an actual behavior change.

The acceptance and broad use of mobile devices (such as smartphones, smart watches, etc.) will further provide opportunities to improve the persuasiveness of forthcoming health behavior change approaches. Particularly, opportunities to collect vast amounts of data, combined with new analytical methods such as deep/machine learning, will enable developers to improve the persuasiveness of their systems. These data can be used to (1) gain deeper knowledge about mental states in real-life situations, (2) get insight into the development, maintenance and course of health

Table 17.2 Persuasive design techniques and proposed[a] corresponding predictors of health behaviour change models

Persuasive design approaches	Technological realization	Psychological dimensions of health behavior change					
		Self-efficacy	Outcome expectancy	Risk perception	Intention-behavior gap	Goal setting	Planning
Self-monitoring and continuous tracking	E.g. EMA[b]	–	–	✓	–	–	–
Visualisation	E.g. tracking chart	✓	✓	✓	✓	✓	✓
Tailored content/feedback	E.g. personalized content	✓	✓	✓	✓	✓	✓
Adaptive feedback/content	E.g. chatbot	✓	✓	✓	✓	✓	✓
Reminders/alerts	E.g. push notifications	–	–	✓	✓	–	✓
Suggestions	E.g. lists	–	✓	–	✓	✓	✓
Recognition of achievements	E.g. rewards	✓	✓	–	✓	✓	✓
Sharing and cooperation	E.g. interactive tools	✓	✓	✓	✓	✓	✓
Social comparison	E.g. leaderboards	✓	✓	✓	✓	✓	✓
Empathy	E.g. emotional responsive android	✓	–	–	–	–	–

[a] Mapping based on expert ratings by the authors HB and EMM of the present chapter
[b] Ecological momentary assessment

conditions, (3) evaluate therapeutic processes, (4) give timely or context triggered just-in-time interventions or to suggest tailored interventions (Brunette et al. 2016; Rathner et al. 2018a, b). This timely feedback, in combination with the experienced social support, reinforce the adoption and maintenance of healthy behaviors (Naslund et al. 2017). Furthermore, the use of wearable sensors can assist in monitoring health and disease management over extended periods of time, as they need little active user input and therefore compliance (e.g. Ben-Zeev et al. 2015; Lanata et al. 2015; Naslund et al. 2017). To make such use of big data sets, deep machine learning will be one promising computational basis (Bengio et al. 2013; Längkvist et al. 2014; Miotto et al. 2018). It is based on an iterative process of computerized pattern recognition and is therefore well suited to analyze big data exploratively. Theories based on the detected patterns can be subsequently tested in confirmatory study designs and therefore may lead to deeper knowledge. Overall, the use of persuasive design to foster health behavior change will improve the likelihood of success in future.

References

Ajzen I (1985) From intentions to actions: a theory of planned behavior. In: Kuhl J, Beckmann J (eds) Action control: From cognition to behavior. Springer, Heidelberg, pp 11–39

Anagnostopoulou E, Bothos E, Magoutas B, Schrammel J, Mentzas G (2018) Persuasive interventions for sustainable travel choices leveraging users' personality and mobility type. In: Ham J, Karapanos E, Morita PP, Burns CM (eds) Persuasive technology. Springer, Cham, pp 229–241. https://doi.org/10.1007/978-3-319-01583-5_56

Bandura A (2001) Social cognitive theory: an agentic perspective. Ann Rev Psychol 52:1–26. https://doi.org/10.1146/annurev.psych.52.1.1

Baumeister H, Krämer L, Brockhaus B (2008) Grundlagen psychologischer Interventionen zur Änderung des Gesundheitsverhaltens. Klinische Verhaltensmedizin und Rehabilitation 82:254–264

Baumeister H, Reichler L, Munzinger M, Lin J (2014) The impact of guidance on Internet-based mental health interventions—a systematic review. Internet Interv 1:205–215. https://doi.org/10.1016/j.invent.2014.08.003

Baumel A, Yom-Tov E (2018) Predicting user adherence to behavioral eHealth interventions in the real world: Examining which aspects of intervention design matter most. Trans Behav Med 8(5):793–798. https://doi.org/10.1093/tbm/ibx037

Baumel A, Birnbaum ML, Sucala M (2017) A systematic review and taxonomy of published quality criteria related to the evaluation of user-facing eHealth programs. J Med Syst 41(128). https://doi.org/10.1007/s10916-017-0776-6

Becker MH (1947) The health belief model and personal health behavior. Slack, Thorofare, NJ

Bendig E, Bauereiß N, Ebert DD, Snoek F, Andersson G, Baumeister H (2018) Internet-based interventions in chronic somatic disease. Deutsches Aerzteblatt Int 115:659–665. https://doi.org/10.3238/arztebl.2018.0659

Bengio Y, Courville A, Vincent P (2013) Representation learning: a review and new perspectives. IEEE Trans Pattern Anal Mach Intell 35:1798–1828. https://doi.org/10.1109/TPAMI.2013.50

Ben-Zeev D, Scherer EA, Wang R, Xie H, Campbell AT (2015) Next-generation psychiatric assessment: using smartphone sensors to monitor behavior and mental health. Psychiatry Rehabil J 38(3):218–226. https://doi.org/10.1037/prj0000130

Bogg T, Roberts BW (2004) Conscientiousness and health-related behaviors: a meta-analysis of the leading behavioral contributors to mortality. Psychol Bull 130:887–919. https://doi.org/10.1037/0033-2909.130.6.887

Brunette MF, Ferron JC, Gottlieb J, Devitt T, Rotondi A (2016) Development and usability testing of a web-based smoking cessation treatment for smokers with schizophrenia. Internet Interv 4:113–119. https://doi.org/10.1016/j.invent.2016.05.003

Chow KKN (2018) Time off: designing lively representations as imaginative triggers for healthy smartphone use. In: Ham J, Karapanos E, Morita PP, Burns CM (eds) Persuasive technology. Springer, Cham, pp 135–146. https://doi.org/10.1007/978-3-319-78978-1_11

Christensen H, Griffiths KM, Farrer L (2009) Adherence in internet interventions for anxiety and depression: Systematic review. J Med Internet Res 11:e13. https://doi.org/10.2196/jmir.1194

Cloninger R, Svrakic D, Przybeck T (1994) A psychobiological model of temperament and character. Arch Gen Psychiatry 50:975–990

Conner M (2008) Initiation and maintenance of health behaviors. Appl Psychol Int Rev 57:42–50. https://doi.org/10.1111/j.1464-0597.2007.00321.x

Day JJ, Sanders MR (2018) Do parents benefit from help when completing a self-guided parenting program online? a randomized controlled trial comparing triple p online with and without telephone support. Behav Ther 49:1020–1038. https://doi.org/10.1016/j.beth.2018.03.002

Domhardt M, Steubl L, Baumeister H (2018) Internet-and mobile-based interventions for mental and somatic conditions in children and adolescents. Zeitschrift für Kinder- und Jugendpsychiatrie und Psychotherapie 1–14. https://doi.org/10.1024/1422-4917/a000625

Domhardt M, Geßlein H, von Rezori RE, Baumeister H (2019) Internet- and mobile-based interventions for anxiety disorders: A meta-analytic review of intervention components. Depress Anxiety 36:213–224. https://doi.org/10.1002/da.22860

Ebert DD, Cuijpers P, Munoz RF, Baumeister H (2017) Prevention of mental health disorders using internet and mobile-based interventions: a narrative review and recommendations for future research. Front Psychiatry 8:116. https://doi.org/10.3389/fpsyt.2017.00116

Ebert DD, Daele T, Nordgreen T, Karekla M, Compare TA, Zarbo C, Brugnera A, Oeverland S, Trebbi G, Jensen KL, Kaehlke F, Baumeister H (2018) Internet and mobile-based psychological interventions: applications, efficacy and potential for improving mental health. a report of the EFPA e-health taskforce. Eur Psychol 23:167–187. https://doi.org/10.1027/1016-9040/a000318

Ferrer RA, Klein WMP (2015) Risk perceptions and health behavior. Curr Opin Psychol 5:85–89. https://doi.org/10.1016/j.copsyc.2015.03.012

Fogg BJ (1998) Persuasive computers: perspectives and research directions. In: Proceedings of the SIGCHI conference on human factors in computing systems. ACM Press/Addison-Wesley Publishing Co., pp 225–232

Fry JP, Neff RA (2009) Periodic prompts and reminders in health promotion and health behavior interventions: Systematic review. J Med Internet Res 11(2):e16. https://doi.org/10.2196/jmir.1138

Gollwitzer PM (1999) Implementation intentions: Strong effects of simple plans. Am Psychol 54(7):493–503

Hamari J, Koivisto J, Pakkanen T (2014) Do persuasive technologies persuade? a review of empirical studies. In: International conference on persuasive technology, pp 118–136

Hammelstein P (2006) Sexuelles Kontaktverhalten. In: Renneberg B, Hammelstein P (eds) Gesundheitspsychologie. Springer, Heidelberg, pp 229–244. https://doi.org/10.1007/978-3-540-47632-0_14

Hammelstein P, Pohl J, Reimann S, Roth M (2006) Persönlichkeitsmerkmale. In: Renneberg B, Hammelstein P (eds) Gesundheitspsychologie. Springer, Heidelberg, pp 61–101

Hardcastle SJ, Hancox J, Hattar A, Maxwell-Smith C, Thøgersen-Ntoumani C, Hagger MS (2015) Motivating the unmotivated: how can health behavior be changed in those unwilling to change? Front Psychol 6:1–4. https://doi.org/10.3389/fpsyg.2015.00835

Heckhausen J (2007) Motivation und Handeln. Springer, Berlin

Kaprio J, Pulkkinen L, Rose RJ (2002) Genetic and environmental factors in health-related behaviors: studies on Finnish twins and twin families. Twin Res 5:366–371. https://doi.org/10.1375/136905202320906101

Kauer SD, Reid SC, Crooke AHD, Khor A, Hearps SJC, Jorm AF, Patton G et al (2012) Self-monitoring using mobile phones in the early stages of adolescent depression: randomized controlled trial. J Med Internet Res 14:e67. https://doi.org/10.2196/jmir.1858

Knoll N, Scholz U, Rieckmann N (2005) Einführung in die Gesundheitspsychologie. Reinhardt, München

Kok G, Schaalma H, Ruiter RAC, Van Empelen P, Brug J (2004) Intervention mapping: a protocol for applying health psychology theory to prevention programmes. J Health Psychol 9:85–98. https://doi.org/10.1177/1359105304038379

Lally P, van Jaarsveld CHM, Potts HWW, Wardle J (2010) How are habits formed: modelling habit formation in the real world. Eur J Soc Psychol 40:998–1009. https://doi.org/10.1002/ejsp.674

Lanata A, Valenza G, Nardelli M, Gentili C, Scilingo EP (2015) Complexity index from a personalized wearable monitoring system for assessing remission in mental health. IEEE J Biomed Health Inform 19:132–139. https://doi.org/10.1109/JBHI.2014.2360711

Längkvist M, Karlsson L, Loutfi A (2014) A review of unsupervised feature learning and deep learning for time-series modeling. Pattern Recognit Lett 42:11–24. https://doi.org/10.1016/J.PATREC.2014.01.008

Lin J, Paganini S, Sander L, Lüking M, Ebert DD, Buhrman M, Baumeister H et al (2017a) An internet-based intervention for chronic pain: a three-arm randomized controlled study of effectiveness of guided and unguided acceptance and commitment therapy. Deutsches Aerzteblatt Int 114:681–688. https://doi.org/10.3238/arztebl.2017.0681

Lin J, Sander L, Paganini S, Schlicker S, Ebert D, Berking M, Baumeister H et al (2017b) Effectiveness and cost-effectiveness of a guided internet- and mobile-based depression intervention for individuals with chronic back pain: protocol of a multi-centre randomised controlled trial. BMJ Open 7:e015226. https://doi.org/10.1136/bmjopen-2016-015226

Ludden G, Van Rompay T, Van Gemert-Pijnen J (2015) How to increase reach and adherence of web-based interventions: A design research viewpoint. J Med Internet Res 17:e172. https://doi.org/10.2196/jmir.4201

McDade-Montez E, Cvengros J, Christensen A (2007) Persönlichkeitseigenschaften und Unterschiede. In: Kerr J, Weitkunat R, Moretti M (eds) ABC der Verhaltensänderung: Der Leitfaden für erfolgreiche Prävention und Gesundheitsförderung. Urban & Fischer, München, pp 60–74

Miotto R, Danieletto M, Scelza JR, Kidd BA, Dudley JT (2018) Reflecting health: smart mirrors for personalized medicine. Npj Digit Med 1(62). https://doi.org/10.1038/s41746-018-0068-7

Montag C, Baumeister H, Kannen C, Sariyska R, Meßner E-M, Brand M (2019) Concept, possibilities and pilot-testing of a new smartphone application for the social and life sciences to study human behavior including validation data from personality psychology. Multidiscip Sci J 2:102–115. https://doi.org/10.3390/j2020008

Muench F, Baumel A (2017) More than a text message: dismantling digital triggers to curate behavior change in patient-centered health interventions. J Med Internet Res 19:e147. https://doi.org/10.2196/jmir.7463

Mylonopoulou V, Väyrynen K, Stibe A, Isomursu M (2018) Rationale behind socially influencing design choices for health behavior change. In: Ham J, Karapanos E, Morita PP, Burns CM (eds) Persuasive technology. Springer Cham, pp 147–159 https://doi.org/10.1007/978-3-319-78978-1_12

Naslund JA, Kim SJ, Aschbrenner KA, Mcculloch LJ, Brunette MF, Dallery J, Marsch LA et al (2017) Systematic review of social media interventions for smoking cessation. Addict Behav 73:81–93. https://doi.org/10.1016/j.addbeh.2017.05.002

Oinas-Kukkonen H, Harjumaa M (2009) Persuasive systems design: key issues, process model, and system features. Commun Assoc Inform Syst 24:485–500. https://doi.org/10.17705/1CAIS.02428

Orji R, Moffatt K (2018) Persuasive technology for health and wellness: state-of-the-art and emerging trends. Health Inform J 24:66–91. https://doi.org/10.1177/1460458216650979

Perski O, Blandford A, West R, Michie S (2017) Conceptualising engagement with digital behaviour change interventions: a systematic review using principles from critical interpretive synthesis. Trans Behav Med 7:254–267. https://doi.org/10.1007/s13142-016-0453-1

Prochaska JO, Velicer WF (1997) The transtheoretical model of health behavior change. Am J Health Promot 12:38–48. https://doi.org/10.4278/0890-1171-12.1.38

Rathner E-M, Djamali J, Terhorst Y, Schuller B, Cummins N, Salamon G, Baumeister H et al (2018a) How did you like 2017? detection of language markers of depression and narcissism in personal narratives. Interspeech 3388–3392. https://doi.org/10.21437/Interspeech.2018-2040

Rathner E-M, Terhorst Y, Cummins N, Schuller B, Baumeister H (2018b) State of mind: classification through self-reported affect and word use in speech. Interspeech 2018:267–271. https://doi.org/10.21437/Interspeech.2018-2043

Reddy V, Bushree B, Chong M, Law M, Thirani M, Yan M, Joshi A et al (2018) Influencing participant behavior through a notification-based recommendation system. In: Ham J, Karapanos E, Morita PP, Burns CM (eds) Persuasive technology. Springer, Cham, pp 113–119 https://doi.org/10.1007/978-3-319-78978-1_9

Renneberg B, Hammelstein P (2006) Gesundheitspsychologie. Springer Medizin, Heidelberg

Riley WT, Rivera DE, Atienza AA, Nilsen W, Allison SM, Mermelstein R (2011) Health behavior models in the age of mobile interventions: are our theories up to the task? Trans Behav Med 1:53–71. https://doi.org/10.1007/s13142-011-0021-7

Roberts BW, Walton KE, Bogg T (2005) Conscientiousness and health across the life course. Rev Gen Psychol 9:156–168. https://doi.org/10.1037/1089-2680.9.2.156

Ryan RM, Lynch MF, Vansteenkiste M, Deci EL (2011) Motivation and autonomy in counseling, psychotherapy, and behavior change: A look at theory and practice. The Counseling Psychologist 39:193–260. https://doi.org/10.1177/0011000009359313

Sander L, Paganini S, Lin J, Schlicker S, Ebert DD, Buntrock C, Baumeister H (2017) Effectiveness and cost-effectiveness of a guided internet- and mobile-based intervention for the indicated prevention of major depression in patients with chronic back pain-study protocol of the PROD-BP multicenter pragmatic RCT. BMC Psychiatry 17:1–13. https://doi.org/10.1186/s12888-017-1193-6

Sariyska R, Reuter M, Bey K, Sha P, Li M, Chen YF, Montag C et al (2014) Self-esteem, personality and internet addiction: a cross-cultural comparison study. Pers Individ Differ 61–62:28–33. https://doi.org/10.1016/j.paid.2014.01.001

Schwarzer R (2004) Psychologie des Gesundheitsverhaltens. Einführung in die Gesundheitspsychologie. Hogrefe, Göttingen

Schwarzer R (2008) Modeling health behavior change: how to predict and modify the adoption and maintenance of health behaviors. Appl Psychol Int Rev 57(1):1–29. https://doi.org/10.1111/j.1464-0597.2007.00325.x

Sheeran P, Webb TL (2016) The intention-behaviour gap. Soc Pers Psychol Compass 10:503–518. https://doi.org/10.1111/spc3.12265

Sheeran P, Maki A, Montanaro E, Avishai-Yitshak A, Bryan A, Klein WMP, Rothman AJ et al (2016) The impact of changing attitudes, norms, or self-efficacy on health intentions and behavior: a meta-analysis. Health Psychol 35:1178–1188

Singh K, Drouin K, Newmark LP, Jae HL, Faxvaag A, Rozenblum R, Bates DW et al (2016) Many mobile health apps target high-need, high-cost populations, but gaps remain. Health Affairs 35:2310–2318. https://doi.org/10.1377/hlthaff.2016.0578

Sniehotta FF, Schwarzer R, Scholz U, Schüz B (2005) Action planning and coping planning for long-term lifestyle change: theory and assessment. Eur J So Psychol 35(54):565–576. https://doi.org/10.1002/ejsp.258

Spelt H, Westerink J, Ham J, IJsselsteijn W (2018) Cardiovascular reactions during exposure to persuasion principles. In: Ham J, Karapanos E, Morita PP, Burns CM (eds) Persuasive technology. Springer, Cham, pp 267–278 https://doi.org/10.1007/978-3-319-78978-1_22

Sutton S (2008) How does the Health Action Process Approach (HAPA) bridge the intention-behavior gap? An examination of the model's causal structure. Appl Psychol Int Rev 57:66–74. https://doi.org/10.1111/j.1464-0597.2007.00326.x

Tikka P, Laitinen M, Manninen I, Oinas-Kukkonen H (2018) Reflection trough gaming: reinforcing health message response through gamified rehearsal. In: Ham J, Karapanos E, Morita PP, Burns CM (eds) Persuasive technology. Springer, Cham, pp 200–212 https://doi.org/10.1007/978-3-319-01583-5_56

Van Ballegooijen W, Cuijpers P, Van Straten A, Karyotaki E, Andersson G, Smit JH, Riper H (2014) Adherence to internet-based and face-to-face cognitive behavioural therapy for depression: a meta-analysis. PLoS ONE 9:e100674. https://doi.org/10.1371/journal.pone.0100674

Venkatesh V, Morris MG, Davis GB, Davis FD (2003) User acceptance of information technology: toward a unified view. MIS Q 27:425–478. https://doi.org/10.2307/30036540

Wais-Zechmann B, Gattol V, Neureiter K, Orji R, Tscheligi M (2018) Persuasive technology to support chronic health conditions: investigating the optimal persuasive strategies for persons with COPD. In: Ham, J Karapanos E, Morita PP, Burns CM (eds) Persuasive technology. Springer, Cham, pp 255–266 https://doi.org/10.1007/978-3-319-78978-1_21

Wangberg SC, Bergmo TS, Johnsen J-AK (2008) Adherence in internet-based interventions. Patient Preference Adherence 2:57–65. Retrieved from https://www.scopus.com/inward/record.uri?eid=2-s2.0-58449124326&partnerID=40&md5=a628ca7e5169cdfbb186e6c18a70a5b3

Wunsch M, Stibe A, Millonig A, Seer S, Dai C, Schechtner K, Chin RCC (2015) What makes you bike? exploring persuasive strategies to encourage low-energy mobility. In: MacTavish T, Basapur S (eds) Persuasive technology. Springer, Cham, pp 53–64 https://doi.org/10.1007/978-3-319-20306-5

Zhang C-Q, Zhang R, Schwarzer R, Hagger MS (2019) A meta-analysis of the health action process approach. Health Psychol https://doi.org/10.31234/osf.io/4pc27

Zhu J (2004) Understanding volition. Philoso Psychol 17(2):247–273. https://doi.org/10.1080/0951508042000239066

Chapter 18
Optimizing mHealth Interventions with a Bandit

Mashfiqui Rabbi, Predrag Klasnja, Tanzeem Choudhury, Ambuj Tewari and Susan Murphy

Abstract Mobile health (mHealth) interventions can improve health outcomes by intervening in the moment of need or in the right life circumstance. mHealth interventions are now technologically feasible because current off-the-shelf mobile phones can acquire and process data in real time to deliver relevant interventions in the moment. Learning which intervention to provide in the moment, however, is an optimization problem. This book chapter describes one algorithmic approach, a "bandit algorithm," to optimize mHealth interventions. Bandit algorithms are well-studied and are commonly used in online recommendations (e.g., Google's ad placement, or news recommendations). Below, we walk through simulated and real-world examples to demonstrate how bandit algorithms can be used to personalize and contextualize mHealth interventions. We conclude by discussing challenges in developing bandit-based mhealth interventions.

18.1 Introduction

Before mHealth, the standard of care was periodic visits to a clinician's office, interspersed with little to no patient support in between visits. At the clinician's office,

M. Rabbi (✉)
Department of Statistics, Harvard University, 1 Oxford St. #316, Cambridge, MA 02138, USA
e-mail: mrabbi@fas.harvard.edu

P. Klasnja
School of Information, University of Michigan, Ann Arbor, USA

T. Choudhury
Department of Information Science, Cornell University, Ithaca, USA

A. Tewari
Department of Statistics, University of Michigan, Ann Arbor, USA

S. Murphy
Department of Statistics and Department of Computer Science, Harvard University, Cambridge, USA

© Springer Nature Switzerland AG 2019
H. Baumeister and C. Montag (eds.), *Digital Phenotyping and Mobile Sensing*,
Studies in Neuroscience, Psychology and Behavioral Economics,
https://doi.org/10.1007/978-3-030-31620-4_18

data is collected to describe the patient's state at that visit time and self-report data about the patient's state prior to the current visit time is collected through an error-prone mechanism of recalling past events. The mHealth model has enabled significant progress in situ data collection between clinic visits; phone sensors can now capture personal data at a millisecond level, and improvement in user interfaces has reduced the burden of self-report information (Kubiak and Smyth 2019). mHealth interventions using persuasive design features are promising approaches for improving patients health (Baumeister et al. 2019; Messner et al. 2019). However providing effective interventions personalized to the patient between patient visits remains challenging.

Two key components of intervening at the right time are personalization and contextualization. Personalization is the process of matching an individual's preferences and lifestyle. e.g., a physical activity intervention can say, "You walked 10 times in the last week near your office. Don't forget to take small walks near your office today." Such personalization can lower barriers to acting on the suggestion (Hochbaum et al. 1952). Contextualization takes personalization one step further by delivering interventions at moments of need or at an opportune moment when the intervention is easy to follow (Fogg 2009). e.g., when a participant reaches the office, a push notification with the earlier walking suggestion can be sent, or, just after a high risk teen reports high stress, a SMS can be sent with ideas to reduce stress.

Contextualization and personalization are complex problems because different people may prefer different interventions and these preferences may vary by context. Fortunately, similar problems have been solved before. When Google places ads or Netflix suggests movies, they adapt their recommendation based on user preferences and characteristics, utilizing bandit algorithms. Here we describe how to repurpose bandit algorithms to personalize and contextualize mHealth interventions. We will start with a simple example, where we personalize a daily list of physical activity suggestions to an individual. We will then extend this simple example to account for contextual factors (e.g., weather). We conclude with a real-world example and discuss future challenges in developing personalized/contextualized interventions with bandit algorithms.

18.2 Background

Bandit algorithms: "Bandit algorithms" are so called because they were first devised for the situation of a gambler playing one-armed bandits (slot machines with a long arm on the side instead of a push button). Each time the gambler picks a slot machine, he/she receives a reward. The bandit problem is to learn how to best sequentially select slot machines so as to maximize total rewards. The fundamental issue of bandit problems is the *exploitation-exploration* tradeoff; here exploitation means re-using highly rewarding slot machines from the past and exploration means trying new or less-used slot machines to gather more information. While exploration may yield less short-term payoff, an exploitation-only approach may miss a highly rewarding slot machine. Researchers have proposed solutions to the bandit's exploit-explore tradeoff across many areas. In particular, once the relevance of bandit algorithms to internet advertising was understood, there was a flurry of work (Bubeck and Cesa-Bianchi 2012). Nowadays, bandit algorithms are theoretically well understood, and their benefits have been empirically demonstrated (Bubeck and Cesa-Bianchi 2012; Chapelle et al. 2012).

An important class of bandit problems is the contextual bandit problem that considers additional contextual information in selecting the slot machine (Woodroofe 1979). Contextual bandit problems provide a natural model for developing mobile health interventions. In this model, the context is the information about the individual's current circumstances, the slot machines correspond to the different intervention options, and the rewards are near-time, proximal, outcomes (Nahum-Shani et al. 2017). In this setup, optimizing mHealth intervention delivery is the act of learning the intervention option that will result in the best proximal outcome in a given circumstance. This is same as solving the contextual bandit problem.

18.3 Optimizing Intervention with a Bandit Algorithm

We will use two simulated examples to explain how bandits can be used to optimize an mHealth intervention for an individual. In Sect. 18.4, we will discuss another real-world mobile application that builds on the ideas introduced in the first two simple examples.

In our first example, the bandit algorithm will be used to select an optimal set of five physical activity suggestions, for an individual, from a set of ten suggestions. A set of five suggestions is optimal if the set leads to the highest level of daily activity for that individual. The second example extends the first by finding a set of five suggestions for each of several contexts. Contextualizing suggestions can be helpful because the same suggestion may be more actionable in certain contexts (e.g., good weather or day of the week).

18.3.1 Personalizing Suggestions for an Individual

Consider a scenario in which Jane's health plan gives her a physical activity tracker and a smartphone app. Jane's health plan has found that the ten activity suggestions from Table 18.1 often work for many less-active people to increase their activity. Note that the order of suggestions in Table 18.1 does not imply any specific ranking. It is unlikely, however, that every individual will be able to follow or prefer to follow all the 10 suggestions equally and there will be inter-personal variability in which suggestions are followed and to what degree. Thus, we set the goal of learning the five suggestions with the highest chance of maximizing Jane's activity. We use the bandit algorithm, which is running as part of Jane's smartphone app, to achieve this

Table 18.1 List of 10
suggestions

1. Walk 30 min
2. Add intervals: walk 5 min, walk very fast for 5 min, repeat 3 times
3. Take the stairs instead of the elevator whenever possible
4. Go for a walk with a friend or your dog
5. Swim a lap, rest 1 min, repeat 10 times
6. Attend a fitness class at your gym
7. Try some of the strength training and bodyweight exercises illustrated by the fitness app on your phone
8. Do yoga
9. Park at the far end of the parking lot to walk farther
10. Do yardwork for at least 10 min

goal. Each morning, the app issues a set of 5 suggestions. The app then monitors Jane's activities throughout the day and uses that information to choose 5 suggestions for the following day.

Formally, we will refer to each set of five activity suggestions as an *intervention option* or *action*. This intervention option or action is the particular choice of the five suggestions. On the morning of day t, the app suggests to Jane the action A_t, where $A_t = [S_{t1}, S_{t2}, S_{t3}, \ldots, S_{t10}]^T$ is a 10×1 vector of binary variables. S_{ti} has a value of 1 if the i-th suggestion from Table 18.1 is shown to Jane on day t, and 0 otherwise. Thus A_t will have 5 entries equal to 1 and 5 entries equal to 0. Further, let Y_t denote the number of active minutes for Jane on day t, which might be called the *proximal outcome* or *reward* of action A_t.

Consider the following linear regression model for the mean of the daily active minutes Y_t on day t in terms of the suggestions:

$$E[Y_t|A_t] = \sum_{i=1}^{10} \beta_i S_{ti}$$
$$= \beta^T A_t \tag{18.1}$$

where the second equality is written more compactly by using vector notation, $\beta = [\beta_1, \beta_2, \ldots, \beta_{10}]^T$. Here $\beta_1, \beta_2, \beta_3, \ldots, \beta_{10}$ respectively represent suggestion $1, 2, 3, \ldots, 10$ s contribution to Jane's number of active minutes. Therefore, Eq. 18.1. has the following simple interpretation: Y_t, the number of daily active minutes, is the sum of the effects of the 5 activity suggestions provided on day t (i.e., suggestions for which $S_{ti} = 1$).

Formally, our goal is to discover the best action $A_t = a^*$ that is, the set of 5 suggestions that makes Jane most active (that results in the highest mean daily active minutes). We can formally write this goal as: given β, determine the action a^* for which

$$\beta^T a^* \geq \beta^T a \tag{18.2}$$

where a is a combination of 5 suggestions from Table 18.1. β is, however, unknown. We can estimate Jane's a^* by running experiments in the following way: at the start of a day t, the app selects action A_t (in other words, it delivers to Jane a combination of 5 suggestions from Table 18.1). The tracker then counts the number of minutes Jane is active on the day (note that this number is the proximal outcome Y_t). If the 5 suggestions are useful, then Jane will be more active that day and Y_t will be high compared to other days with a different set of 5 suggestions. Now, the question is: how to select the 5 suggestions each day? One simple approach is to select 5 suggestions out of 10 with equal probability. But such a uniform selection strategy will select more useful and less useful suggestions equally. A more sophisticated approach is to use the information already available from the past experiments to select future suggestions that will both yield additional information about a^* and give as few less useful suggestions as possible. Note that here we face the same exploit-explore tradeoff faced by the classic bandit setting's gambler—i.e., how to balance exploiting suggestions that seemed useful in the past with exploring less frequently issued suggestions.

An effective approach to delivering less useful suggestions as little as possible is "optimism in the face of uncertainty" epitomized by the Upper Confidence Bound (UCB) technique (Auer et al. 2002; Li et al. 2010). Bandit algorithms based on the UCB have been well studied and possess guarantees of minimizing the number of less useful suggestions. The key intuition behind the UCB idea is the following: First, for each choice of action a_t, a confidence interval is constructed for the linear combination $\beta^T a_t$. Recall this linear combination represents $E[Y_t | A_t = a_t]$, the expected proximal outcome after receiving action, a_t. Then the UCB bandit algorithm selects the action with the highest upper confidence limit. Note that the upper confidence limit for $\beta^T a_t$ can be high for either of two reasons: (1) either $\beta^T a_t$ is large and thus a_t is a good action to make Jane active, or (2) the confidence interval is very wide with a high upper limit, indicating that there is much uncertainty about the value of $\beta^T a_t$. Using the upper confidence limit represents UCB's optimism; UCB is optimistic that actions with high upper confidence limits will be the best actions, even though a larger upper confidence limit can mean more uncertainty. However, if an action with high upper confidence is indeed not the optimal action, then selecting the action will reduce the uncertainty about the effect of this action. This will help UCB realize that the action is indeed not useful.

How does UCB choose an action using the upper confidence interval? By following these two steps. The first step involves using Eq. 18.1 to estimate β assuming homogeneous error variance. We might use ridge regression to estimate β because ridge regression regularizes to avoid overfitting, especially when Jane has just begun to use the app and we have less data (Li et al. 2010; Bishop 2007). In this case the estimator of β, denoted by $\widehat{\beta}_t$, after t days of using the bandit algorithm is:

$$\widehat{\beta}_t = \widehat{\Sigma}_t^{-1} \left(\sum_{u=1}^t A_u Y_u \right) \tag{18.3}$$

where $\widehat{\Sigma}_t^{-1} = \sum_{u=1}^{t} \left(A_u A_u^T \right) + I_{10}$ and I_{10} is an 10×10 identity matrix. Equation 18.3 is the standard solution for ridge regression. The second step is to construct an upper confidence limit for $\beta^T a$ for each possible action a; the upper confidence limit on day t for action a is given by $\widehat{\beta}_t^T a + \alpha \sqrt{a^T \widehat{\Sigma}_t^{-1} a}$, where α is an appropriate critical value. Note, since we assumed homogeneous error variance, $\widehat{\Sigma}_t^{-1}$ is proportional to the covariance for $\widehat{\beta}_t$, and $a^T \widehat{\Sigma}_t^{-1} a$ is the covariance of $\beta^T a$. Thus, $\sqrt{a^T \widehat{\Sigma}_t^{-1} a}$ represents standard deviation of $\beta^T a$ and the upper confidence limit of $\beta^T a$ has an interpretable form, which is simply the current estimate, $\widehat{\beta}_t^T a$, plus its standard deviation multiplied up to a constant factor α. Then, to choose the UCB action for day $t + 1$, we calculate the a_{t+1} for which

$$\widehat{\beta}_t^T a_{t+1} + \alpha \sqrt{\alpha_{t+1}^T \widehat{\Sigma}_t^{-1} \alpha_{t+1}} \geq \widehat{\beta}_t^T a + \alpha \sqrt{a^T \widehat{\Sigma}_t^{-1} a} \tag{18.4}$$

for all actions a. i.e., a_{t+1} is selected to maximize the upper confidence limit on the mean of Y_{t+1}. This approach possesses strong guarantees to minimize the number of less useful suggestions (Li et al. 2010; Auer 2002).

Here we summarize how the UCB bandit algorithm works on Jane's smartphone. First there is an "exploration phase" to allow the UCB algorithm to form preliminary estimates of β. This phase lasts for a number of days, say t_0 days, during which each morning the UCB bandit algorithm randomly selects an action, that is, uniformly selects five activity suggestions from the 10, and delivers these suggestions to Jane in the application. Then at the end of day t_0, the UCB bandit uses an incremental calculation to form $\widehat{\beta}_{t_0}$ and $\widehat{\Sigma}_{t_0}$ based on the selected action, Jane's activity minutes, Y_{t_0}, for that day and the prior day's $\widehat{\beta}_{t_0-1}$ and $\widehat{\Sigma}_{t_0-1}$. Next the UCB algorithm calculates the upper confidence limit for each action and selects the action a_{t_0+1} with the highest upper confidence limit. On the next morning, Jane is provided the five suggestions as specified by a_{t_0+1}. The UCB algorithm repeats the process by estimating new $\widehat{\beta}_{t_0+1}$ $\widehat{\Sigma}_{t_0+1}$ and an updated set of 5 suggestions are chosen for the next day and so on.

18.3.2 A Simulation Example

In this section, we use a simulated example to demonstrate how a UCB bandit algorithm can personalize suggestions for Jane. We assume the following simple model of how Jane responds to the suggestions: When Jane sees a suggestion, she follows it with probability p or does not follow it with probability $1 - p$. If Jane follows the suggestion, she spends D minutes following it on a particular day. We assume D is random and normally distributed, because Jane may not spend the same amount of time each time she follows the same suggestion. In Table 18.2, we created an artificial example scenario with p and D values for different suggestions. The D values are written as mean $\pm$ standard deviation. We also show the expected number

Table 18.2 A simulated scenario for Jane where p represents the probability of following a suggestion when Jane sees it, and if the suggestion is followed, "Duration" represents the number of daily minutes spent following the suggestion. Finally, p and "Duration" are used to compute the expected value $pE[D]$, which also represents β values for the suggestion

Suggestions	p	Duration, D (min)	Expected duration $pE[D]$ (min)
1. Walk 30 min	1	15 ± 4	15.0
2. Add intervals: walk 5 min, walk very fast 5 min, repeat 3 times	$\frac{1}{90}$	21 ± 5	0.4
3. Take the stairs instead of the elevator whenever possible	$\frac{5}{7}$	7.5 ± 2	5.2
4. Go with a friend or your dog for a walk	$\frac{6}{7}$	22 ± 10	18.9
5. Swim a lap, rest for 1 min, repeat 10 times	0	–	–
6. Attend a fitness class at your gym	$\frac{1}{14}$	31 ± 5	2.2
7. Try some of the strength training and bodyweight exercises illustrated by the fitness app on your phone	0	–	–
8. Yoga	$\frac{4}{7}$	18 ± 3	10.3
9. Park at the far end of the parking lot to walk further	$\frac{4}{7}$	11 ± 2	6.3
10. Do yardwork for at least 10 min	$\frac{3}{14}$	24 ± 5	5.1

of activity minutes that Jane spends following a suggestion when she sees it. This expected number is $p \times E[D] + (1 - p) \times 0 = pE[D]$. These expected minutes are also β values in Eq. 18.1. Note that β values are unknown in real world setting. We use known β values in a simulated example to show how the UCB algorithm finds the suggestions with higher β values.

With the above setup, we run the simulation in two stages. In the first stage, suggestions are included with equal probability in the five suggestions on each of the first fourteen days. This initial "exploration phase" helps to form an initial estimate of β. In the second stage, we run the UCB bandit algorithm: on each day, we compute $\hat{\beta}_t$, according to Eq. 18.3, and choose an action using Eq. 18.4. We run these simulation for 56 days, or 8 weeks. We run 200 instances of the simulation to account for randomness in the problem. One source of this randomness comes from the exploration phase, where the app generates non-identical sequences of random suggestions based on when Jane starts using the app. We deal with this randomness by resetting the randomization seed after each simulation run. Another source of randomness comes from the within-person variability of how Jane responds to the suggestions. We create a second stream of random numbers to simulate how Jane responds to the suggestions. The seed of this second stream remains unchanged after each simulation run;

we do not reset this seed because, doing so will add the randomness of resetting the seeds to the within-person variability.

Table 18.3 shows the results, where we report the mean of the β estimates. At the top, we list the actual β values. We then list in each row how many times a suggestion is issued by UCB over a two week period. We use boldface for the top five suggestions (1st, 3rd, 4th, 8th, 9th in Table 18.1). The simulation shows that after the two-week exploration phase, UCB chooses the top (boldfaced) suggestions more times than the less useful ones. Since a suggestion can be picked only once a day, the top suggestions 1, 2, and 8 from Table 18.3 are picked nearly every day after the exploration phase (11–14 days between week 3–4, 5–6, and 7–8). However, suggestions 3, 9, and 10 all have similar β values. As a result, UCB is often uncertain among them and chooses the 10th suggestion sometimes wrongly, since it is not in the top five suggestions.

18.4   Optimizing Interventions for Different Contexts

In the earlier section, we discussed an example of personalizing suggestions with the UCB algorithm. Our goal was to demonstrate the inner workings of a bandit algorithm in a simple setting. Here we discuss extending the prior example to a more realistic setting where we tailor suggestion based on users' context. Indeed, context can determine whether, and the degree to which, certain suggestions are actionable. For example, Jane may only be able to act on the yardwork suggestion on the weekend, or she may appreciate and act on the reminder to take her dog for a walk when the weather is good. By adapting suggestions to different contexts, we hope to enhance her activity level. Fortunately, we can contextualize suggestions by re-purposing the bandit technique described already. We briefly describe one way to do so below.

For clarity, we will first consider a very simple context involving only the weather and day of the week. For these two contexts, there are two states (i) weekend or weekday, (ii) good or bad weather, where we consider the whole day as bad weather if only part is. Thus, each day belongs to one of four different context combinations (see Table 18.4). Note this simple characterization of only 4 contexts is to convey the idea of contextualization rather than actually to realistically handle a large number of contexts.

For these four context combinations, the task of contextualizing suggestions boils down to optimizing the suggestions for each of the four. An intuitive approach is to use 4 different bandit algorithms, one for each context combination. Depending on the context on day t, the corresponding bandit would be activated for optimizing suggestions for that context. Recall that an action is a set of five activity suggestions from the 10 in Table 18.1. Each of the four different bandit algorithms uses a model such as Eq. 18.1. but with different βs due to the different contexts. We represent this difference by sub-scripting β as β_k for the k-th ($k = 1, 2, 3, 4$) context. So, the goal is to learn the optimal action a_k^* that maximizes the average number of minutes

Table 18.3 Number of times suggestions are picked by the app within each of the two-week intervals. $\widehat{N}$ denotes the number of days the app selects a suggestion in the time frame mentioned within parenthesis. Note, the number of times a suggestion can be selected during a two-week period is at most 14 (i.e., $\widehat{N} \leq 14$).

Suggestions	1	2	3	4	5	6	7	8	9	10
β	**15.0**	0.4	**5.2**	**18.9**	0.0	2.2	0.0	**10.3**	**6.3**	5.1
$\widehat{N}$ (week 1–2)	**7.1**	7.2	**7.0**	**7.0**	6.8	6.9	6.9	**6.8**	**7.1**	7.1
$\widehat{N}$ (week 3–4)	**12.4**	3.9	**6.3**	**13.4**	2.8	4.5	2.5	**9.6**	**7.8**	6.5
$\widehat{N}$ (week 5–6)	**12.8**	3.5	**6.3**	**13.7**	2.6	4.3	1.7	**10.1**	**8.1**	6.7
$\widehat{N}$ (week 7–8)	**13.1**	3.4	**6.4**	**13.8**	2.4	4.3	1.6	**10.1**	**7.8**	6.8

Table 18.4 Different types of contexts

	Context
1	Bad weather, weekend
2	Bad weather, weekday
3	Good weather, weekend
4	Good weather, weekday

active for Jane in context k. That is, for $k = 1, 2, 3, 4$ the goal is to learn the action a_k^* which satisfies

$$\beta_k^T a_k^* \geq \beta_k^T a_k$$

Again, one UCB bandit algorithm can be run per context to learn the optimal five suggestions for that context.

Note that using a separate bandit algorithm for each context is not a feasible approach in a real-world setting; there are too many possible contexts. It would take the bandit algorithm many days to obtain good estimates of the β_k parameters. However, we can use a few tricks to handle large number of contexts. First, we may know a priori that some suggestions are equally actionable across different contexts and some suggestions are not at all actionable in certain contexts. If the suggestions are equally actionable across contexts, we can use the same β_k parameter values for these contexts. And if a suggestion is not actionable in a given context we can set its parameter in β_k to zero. Second, we can pool information across people. For example, some suggestions, such as yardwork, are more actionable on weekends for most people. Thus, we don't need to find β_k for each user individually. Pooling information, however, requires a Bayesian approach where for a new user, initially β_k is pooled from prior users and once some data from the user is available, β_k is then adapted to make more user-specific changes. Bayesian approaches to bandit algorithms are beyond the scope of this chapter; but the techniques are along the same lines as UCB (Chapelle and Li 2011).

18.5   A Real-World Example

Earlier, we gave two simple examples of how the UCB bandit algorithm can personalize and contextualize mobile health interventions. Real-world examples, however, are more complicated, with many potential suggestions and many contexts. Below we discuss an mHealth app called MyBehavior that has been deployed multiple times in real world studies (Rabbi et al. 2018; Rabbi et al. 2015). MyBehavior utilizes phone sensor data to design unique suggestions for an individual and subsequently uses a bandit algorithm to find the activity suggestions that maximize chances of daily calorie burns. Like the example in Sect. 18.3, MyBehavior issues the suggestions once each morning. The number of suggestions, however, is higher than in Table 18.1

because the suggestions in MyBehavior closely match an individual's routine behaviors, and routine behaviors are dynamic. In the following, we briefly discuss how MyBehavior uses the bandit algorithm. More information on this can be found in (Rabbi et al. 2017).

18.5.1   MyBehavior: Optimizing Individualized Suggestions to Promote More Physical Activity

The following discussion of MyBehavior first covers how unique suggestions are created for each individual. We then briefly discuss how a bandit algorithm is used to find optimal activity suggestions that have the highest chance of maximizing an individual's daily calorie burn.

The MyBehavior app tracks an individual's physical activity and location every minute. The detected physical activities include walking, running, driving, and being stationary. The app then analyzes the location-tagged activity data to find patterns that are representative of the user's behaviors. Figure 18.1 shows several examples of behaviors found by MyBehavior. Figure 18.1a and b respectively contain places where a user stayed stationary and a location where the user frequently walked. Figure 18.1c shows similar walking behaviors from another user. MyBehavior uses these behavioral patterns to generate suggestions that are unique to each individual. For example, one intervention may suggest an activity goal at specific locations that the user regularly goes to. Such tailoring makes feedback more compelling, since a user's familiarity with the location enhances adherence (Fogg 2009).

(a)   **(b)**   **(c)**

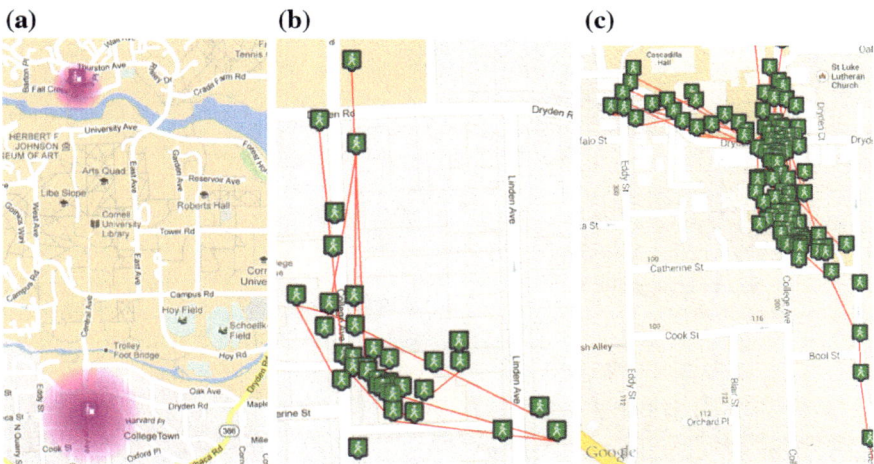

Fig. 18.1 Visualization of a user's movements over a week **a** heatmap showing the locations where the user is stationary everyday **b** location traces of frequent walks for the user **c** location traces of frequent walks for another user

Specifically, MyBehavior creates three kinds of uniquely individualized suggestions: (i) for stationary behaviors, MyBehavior pinpoints the locations where the user tends to be stationary and suggests taking small walking breaks every hour in these locations. (ii) for walking behaviors, MyBehavior locates the different places the user usually walks and suggests continuing to walk in those locations (iii) for other behaviors, e.g., participation in yoga class or gym exercises, MyBehavior simply reminds the user to keep up the good work. Figure 18.2 shows several screen shots of the MyBehavior app, where Figs. 18.2a–c are suggestions for three separate users. Since MyBehavior suggestions are tailored to the user, the first suggestion at the top of each screen shot is to walk, but the locations are different. Also, the first and third users receive a gym weight training exercise suggestion that the second user does not.

Now, how does MyBehavior decide which suggestions to give? MyBehavior uses a bandit algorithm like that in Sect. 18.3's first example, where suggestions are issued once a day. But MyBehavior can offer many more suggestions than Table 18.1 contains, depending on the variety of locations in which a user might be sedentary or active, etc. Fortunately, the bandit algorithm can still efficiently adapt to these high numbers of tailored suggestions. Rabbi et al. (2017) details how this optimization works, but the key intuitions are the following: (i) Most human behaviors are highly repetitive and routine and occur in the same locations. Routine behaviors and locations will be detected early and thus included soon in the individual's list of suggestions. (ii) The suggestions relating to routine behaviors and locations are more likely to be followed than suggestions of non-routine behaviors in non-routine

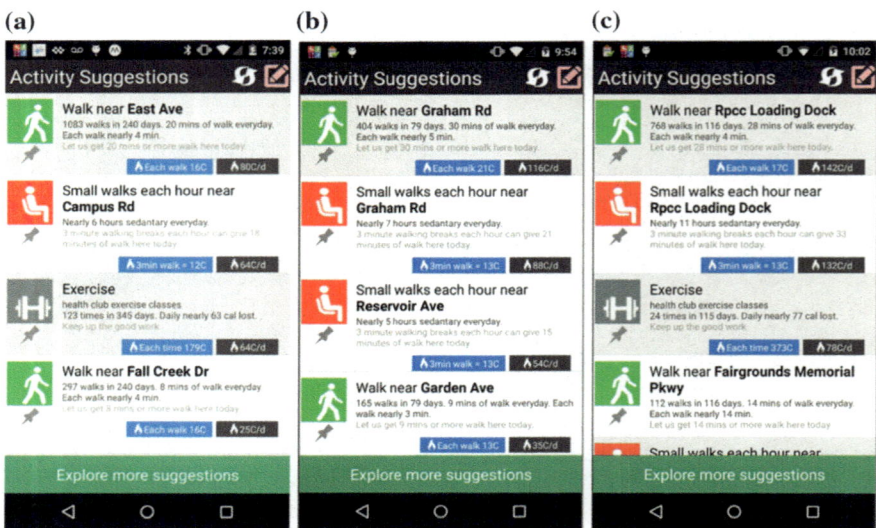

Fig. 18.2 MyBehavior app screenshots for three different users. Figures 18.1 and 18.2 have been reproduced from Rabbi et al. (2015) with appropriate permission from the authors

locations. Thus, the bandit will learn about the effects of these suggestions more quickly and these suggestions will likely remain effective if the user's routine does not change.

18.6 Discussion

In the last two sections, we discussed several examples of how bandit algorithms can optimize mobile health interventions. The bandit algorithm balances experimenting with different activity suggestions and selecting activity suggestions that currently appear most useful. This balancing act ensures that the algorithm acquires necessary information while maintaining an engaging user experience by providing as few less-useful suggestions as possible. While we showed that bandit algorithms can be useful to personalize and contextualize suggestions, there are additional real-world complexities that pose new challenges for bandit algorithms to address:

Ignoring delayed effects: In bandit algorithms, the optimal action is the action that maximizes the immediate reward (proximal outcome). In other words, bandit algorithms ignore the potential impact of the action on future context and future proximal outcomes. Some actions, however, can have long-term negative effects even if the short-term effect is positive. e.g., delivering an office walking suggestion may increase a user's current activity level, but the user might become bored after repeating the office walk several days, thus future suggestions may be less effective. In these cases, other algorithms that explicitly allow past actions to impact future outcomes (Sutton and Barto 1998) might be used. Precisely, the outcome of these algorithms are $Y_t + V(X_{t+1})$, where $V(X_{t+1})$ is the prediction of the impact of the actions on future proximal outcomes given the context X_{t+1} at the time $t+1$ (a bandit algorithm acts as if $V(X_{t+1}) = 0$). These algorithms tend to learn more slowly than bandit algorithms, since we need additional data to form the prediction $V(X_{t+1})$. We conjecture that the noisier the data is, the harder it will be to form high quality predictions of $V(X_{t+1})$ and thus as a result, bandit algorithms may still be preferable.

Non-stationarity: Most bandit algorithms assume "stationary" settings; i.e., the responsivity of a user in a given context to an action does not change with time. This assumption can be violated in real-word settings; in MyBehavior, for example, we observed that many suggestions become ineffective when people switched job and moved from one location to another. Such changes over time are often referred to as "non-stationarity." Other types of non-stationarity can be caused by life events such as a significant other's illness or aging. Bandit algorithms are typically slow to adapt to non-stationarity. Speeding up this process is a critical direction for future bandit research.

Dealing with less data: In real world applications, where the number of contexts and actions are many, bandit algorithms will need a lot of burdensome experimentation to find the optimal action for a given context. One way around this is to use a "warm start." A warm start set of decision rules that link the context to the action can be constructed using data from micro-randomized trials (Klasnja et al. 2015) involving similar individuals. Recently Lei et al. (2014) developed a bandit algorithm that can employ a warm start. However, we still need to test whether, and in which settings, warm starts will sufficiently speed up learning.

Adverse effects: Since mHealth interventions are generally behavioral, the risk of personal harm is often minimal. Nonetheless, there could be potential iatrogenic effect because phones cannot capture every piece of contextual information and bandit algorithms ignore the long-term effects of interventions. Since bandit algorithms don't take interventions' long-term effects into account, the algorithm may notify or otherwise deliver interventions too much and thus cause annoyance and reduce app engagement. Future work needs to investigate how to account for such long-term adverse effects. Furthermore, current phone sensors cannot automatically capture critical contextual information such as a user's health risks, preferences, barriers, emotional states, etc. Incomplete information may cause the algorithm to provide less appealing (e.g., not suggesting an activity that a user likes but didn't do often in the past) and inappropriate suggestions (e.g., asking someone who is injured to walk). Providing human control over the suggestion generation process can mitigate these problems; e.g., a user can delete inappropriate suggestions and prioritize the suggestions that are more appealing (Rabbi et al. 2015).

Acknowledgements This work has been supported by NIDA P50 DA039838 (PI Linda Collins), NIAAA R01 AA023187 (PI S. Murphy), NHLBI/NIA R01 HL125440 (PI: PK), NIBIB U54EB020404 (PI: SK). A. Tewari acknowledges the support of a Sloan Research Fellowship and an NSF CAREER grant IIS-1452099.

References

Auer P (2002) Using confidence bounds for exploitation-exploration trade-offs. J Mach Learn Res 3(Nov):397–422

Auer P, Cesa-Bianchi N, Fischer P (2002) Finite-time analysis of the multiarmed bandit problem. Mach Learn 47(2–3):235–256

Baumeister H, Kraft R, Baumel A, Pryss R, Messner E-M (2019) Persuasive e-health design for behavior change. In: Baumeister H, Montag C (eds) Mobile sensing and digital phenotyping: new developments in psychoinformatics. Springer, Berlin, pp x–x

Bishop CM (2007) Pattern recognition and machine learning. Springer

Bubeck S, Cesa-Bianchi N (2012) Regret analysis of stochastic and nonstochastic multi-armed bandit problems. Found Trends® Mach Learn 5(1):1–122

Chapelle O, Li L (2011) An empirical evaluation of thompson sampling. In: Advances in neural information processing systems, pp 2249–2257

Chapelle O, Joachims T, Radlinski F, Yue Y (2012) Large-scale validation and analysis of interleaved search evaluation. ACM Trans Inf Syst (TOIS) 30(1):6

Fogg BJ (2009) A behavior model for persuasive design. In: Proceedings of the 4th international conference on persuasive technology, ACM, vol 40

Hochbaum G, Rosenstock I, Kegels S (1952) Health belief model. United States Public Health Service

Klasnja P, Hekler EB, Shiffman S, Boruvka A, Almirall D, Tewari A, Murphy SA (2015) Microrandomized trials: an experimental design for developing just-in-time adaptive interventions. Health Psychol 34(S):1220

Kubiak T, Smyth JM (2019) Connecting domains—ecological momentary assessment in a mobile sensing framework. In: Baumeister H, Montag C (eds) Mobile sensing and digital phenotyping: new developments in psychoinformatics. Springer, Berlin, pp x–x

Lei, H., Tewari, A., & Murphy, S. (2014) An actor-critic contextual bandit algorithm for personalized interventions using mobile devices. Advances in Neural Information Processing Systems, 27

Li L, Chu W, Langford J, Schapire RE (2010) A contextual-bandit approach to personalized news article recommendation. In: Proceedings of the 19th international conference on World wide web, ACM, pp 661–670

Messner E-M, Probst T, O'Rourke T, Baumeister H., Stoyanov S (2019) mHealth applications: potentials, limitations, current quality and future directions. In: Baumeister H, Montag C (eds) Mobile sensing and digital phenotyping: new developments in psychoinformatics. Springer, Berlin, pp x–x

Nahum-Shani I, Smith SN, Spring BJ, Collins LM, Witkiewitz K, Tewari A, Murphy SA (2017) Just-in-time adaptive interventions (JITAIs) in mobile health: key components and design principles for ongoing health behavior support. Ann Behav Med 52(6):446–462

Rabbi M, Aung MH, Zhang M, Choudhury T (2015) MyBehavior: automatic personalized health feedback from user behaviors and preferences using smartphones. In: Proceedings of the 2015 ACM international joint conference on pervasive and ubiquitous computing, pp 707–718

Rabbi M, Aung MH, Choudhury T (2017) Towards health recommendation systems: an approach for providing automated personalized health feedback from mobile data. Mobile health. Springer, Cham, pp 519–542

Rabbi M, Aung MS, Gay G, Reid MC, Choudhury T (2018) Feasibility and acceptability of mobile phone-based auto-personalized physical activity recommendations for chronic pain self-management: pilot study on adults. J Med Internet Res 20(10):e10147

Sutton RS, Barto AG (1998) Reinforcement learning: an introduction. MIT Press

Woodroofe M (1979) A one-armed bandit problem with a concomitant variable. J Am Stat Assoc 74(368):799–806

Printed by Printforce, the Netherlands